Engineering Practical Rope Rescue Systems

Michael G. Brown

Australia • Brazil • Japan • Korea • Mexico • Singapore • Spain • United Kingdom • United States

DELMAR
CENGAGE Learning™

Engineering Practical Rope Rescue Rystems
Michael G. Brown

Business Unit Director: Alar Elken

Executive Editor: Sandy Clark

Acquisitions Editor: Mark Huth

Developmental Editor: Jeanne Mesick

Editorial Assistant: Dawn Daugherty

Executive Marketing Manager: Maura Theriault

Marketing Coordinator: Kasey Young

Executive Production Manager: Mary Ellen Black

Project Editor: Barbara L. Diaz

Art Director: Rachel Baker

Channel Manager: Mona Caron

For product information and technology assistance, contact us at
Cengage Learning Customer & Sales Support, 1-800-354-9706

For permission to use material from this text or product,
submit all requests online at **cengage.com/permissions**
Further permissions questions can be emailed to
permissionrequest@cengage.com

Library of Congress Control Number: 99-056517

ISBN-13: 978-0-7668-0197-4

ISBN-10: 0-7668-0197-7

Delmar
Executive Woods
5 Maxwell Drive
Clifton Park, NY 12065
USA

Cengage Learning is a leading provider of customized learning solutions with office locations around the globe, including Singapore, the United Kingdom, Australia, Mexico, Brazil, and Japan. Locate your local office at: **international.cengage.com/region**

Cengage Learning products are represented in Canada by Nelson Education, Ltd.

For your lifelong learning solutions, visit **delmar.cengage.com**

Visit our corporate website at **www.cengage.com**

Notice to the Reader
Publisher does not warrant or guarantee any of the products described herein or perform any independent analysis in connection with any of the product information contained herein. Publisher does not assume, and expressly disclaims, any obligation to obtain and include information other than that provided to it by the manufacturer. The reader is expressly warned to consider and adopt all safety precautions that might be indicated by the activities described herein and to avoid all potential hazards. By following the instructions contained herein, the reader willingly assumes all risks in connection with such instructions. The publisher makes no representations or warranties of any kind, including but not limited to, the warranties of fitness for particular purpose or merchantability, nor are any such representations implied with respect to the material set forth herein, and the publisher takes no responsibility with respect to such material. The publisher shall not be liable for any special, consequential, or exemplary damages resulting, in whole or part, from the readers' use of, or reliance upon, this material.

Printed in the United States of America
8 9 10 11 12 13 11 10 09 08

FOREWORD

Engineering Practical Rope Rescue Systems is the book we have been waiting for—a quality technical rope rescue text written by someone who actually does it as a profession. Mike Brown is eminently qualified to write the definitive manual on technical rope rescue. A professional fireman for twenty years and a rope rescue training officer for eighteen years, he speaks and writes from experience. He has been there from the beginning of the modern rope rescue techniques evolution. His experience in the field covers everything from shipboard rescue to the Oklahoma City bombing incident. It is rare when an individual with the level of expertise that Brown has is able to communicate that knowledge with the clarity and enjoyable style that we see in this book. It is destined to be the standard by which all other rope rescue texts are compared. Brown makes no attempt to force the reader to use only a system that he prefers, but presents systems that have worked in the past. He also makes clear that individual preferences are fine as long as they are based on sound principles.

That this book is written from the professional rescuer perspective is apparent from the beginning when he discusses the relevant NFPA standards and the OSHA and ANSI standards as they pertain to rope rescue. One of the many outstanding features of this book is the effort by Brown to help the reader comprehend how the natural laws of physics, such as friction and gravity, control the way systems are designed and operated. A significant effort is made to help people understand the materials that are used to manufacture the products used in rope systems. Everything from the metallurgy of aluminum to the manufacturing process of rope is explained to help the reader make informed decisions. The reader is led through the process of team development and training, which is a welcome departure from other texts that do not prepare the reader for the significant commitment of time and effort involved in becoming proficient in technical rope rescue techniques. The text covers all the usual techniques of rappeling, lowering systems, ascending systems, hauling systems, patient handling, etc. Of special importance, and again indicative of the background of the author, is the addressing of personal escape systems, which all firefighters need to know. This text covers techniques from the simplest rappel to the most accurate discussion of traverse systems yet published.

Another area in which this book excels is in discussing actual engineering of systems. Very important principles are discussed clearly and simply. Brown has done the research necessary to prevent the unfortunate habit we have seen in other texts of author's repeating inaccurate information because they did not recognize the mistakes in earlier books. The application of simple machines such as levers and pulleys and how they affect sophisticated rigging systems is clearly explained. The importance of elevated leading edge systems, and how they might affect sophisticated rigging systems is clearly explained. The importance of elevated leading edge systems, and how they are stabilized, and the use of angled rope systems for mass evacuation is detailed for the reader. For once, even the layman can understand force vectors.

The author has done an outstanding job in this book of presenting many highly technical and misunderstood principles in an entertaining and understandable manner. Actual field incidents are used to help illustrate many important points. This book is the best reference for the technical rope rescuer. It is intended for use by those who risk their lives in the pursuit of saving others. Such people deserve the best and most systems accurate information to assist them in their endeavors. *Engineering Practical Rope Rescue Systems* provides that information.

Kyle Isenhart
Rescue Systems Inc.

PREFACE

The topic of rope rescue, it would seem, is finite. There is an exciting beginning, a meaty intermediate level, and then an advanced and final level. Nothing could be further from the truth. Its study is like both the Hubbell Space Telescope and a powerful electron microscope—the more you look both inward and outward, the more you see. Collecting its information is both exciting and endless. This book is a humble attempt to collate practical rope rescue information in a readable, informative, and entertaining format.

Surprisingly, there was much discussion regarding the title, *Engineering Practical Rope Rescue Systems.* It seems there is always a target audience, people who are "supposed" to want to buy the book, people who have a need or use for the work. How can a single compendium of rope talk span all of the rope-type target people so that everyone gets something out of it? To me, *engineering* is what we do. We identify a need, we design a system to satisfy that need, we identify the tools needed for accomplishing the task, and then we put it all together and make it work. It's kind of like building an erector set. We engineer systems—complementary groups of components—to accomplish lifesaving tasks in extremely dangerous environments. We are rope-system engineers. The discussion, though, was that the word "engineer" was too steeped in academia and might scare off droves of would-be users of the book—that practitioners of rope rescue (and basically any fire and rescue-oriented discipline) are less concerned about engineering than about tying the rope to a tree and chucking it over the side of a cliff. Realistically, however, I also wanted this book to be practical, and not a collection of minutiae and rope facts. It had to be user friendly, practical enough to use as a text, and good enough to inspire a wide variety of rope-users to develop their talents and abilities. I think "practical" balanced the word "engineering" enough so my publisher could get some sleep at night. And finally, *Engineering Practical Rope Rescue Systems* is intended to emphasize that no single piece of the equation can perform high-angle rescue by itself. Gravity, the human element, and friction are the three primary tools. All those neat pieces of equipment are the secondary tools that must somehow be formed into a harmonious and fluid *system* that safely transports people to a more secure place. Titles, like books themselves, can scare you away, or they can grab you and not let you go. I hope this one is the latter.

There is a decided fire and urban/industrial rescue slant to this book, as that is my primary background. However, the principles discussed throughout this book have been specifically designed to examine and explain generic rope systems and equipment for maximum benefit by any rescue team that chooses to use the book as a reference source. Gravity does its thing everywhere people need rescue by rope, and rope cuts just the same on sharp edges in caves, on mountains, and in the petroleum plant. There are myriad different influences that make each rescue environment a little bit different. It is up to you and your team to interpret local environmental factors and apply them to the generic techniques discussed herein. Additionally, nothing in the book is intended to alter local protocol or to challenge the Authority Having Jurisdiction (AHJ) where you live and work. Perhaps some concepts are logical enough to have a positive influence on local protocol discussion and may become useful in upgrading or updating the services provided by your rescue team.

Wonderfully, rope science is a continuous source of debate, ranging from discussion to fisticuffs. No one, and I mean *no one,* agrees with *anyone* about how to do *anything* in rope rescue *all* the time. There are simply too many combinations of equipment, terrain, and technique. Writing a rope rescue book immediately and publicly shares one's own little view on the topic. If you are consumed with its study, as I am, it is a personal thing to which you automatically develop some defenses. Some people will attempt to destroy certain things they do not like about me or the book, as they do everything that requires a different perspective. That is our tradition. It is the dark part of our culture. The ceaseless questioning is good—it keeps the business alive and dynamic. But we are a personal and controversial bunch. Collectively, we wear each other out for the simplest differences of opinion. Under the guise of safety, we ridicule techniques and actions that are very similar to our own. And yet there is a common bond that we hate to admit. We are all rope dogs. Most of the fire and rescue community looks at us as if we are from another dimension. We are from Planet Rope. We clank, we are cocky, and we are an egotistical group of varmints who think we can defy basic laws of nature. But I would defend the goofiest rope dog in a second over his or her right to practice the trade, to learn, to get better, and to argue until the cows come home.

This book is intended to be a reference of rope ideas, information, technique, and history. It includes methods that I have grown to accept as safe, reliable, efficient, easy to teach and learn, and effective. There are many ways to accomplish these same tasks, and you may prefer some over the ones I have suggested. Please use this book as a source where you are comfortable. Do not use techniques in this book that make you uncomfortable. Likewise, do not expect future editions of this book to be just like this one. I learn new things every day, and some things may change my opinions and what I present in future revisions. This is a highly dynamic discipline. I think that is part of why rope is my passion, and I would wager it is a good part of why you are reading this today.

Intentionally, there are portions of this book that are a bit involved. There are other good rope rescue books on the market that address the basics. But I think there is a vacuum where books that look at details are concerned. Some of those details are as follows.

A detailed discussion of national standards, laws, and trends has been included to give the reader some depth and background on how we got to where we are today. I have painfully weeded out much of the bureaucratic mumbo-jumbo and carefully paraphrased the important things that you should probably be familiar with.

Personal Emergency Escape Rope Systems (PEERS) has been included because it is a timely subject. NFPA 1983 (see Chapter 1) now allows for the use of small but strong escape rope. I think we are seeing the birth of a new and lifesaving discipline that will probably be commonplace in the fire service in another five or ten years.

The section on the metallurgical properties of our hardware (Chapter 3) has not been included for the beginner. It has been included for those intermediate and advanced rope dogs who are thirsty for more. There is a feeling of confidence in understanding the material of the hardware on which we hang our lives.

The section on mechanical advantage (Chapter 8) does not *just* discuss pulleys, our primary mechanical advantage tool. It offers explanations of all types of mechanical advantage (inclined planes, levers, and wheels) to give the reader a perspective on simple machines in general and how they might be adapted to benefit us as technical rescuers.

Another controversial discussion takes place in Chapter 9 on building tensioned rope systems, or highlines, if you prefer. There is a school of thought (contrary to suggestions in NFPA 1006 proposed, Chapters 4–8) suggesting highlines are too complicated and dangerous for trained firefighters and rescuers to safely engineer, build, and operate. Some believe that highlines, the pinnacle of our skills, the accumulation of all equipment and technique, the maximum multiplication of forces, are just too dangerous and complicated for most people to understand, especially from a book. They must be kept secret, shrouded in their own complicity, for use only by rope rescue's elite. "IN THE NAME OF SAFETY, NEVER BUILD A HIGHLINE," they shout from the mountain tops. I cannot disagree more. The highline information is included for those who care to explore it, and who are excited by the endless possibilities.

An exciting and timely discussion revolves around the use of load releasing hitches (LRH). The British Columbia version of LRH is a very popular, highly marketed, and reliable tool. After years of using numerous load releasing techniques, I have settled on the versatility and utility of the Hokie Hitch LRH. Everything has its pros and cons. The BC-LRH is a slightly better shock absorber and has greater reach, but it can allow the rope to be released completely if it is improperly handled and can, at least, cause shock to occur to other portions of a system. The Hokie Hitch (HH-LRH) has an automatic catch feature that prevents a complete, potentially hazardous release. The discussion is analogous to favoring a Ford over a Chevrolet. They are both great cars—you just grow accustomed to one or the other. My experience has shown that students and real-time rescuers can easily be taught to operate the HH-LRH with a margin of safety that is uniquely confidence building.

Ultimately, the goal of this book is to help make rope rescue a realistic, valuable, and enjoyable experience. Rope rescue is most of all serious business, but it is about the most fun business you can get into. Enjoy, and be eternally safe.

Michael G. Brown
e-mail: ropebug©aol.com

ACKNOWLEDGMENTS

I am sitting in an insanely nervous airport full of angry would-be passengers on a cold January Saturday afternoon in Charlotte, North Carolina. Wintry weather across the Northeast has had a ripple effect on the South and, due to snow in New York, I am inhabiting, for a while at least, a place meant only for short visits. By Monday I should be in Sarasota, lapping up the warm subtropical sun, polishing my farmer's tan, and enjoying my passion—teaching technical rescue, rope rescue to fire and rescue people.

Sitting in a noisy airport full of people gives me a chance to reflect on and be thankful for the many people who have influenced me and, therefore, influenced the outcome of this book. After twenty-three years of learning, performing, and teaching fire and rescue skills, it seems that I am probably indebted to at least as many people as are in this beehive of an airport.

It is impossible to write a book without hundreds of influences. It is equally impossible to properly acknowledge every person to whom I owe a debt of gratitude for helping me in some way to compile this collection of information. To those I have neglected to thank personally, I apologize. Your contributions to this work are no less important, and I am indeed grateful.

I am grateful to my family; my mother, Margaret; my dad, J. Gordon (deceased); sisters Karen and Sherry; children Katie and Garrett; and Ted, John, Paul, and Rett-Ro (the yellow dog); all of whom have had some piece of their lives affected by this work. Their support, both directly and indirectly, has given me the strength to carry this book to conclusion. I am, however, most grateful to my wife, Lonni for demonstrating the incredible will to live and the indomitable power of the human spirit. She was sick throughout most of the time I spent on the computer piecing together my thoughts. I am thankful to God for making her better as this book goes to press. More eloquent methods of describing her strength in relation to my meager efforts at writing this book are beyond my abilities as a writer. Next to raising my children, the singular most important, large, and involved piece of business I've ever attempted is writing this book. But it is the smallest effort measurable by the most sensitive instrument compared to the war she has fought and won with cancer. She has placed me in a kind of perpetual reality check to which I am forever indebted. Lonni, I will love you forever.

Regarding any perceived abilities that I may have in understanding and communicating the art of rope rescue, I owe equally to my great friends and mentors, Dean Paderick and Kyle Isenhart. Dean had the foresight to become the catalyst behind one of the first and finest special operations teaching forces in the country, Virginia's Department of Fire Programs, Heavy and Tactical Rescue Team. Dean recognized the need for state-of-the-art technical rescue training for the Commonwealth. He also saw the need for an avenue of professional energy direction for dozens of smart, energetic, and talented firefighters and rescuers with no place to go and no course to teach. There is no way the Commonwealth, Virginia's firefighters, and countless other rescuers, including me, can ever come close to paying Dean back for his patience and his service.

Equally, I have no hope of repaying Kyle Isenhart for his incalculable technical expertise and infinite patience in helping me to hammer out the myriad details involved with writing this book. Thanks to Marianne Lorraine Isenhart (The Inspirator) for dogging Kyle to finish the foreword for the book on time (and more). Whenever I bogged down, lost my focus, or did not have a clue about what I was trying to say, which was often, I called Kyle on the *Rescue Systems Incorporated* toll-free line and, regardless of the topic, he made it become clear almost instantly. Kyle is completely without peer in understanding the big picture about using a strong, flexible linear medium and its associated tools to affect human rescue. A polymer chemist by trade, an educated engineer by design, a guano-stomping speleotech since birth, and the absent-minded professor in appearance, Kyle has the unique ability to see the practical application of nearly everything that is rope. While other knowledgeable people, in the name of safety, treadmill endlessly in the minutiae of some world-saving technique or some life-inducing piece of equipment, Kyle has championed the cause of rational thinking. He has a practical approach based on understanding the physics and interdynamics of interrelated components of systems. "All of this was around in some form or the other long before we got here," one could hear him say. "Build the system, resist overengineering the obvious, trust your equipment, and lose the little stuff," meaning the endless tiny technical details that some people insist are essential to build a rope system. His selfless style has been the contribution that has kept rope rescue in the realm of the everyday person—an art to be enjoyed by everyone who cares enough to learn and practice endlessly. It's not some mysterious, amorphous, ambiguous techno-babble to be used only by superhuman rescuers with a special comm-line to the rescue gods. Much of the thinking that went into this book will be familiar to those of you fortunate enough to have spent some time with Kyle. When it does, you will smile and think, "Man, that makes so much sense." For that, I say thanks to Kyle Isenhart.

I am thankful also to my friends and business partners in *Spec. Rescue International*—Dean Paderick (again), Chase Sargent,

Mike Brown and his son, Garrett.

Buddy Martinette, Jon Rigolo, Andy Speier, and Cindy Hendrick, the office boss. Their influence has been always positive, whether it was to be available for discussion, to go jump on a rope system somewhere, to review a particular paragraph or technique, or just to be infinitely patient with my silly self.

To the Tidewater Regional Technical Rescue Team, Virginia Beach Fire Department Technical Rescue Team, FEMA's USAR VA TF-2 (sometimes one and the same), and too many others to name who sometime worked rescue calls by my side, wondering if we were going to live through it all, and at other times were quietly patient with my quirky manners, I am deeply indebted.

I owe much thanks to my good friend Martin Grube who was responsible for a majority of the photographs in this book. Fortunately for me, Martin has a keen eye and always made himself available in a flash whenever we needed to take some more shots.

Thanks go to Virginia's Department of Fire Programs' Heavy and Tactical Rescue Team, Joe Thomas, Dean Paderick (again), Charlie King, Mark Light, Harold Chrimes, Charlie Eisele, Chase Sargent (again), Buddy Martinette (again), Terry Kisner, Willie Rice, David Layman, Steve Parrott, Alan Austin, Jerry and Pam Woodson, Perry Ehle, Jimbo Meagher, Joel Kanasky, John Burruss, Duane Krohn, Keene Black, Benji Barksdale, Dan McMaster, Phil Perry, Joey Stump, Mike Lewis, Vince Holt, Byron Andrews, Hoyle Green, Tony Carroll, Steve Cover, Darryl Klopp, Chilhowie Chuck Swecker, Dennis Keane, Terry McAndrews, Scott Prentice, Leon Dextrauder, Warren Whitley, Gary Rock, Jon Rigolo (again), Mackie Tabor, Paul Gleaton, Tom Bahr, Mark Cumashot, Ralph Trowbridge, Mike Woods, John Shahan, Robert Sayles, and Gene Wall.

My firefighting shift at VBFD Co. 16-C, Battalion Chief G. W. Painter, firefighters Dean DaSilva, Mike McAndrews, Terry Payne, Tina Ryner, Skip Frey, and David Foxwell have each been inspirational in their own way. To Tidewater Regional Fire Academy Instructors of yesterday and today Kurt Southall, Mark Piland, Jim Milligan, Bob Anderson, Martin Grube, Larry McAndrews, "Hacksaw" Sam Reynolds, Don Brown, Kurt Nellis, Glenn Dunn (who taught me how to untie a bowline), and "Bulldog" Bennett, whose cumulative contributions are beyond measure, I say thank you.

If you have spent any time studying rope rescue, and unless your head has been completely buried in the sand, you have had the opportunity to learn something from Steve Hudson of *Pidgeon Mountain Industries (PMI),* La Fayette, Georgia. Steve taught me the true definition of a descent control device (a table) and has probably forgotten more about ropes and cordage and using them than I will ever know. Steve was a powerful influence for the book because he told me I would never write a book and, maybe, if it were not for his support, I would not have. I am grateful. Thanks also go to Arnor Larsen, Rock Thompson, Richie Wright, Tom Vines, Bruce Smith, Tim Gallegher, Hoyt Harvel, Steve Fleming, Alim Shariff, Troy Thornton, Ed Waggoner, Tommy Carr, and Mark Hill; and *Seattle Manufacturing Company's* Margaret Brown, Garin Wallace, and Matt Gilbert; *Sterling Rope's* Carolyn Brodsky, Gene LagoMarsino, John Juraschek, Paul Niland; and *SkedCo's* Bud Calkin. I am deeply indebted also to friends and business associates Rich Hansen, Bob Tanenholz, J.B. Anderson, Bill Manning, Diane Feldman, Henry Dineen, Bill Thorn, Steve Jellie, Don Rodgers, John Elwood, Verne Riggall, Jimmy Bridgeman (Bigum), Ron Zawlocki, and Mike Zito; Wellington Puritan's Terry McMichael; Blue Water's Dick and Glenn Newell, John Weinel, Jerry Smith, Reed Thorne, John McNulty, Bill Taggant, Jim Frank, Scott Milsap, Brad Womack and Micki Liski; John Yates, Tetter Robinson, Jerry Olinger, Herman Heikenin, Doc. David Cash, Steve Pyle, Paul Willie, Harry Diezel, Tom Leighton, Rick Johnson, Steve Overton, Pete Peterson, Greg Ettel, Bill Dadmun, Dog Ferguson, Bill Densmore, and every student I've ever had. Thanks for helping me get a little better handle on the wonderful world of rope rescue.

The author and Delmar Publishers also gratefully acknowledge the reviewers who participated in this project. Their comments, suggestions, and insight were invaluable. Our thanks to:

Thomas Welle
Red Rocks Community College
Lakewood, CO

Richard P. Kasko
Lamar University
Beaumont, TX

Joseph De Francisco
Madison County Fire & Rescue Service
Wampsville, NY

Mike Sudol
Chittenango Fire Department
Chittenango, NY

Robert Cobb
Jersey City Fire Department
Jersey City, NJ

Steve Fleming
NASAR
Fort Collins, CO

Jerry Slaughter, NREMT-P
HEMSI High Angle Rough Terrain (HART) Team
Huntsville Cave Rescue Unit, Inc.
Huntsville, AL

CONTENTS

CHAPTER

1 Introduction to Rope Rescue

OBJECTIVES

Upon completion of this chapter, you should be able to:

- explain the three most important tools and how they interact to affect safe high level rescues.
- discuss why NFPA 1983 was developed and how it currently affects rope rescue.
- define the leveling process of NFPA 1670 and how it may be applied to your fire or rescue organization.
- list the requirements for a rope rescue provider from NFPA 1006.
- compare the measurement methods of pounds force and kilonewtons.

INTRODUCTION

Gravity

▶ **gravity**

The gravitational attraction of the mass of the Earth, the moon, or a planet of bodies at or near its surface.

Where would we be without gravity? It is the very foundation of everything physical. It is a force that is simple and at the same time, extraordinarily complex. It is basic and easily understood, but it is also advanced and perplexing—a paradox of sorts. On the simple side, *Merriam-Webster's Collegiate Dictionary* defines **gravity** as a quality of having weight, or the gravitational attraction of the mass of the Earth, the moon, or a planet of bodies at or near its surface. More specifically, we understand gravity to be the force between two objects of mass, where mass is commonly measured by the amount of weight or matter contained in an object. Gravity is also highly relative and diverse. A plumb bob will lean toward a nearby mountain, because the weight of the mountain is much greater than the weight of the plumb bob and string (Figure 1–1). However, the mountain is but a speck compared with the mass of the Earth, which is so great that it holds the moon, a kind of huge plumb bob, locked into a gravitational orbit because of its relative proximity. Imagine, too, the collapse of a star far out in space. When its nuclear fuel tank, hydrogen, runs dry, fusion is no longer generated to counter the effects of the tremendous mass of the star. The star simply collapses in on itself, the victim of its own gravity. In fact, it is surmised that the star collapses so completely as to occupy about the space of a pin point but with its original mass and gravitational pull. It becomes a black hole, sucking in everything within its massive gravitational pull and compounding its mass. Even light cannot escape.

Back on Earth, gravity seems simpler, almost friendly. Soil stays on the ground, water stays relatively put. Your food stays on the plate, your dog enjoys chasing the ball, which *always* comes back down. Your house, your fire station, that bridge all rely on gravity and the interrelational friction of all their parts to stay together. A construction engineer once said to me, "It's all one big balancing act."

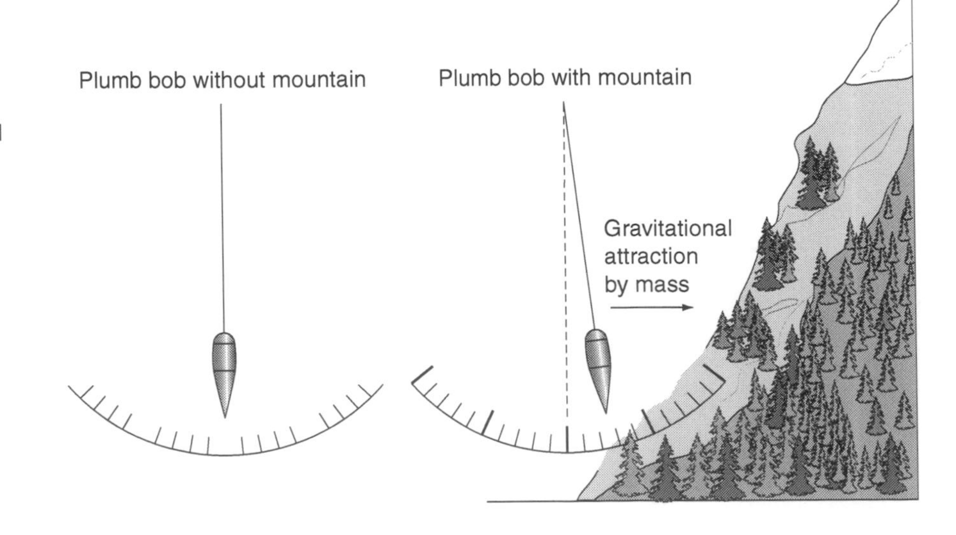

FIGURE 1–1

Mass is attractive. The greater the mass of an object, the greater its attraction on surrounding objects. The Earth's mass attracts us toward its center of gravity.

NOTE: Gravity is our finest and most important tool. In rope rescue it is the engine in our transportation system.

In fact, we are so completely reliant on gravity that, without it, we would immediately cease to live. The consumption of life-sustaining water, air, and food would be the very least of our problems. Our atmosphere would rush into space in an instant, as would our water and every edible thing on the planet that was not bolted down. Within minutes, the earth itself would start to disintegrate.

Human Factors

Our next most important tools are the human factors. These are our abilities as humans to understand the interactions and simple physics of the natural and synthetic objects in our midst, some of which are obstacles and some tools, *and* the ingenuity required to problem solve and accomplish our goals. As humans, we also develop life-preserving anxieties, that is, feelings of nervousness or apprehension when we are around something that makes us uncomfortable or that could harm us. These emotions keep us from venturing too close to the edge or too far into the cave (Figure 1–2), and even keep us from jay-walking on a busy city street. Falling is one of our first *learned* anxieties. It hurts, one thinks, to fall. The fall can be kind of fun, but whatever we run into on the way down, whatever is between us and the center of the Earth, had better be plenty soft or it's going to hurt when we meet it. As rescuers, our job is to save people from unusual predicaments. We do not always have the option of *avoiding* heights from which we might fall. In fact, we must also learn to work with a certain level of fear, and appreciate the accomplishment, but we must control our anxieties and not allow *them* to control *us.*

Other human factors are the proficiency skills—the efficiency with which we work and how we interact with members of our

FIGURE 1–2
The fear of falling is a natural, self-preserving instinct.

team. Specifically, our individual proficiencies are a measure of how skillful and knowledgeable we become at our jobs. For example, how much more rope does it take to tie a figure eight on a bight than to tie a bowline? How quickly can you make a münter hitch on a carabiner and bale out of a burning building? When you hand a carabiner to your partner, is the gate open, closed, or locked? Is your load sharing bridle dangerously long? When does a 3/8-inch Hilti bolt with 8-mm SMC hanger in solid granite pull out when pulled in tension? In shear? Is that a Class I or a Class II lever? Which OSHA standard most closely affects confined space entry? In addition, how do we interact with the other members of our team? Do we understand and effectively merge the individual strengths and weaknesses of all our team members? How do our egos interact? Who are the official and unofficial leaders of the group? When do we have too many people involved in an operation? When are we working so fast that we approach our point of inefficiency? All of these factors are constantly evolving and changing. Mature rescue teams examine their growing pains and attempt to understand and channel the human aspects toward the good of the group and, ultimately, the good of those in peril.

NOTE: The rescue tool box now contains two items. First, our finest tool, gravity, is so helpful that it transports itself to the scene and offers itself to us as our primary engine. Second, we bring with us all of the human factors.

Friction

The next tool in order of importance is friction. Friction is something we cause (intentionally or unintentionally) on the scene. The rope engineer's "balancing act" requires something that temporarily equalizes, controls, and counteracts our primary tool, gravity, in much the same way that the boom, cables, and winch brake on a crane control a load that is being lowered.

► **friction**
The resistance to relative motion between two bodies in contact.

Webster's defines **friction** as the resistance to relative motion between two bodies in contact. Friction, like gravity, can be our best friend or our worst enemy. Good friction balances the effects of gravity as we slide a fixed rope toward the ground. The good friction caused by the rope's contact with an anchored closed end rack is *exactly* equal to the downward pull of gravity on a rescue load. Good friction is the heat that is absorbed into the metal bars of the rack and then radiated and dissipated into the atmosphere. In concert with gravity, it causes our feet to bind with the ground so we can pull on a hauling system. Bad friction causes knots to disintegrate as they are tensioned to their point of failure. It can cut or burn its way through a stationary piece of synthetic fiber being rubbed by a moving piece of synthetic fiber, such as nylon on nylon. Friction causes pulleys to be inefficient, and it heats and wears out bearings. It is the difference between theoretical and actual mechanical advantage. Emotional friction is a side effect of interpersonal relationships, such as those that are developed within the framework of a human team. This friction, like anxieties, can be studied and eventually controlled. Teams can learn to work with a level of friction, can appreciate the accomplishment, and can even enhance team efficiency when it is channeled properly.

NOTE: Our three most important tools—gravity, the human factors, and friction—must be in harmony in order to routinely effect safe, high level rescues.

Cooperation

How does all of this correlate to performing the art of rope rescue? To accomplish rope rescue tasks, we must have a cooperative relationship with all of the tools. The ancillary tools—rope, descent control devices, carabiners, prusiks, pulleys, and the like—are simply the connections for the primary tools.

Understanding these three major tools, gravity, the human element, and friction (Figure 1–3), will help make all the ancillary tools and techniques seem simple, thereby increasing our ultimate survival profile while using ropes as a transportation system to rescue people in peril (Figure 1–4).

PUTTING ROPE RESCUE IN ITS PLACE

Rope rescue systems give us a reliable and safe means of transportation where there is simply no other choice. They provide a

FIGURE 

The safety and efficiency of a rope rescue team increase as the three major tools—gravity, friction, and the human element—are harmoniously combined.

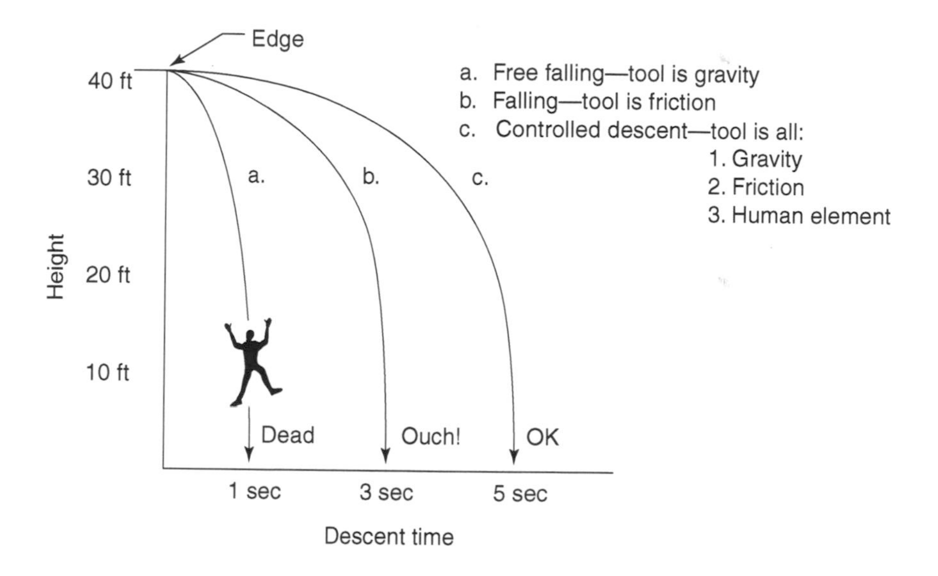

FIGURE 

Photograph of man rappeling. (*Courtesy Craig Aberbach*)

NOTE: Rope rescue should be considered only after all other means of accomplishing the task have been rejected. For example, consider the possibility of using interior stairwells, fire escapes, boats, alternate exits or ladder trucks, or of waiting for one of those options to become available.

synthetic highway with remarkably safe travel parameters. Thanks to advances in chemistry and metallurgy, the understanding of interactive physics, and a devotion to adequate training regimens, rope rescue concepts have become a safe and valuable option for emergency-scene incident managers.

Rope rescue, like all technical rescue disciplines, is a nontraditional activity. Unlike working a cardiac arrest or putting out a house fire, it is not performed daily (see Hazard Curve Paradox in Chapter 2 of this text). In fact, many incident managers are hesitant to use rope rescue teams because they are uninformed as to the relative safety and increased medical stability rope rescue can provide for injured persons over traditional methods of transportation. Some managers recall the horrors of manila rope and refuse even to consider developing alternative transportation systems for their injured citizens. In addition, many simply hope that the once-in-a-lifetime rope rescue will not occur on their watch.

STANDARDS

NFPA 1983

Standard on Fire Service Life Safety Rope and System Components

One of the best tools a progressive emergency services manager, regardless of orientation—urban or wilderness, fire or non-fire—can use to safely initiate a rope or technical rescue program is to align with nationally accepted consensus standards like the National Fire Protection Association (NFPA) **NFPA 1983,** *Standard on Fire Service Life Safety Rope and System Components,* current edition.

SARA Title III opened the floodgate to major changes in the way fire and rescue services performed their duties. Next came Occupational Safety and Health Administration (OSHA) laws dealing with confined space rescue activities (29CFR 1910.146) and respiratory protection laws (29 CFR 1910.134). In the mid-1990s the fire service began to look at possible problems in complying with OSHA's **two in two out rule.** Primarily, staffing levels did not always allow for a fire company to wait around for assistance before entering a burning structure. Essentially, the fire service had to rework the way they responded to fires by either hiring more people (very expensive) or enacting policy that mandated firefighters stand

▶ **two in two out rule**

OSHA's rule that if two firefighters enter a building with an IDLH environment, there must be two firefighters suited up and waiting outside for possible rescue.

Regulation of Fire Service Emergency Procedures

Prior to 1986, there was almost no regulation or control over the way the fire service performed emergency situation mitigation. Excluding some local standard operating procedures, medical protocol, and ubiquitous motor vehicle laws, there just were not many enforcement capabilities nationwide. Fire departments enjoyed the, "Hey it was an emergency, we did the best we could," mentality.

On October 17, 1986, Congress passed into law *The Superfund Amendment and Reauthorization Act* (SARA) of which Title III made local jurisdictions responsible for the monitoring of chemical emergencies and any subsequent cleanup. For the first time, fire departments had to perform in a certain way or risk fines or jail time.

NOTE: And while the NFPA does not carry the weight of law, in most courts in the country, a national consensus standard is easily substituted in the absence of an applicable law.

around waiting for help before they can enter a structure, both alternatives not looked upon highly by the public. The rule states that if two firefighters enter a building with an Immediately Dangerous to Life and Health (IDLH) environment, there must be two firefighters suited up and waiting outside for possible rescue. Backup teams are required for immediate rescue of any employees working in hazardous atmospheres.

While OSHA and the National Institute of Safety and Health (NIOSH) started leveraging their way into the operation of fire and rescue services, other organizations also had an impact on them. The National Fire Protection Association created the only national consensus standards that directly influenced the fire and rescue services, which, in some respects, was the best thing that could have happened to budding rope rescue programs. As a national consensus organization, the NFPA follows guidelines that allow for input by the individuals, agencies, and companies that are most likely to be influenced by a particular standard. More than 100 years old, the NFPA organizes, coordinates, and publishes the results of recommendations by currently 225 committees covering everything from the manufacture of fire boots and breathing apparatus to fire trucks, sprinklers, and rope rescue gear—almost anything fire-related in the country. Ultimately, an NFPA standard can bear the weight of law in lieu of actual legal precedence. For the fire service, this means that if OSHA, NIOSH, or SARA laws, or a local ordinance does not cover what your department is doing, there is probably an NFPA standard that does. Prior to SARA Title III, there simply were no nationally accepted programs or guidance tools, policy, standards or laws, for managers. No guidance, no program was the reality in many places.

A horrible fire-scene accident in New York City was the impetus for the standardization of rope rescue groups across the country.

CASE STUDY

New York City Fire Department Fatalities: The Catalyst (excerpts taken from N.Y.C. Department of Investigation, report #1034/80D, dated December 1980).

At 1830 hours on the evening of 27 June 1980, a gas explosion ripped through Margie Windell's kitchen in apartment number sixty-one at 512 West 151st Street in Manhattan. The resulting fire spread quickly to adjacent apartments on the sixth floor and into apartment number seventy-one directly above the Windell apartment. The building, constructed in the early 1900s, was a brick tenement containing four apartments on each of the eight floors. Shaped like a rectangle, the building had a central air shaft with windows from the apartments, some of which were equipped with nonconnecting fire balconies.

Among the first units to arrive was Ladder Company 28 with an officer and five firefighters, including firefighter Gerard Frisby. Ladder

28 was directed to search through the heavy smoke and heat on the seventh floor for possible patients. At about 1915 hours, firefighter Robert Sexton noticed another firefighter slumped in the window of apartment number seventy-one, which was heavily involved in fire. It was firefighter Frisby from Ladder Company 28. Sexton was training a hose stream into apartment number seventy-one from a fire balcony across the seventh floor air shaft. He could see that Frisby was semi-conscious, and that smoke and flames were beginning to roll across the ceiling, over Frisby's head, and out the window. Firefighters working on the opposite side of the air shaft were directing a hose stream over Frisby's head attempting to slow the advancing fire. Straddling the fire platform and Frisby's window ledge, Sexton grabbed Frisby by his bunker coat and shook him, attempting to establish his level of consciousness. Frisby looked up, his eyes rolled backwards, and he was unable to respond.

The firefighters across the air shaft alerted the firefighters operating on the roof above the fire to the situation. Firefighter William Murphy of Rescue Company 3 was working the roof and moved into position just to the right of the window that Sexton and Frisby were occupying. Other firefighters lowered Murphy over the edge of the roof attached to his 3/8-inch nylon personal safety rope. When Murphy reached the window, Sexton moved back onto the nearby fire balcony to avoid the extreme fire conditions. Murphy stood with both feet on the ledge in front of Frisby's window, took both of Frisby's arms and shook him. Frisby did not respond. Murphy was unable to lift Frisby but stayed with him until further help arrived.

Members located a 1/2-inch rescue rope and hand lowered firefighter Lawrence Fitzpatrick, tied to his ladder belt, over the edge of the parapet. The parapet was lined with smooth, glazed terracotta coping material, presenting a relatively good edge for the lowering operation. Reaching the window, Fitzpatrick, who was known to be the strongest member of Rescue 3, climbed to his left about four feet and reached Frisby. Murphy, seeing the stronger Fitzpatrick had arrived on the rescue rope, swung out of the way. Fitzpatrick leaned into the window, grabbed Frisby in a bear hug and lifted him over the window gate, and then swung free from the ledge. A weight of approximately 440 pounds, comprised of both firefighters and their equipment, was now sharply applied to the rope.

Firefighter Foley, across the air shaft on the roof and Ms. Birdie Hall, a tenant in a seventh floor apartment who had returned despite the building's evacuation, watched the rescue attempt. Although smoke was coming from the window above Frisby, the stricken firefighter and Fitzpatrick were visible. Lighting conditions were relatively good. Sunset was to occur at 2031 hours and the sky was clear. Both Foley and Ms. Hall saw Fitzpatrick take Frisby and swing away from the window in an arc to the right. Suddenly, both men plummeted downwards and out of sight, into the air shaft. Ms. Hall heard a thump but did not look down. Firefighters began shouting that the men had fallen. Firefighters Foley and Schneible of Rescue 3 were the first to reach Frisby and Fitzpatrick. Foley detected what he perceived to be slight signs of life in Fitzpatrick. Both were given

CPR until reaching Columbia Presbyterian Hospital. Neither man was conscious. At the hospital, Frisby was pronounced dead at 2015 hours and Fitzpatrick at 2025 hours.

The lengthy investigation conducted by a special division of then-Mayor Edward Koch's administration concluded in part: "It is our conclusion that two firemen fell to their deaths because the 1/2-inch nylon rope they were using was unable to support their combined weight due to major abrasion and heavy use during the two years and nine months it had been in service by the fire department. We found no evidence to support the hypothesis that the rope was cut by a sharp edge of a piece of broken (edge) coping on the roof parapet." The report went on to highlight managerial problems within the department, primarily in keeping abreast of modern rope technology, and deficiencies in tracking, maintenance, inspection, and recording the use of life safety rope.

NFPA 1983

NFPA 1983

Standard on Fire Service Life Safety Rope and System Components

The New York City fatalities were the catalyst that started NFPA 1983, *Standard on Fire Service Life Safety Rope and System Components.* Updated every five years (like all NFPA standards) NFPA 1983 was first issued on 6 June 1985. It had the effect in many departments of creating a "liability umbrella" for managers wanting technical rescue teams. In essence, managers could say, "You assure that all of your equipment meets NFPA 1983, and you follow the general guidelines and philosophy of safety represented by a normal and prudent person performing the same activities, and you can have a program." Rope rescue programs started sprouting like daisies across the country due in part to NFPA 1983.

Another benefit of NFPA 1983 was that, for the first time, a nationally recognized consensus-standard-making organization was creating some rope-rescue-oriented definitions. Prior to NFPA 1983, the main rope-related organizations in the United States agreed on very little regarding standardizing rope rescue definitions. The primary organizations influencing rope rescue in the 1970s and early 1980s were the Mountain Rescue Association (MRA); the National Speleological Society Vertical Section; National Association of Search and Rescue (NASAR); and to a lesser degree the fire service; rock climbing schools; the European ropers, including the Union of International Alpine Associations (UIAA); and some special warfare military organizations. While there was much good information available, no one could agree on even the most basic definitions. It was hard for knowledgeable persons from different fields to even have a conversation (Figure 1–5) because time was wasted trying to agree on simple, basic definitions. For example, to a rock climber a one-person load might be 200 pounds, to a firefighter it might be 300 pounds, to a rescue guru in France it might be 200 kilograms, and to a caver with loads of lighting and equipment it might be 350 pounds. There simply was no nationally

FIGURE 1–5

NFPA standard 1983 defined some basic rope rescue definitions to help unify the different user groups.

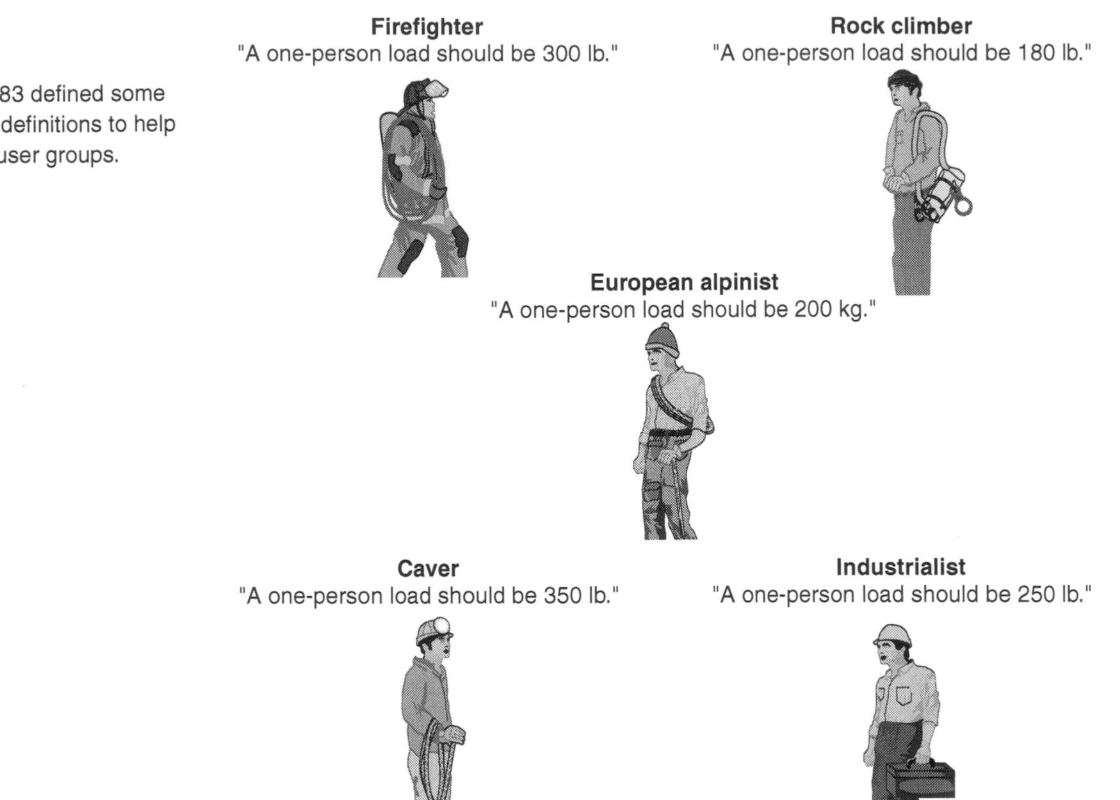

► **ropebotics**

An esoteric term sometimes used to describe rope rescue instructors that are like robots in their delivery and rope use philosophy, highlighted by their inflexible view toward other types of equipment and different techniques.

NOTE: Three hundred pounds was adopted by the original NFPA 1983 committee as a one-person ("one man" at that time) load. This was thought to best represent the average firefighter with all his gear and left some margin for error. A two-person load, therefore, was 600 pounds and a three-person load 900 pounds.

NFPA 1983

Standard on Fire Service Life Safety Rope and System Components

accepted authority to make definitions. NFPA 1983 changed all that when the committee put down hard definitions that are still relevant today to the fire service. These definitions have become accepted by most of the other organizations due to the continuing vacuum and almost constant disagreement at the micro-rope (**ropebotics**) management levels. The NFPA is the only nationally accepted consensus standard, so most non-fire, rope-oriented rescue organizations feel obligated to some degree to use NFPA 1983 (and probably NFPA 1670 and NFPA 1006 in the future) as a reference.

The NFPA 1983 committee went on to boldly answer the biggest question at the time: What should the proper safety margin be for life safety rope? In other words, how much strain (tension) should a rope and other equipment be expected to take and still have some strength left over for a reasonable safety margin, before breaking? This is known as breaking strength. Many laws and standards, the American National Safety Institute (ANSI), NIOSH, and many industrial concerns used 5:1 as an acceptable margin of safety. Specifically, if the piece of equipment was expected to hold 500 pounds during normal operation, it could not break when tested until at least 2,500 pounds force was applied. Some mountain and cave rescue groups expanded this safety margin to 7:1 or even 10:1. For example, if the rescue load was expected to be about 500 pounds, at 10:1 a rope would only have to have a 5,000 pound breaking strength. Realistically, most of the cavers were using 11.1-mm or 7/16-inch (Goldline and Goldline II™) ropes with breaking strengths

FIGURE 

NFPA 1983		Safety margin × 15
One-person load = 300 lb	×	15 = 4,500 lb = 9 mm, or 3/8 in.
Two-person load = 600 lb	×	15 = 9,000 lb = 12.74 mm, or 1/2 in.

NFPA 1983 rattled the nation by defining the safe working load on life safety ropes as 15 to 1. A two-person rope would need to have a minimum breaking strength of at least 9,000 lbf (40kN).

of about 5,500 pounds force, providing a little more than a 10:1 safety margin. The NFPA 1983 committee, still reeling from the effects of the New York City firefighter fatalities, pushed a 15:1 safety margin through the standards-making process. The cavers, mountaineers, and industrial safety people were stunned. They insisted that no one would ever use a 7,500-pound rope (15 × 500 pounds), because it would be too heavy, too expensive, and too hard to tie (a bad or hard hand), and none of the equipment (at the time) would fit. The ultimate heresy was the announcement of the *300-pound,* one-man load. Rescue work always anticipated at least two people on the rope (Figure 1–6), which meant 600 pounds × 15 (600 × 15) or a 9,000-pound force (lbf) breaking strength rope . . . the dreaded half-incher.

The most controversial issue in the standard, and another reflection on the New York City firefighter fatalities, was the recommendation that a rope could only be used once in an emergency, and then it must be destroyed. In fact, this was confusing because NFPA 1983, was supposed to consist of "minimum manufacturers' recommendations" and not *use* standards. To most practitioners of rope rescue, this was expensive as well as ludicrous. Even the rope manufacturers, who stood to make a great deal of money selling new rope every time one was used, considered it an insult to their products. The "use once and destroy" statement was drawing credibility out of NFPA 1983.

NFPA 1983

Standard on Fire Service Life Safety Rope and System Components

A nearly identical second edition (NFPA 1983–90) was issued on 20 July 1990. Still maintaining a good foundation for fire and rescue teams, the 1990 edition committee maintained the "use once and destroy" rope mentality. That job and others were relegated to the third edition NFPA 1983–95 committee.

Summary of NFPA 1983–95

The 1995 edition committee made their edition more practical for rescuers and equipment manufacturers, and they placed a particular emphasis on emergency escape issues. Carrying an effective issue date of 11 August 1995, NFPA 1983–95 is designed, like its predecessors, specifically *not* to be a *use* standard, like NFPA 1500 *Fire Service Life Safety.* NFPA 1983 is a manufacturer's minimum requirements document to allow equipment to bear the NFPA symbol.

NFPA 1500

Fire Service Life Safety

The early 1995 edition committee members convinced the NFPA Standards Council to make a change in the 1990 edition through a Transient Interim Amendment (TIA). This tactic effectively used NFPA 1500 super document, *which is a use standard,* as a kind of legislative springboard. In May 1993, NFPA 1500 effectively overruled the eight-year-old "use once and destroy" portions of NFPA 1983 (NFPA 1500-93, Chapter 5, para. 8). A life safety rope could be used in multiple emergency situations if:

a. Rope has not been visually damaged
b. Rope has not been exposed to heat, direct flame impingement, or abrasion
c. Rope has not been subjected to any impact load
d. Rope has not been exposed to liquids, solids, gases, mists, or vapors of any chemical or other material that can deteriorate rope
e. Rope passes inspection when inspected by a *qualified person* following the manufacturer's inspection procedures before and after use.

In addition to the "use once and destroy" changes, the 1995 edition mandated third-party certification (see Chapter 2—Certification). This meant that a bona-fide testing organization had to certify that products met the specifications outlined in the standard before they could carry the NFPA label. To many rescuers, team fire departments, instructional organizations, and fire service managers, the NFPA symbol on equipment is mandatory. It implies an attempt to *perform* rescues in a safe and prudent manner, which is just one part of the liability umbrella. NFPA 1983 also encourages many behaviors or uses, via its marking system. Chapter 3, for example, details what manufacturers must tell the users to do on the labels that come attached to the equipment. Because of the tremendous amount of information included and the concise definitions, NFPA 1983–95 has become a major reference source for rescue teams across the country, regardless of their orientation. As usual NFPA 1983–95, contrary to what many instructors teach, *does not* tell rescuers how to engineer their systems, where to place what equipment, or what safety margins should be built into what systems.

NFPA 1983

Standard on Fire Service Life Safety Rope and System Components

The standard contains seven chapters and an informational appendix:

Chapter 1—Administration
Chapter 2—Certification
Chapter 3—Product Labeling and Information
Chapter 4—Design and Construction Requirements
Chapter 5—Performance Requirements
Chapter 6—Testing Requirements
Chapter 7—Referenced Publications
Appendix
Index

Chapter 1—Administration. This chapter states that the document is intended for use primarily by the fire service; however, it lends itself for use by "similar emergency service organizations" (section 1–1.1).

The standard also states that the document does not apply to special rescue operations like water rescue, cave rescue, mountain rescue, industrial fall protection, or even recreational uses. The purpose of this language is to allow rescue teams to use the information in the document as reference material, which many non-fire rescue organizations have done while performing legitimate and safe rescues, without being bound to follow the standards word for word.

► **fall factor**

The severity of a fall that is expressed as a ratio calculated by dividing the distance fallen by the length of the rope used to arrest the fall.

The standard also addresses, for the first time, the issue of **fall factors,** which is the severity of a fall that is expressed as a ratio calculated by dividing the distance fallen by the length of the rope used to arrest the fall (Figure 1–7). Whenever possible, it is preferable to perform rope rescues from above or at least anchored above the target, including "operations where personnel are required to work above anchor points or in operations where the fall factor might exceed 0.25" (section 1–1.3). This paragraph was intended to prevent someone from using equipment detailed in NFPA 1983 to attempt rescue by climbing above an anchor, such as in lead climbing on a rock face. Such activity requires a very stretchy rope to absorb some of the energy of a falling climber, whereas the equipment recommended by NFPA 1983 is low stretch, bigger, and somewhat heavier than climbing and recreational gear.

NFPA 1983

Standard on Fire Service Life Safety Rope and System Components

Chapter 1 also includes an extensive list of definitions, many of which will be referenced throughout this book.

Chapter 2—Certification. This chapter discusses in detail the requirements of equipment manufacturers to have their products tested by a third party. The first party is the NFPA, the second party is the manufacturer, and the third party is the testing organization. As with all NFPA standards that deal with manufacturers of equipment, a third party, such as Underwriters Laboratories (UL), Factory Mutual (FM), or Safety Equipment Institute (SEI) must be hired to "certify compliance with the standard." This requirement is new to the 1995 edition and was mandated by the Standards Council of the NFPA, not the committee. The equipment manufacturers objected

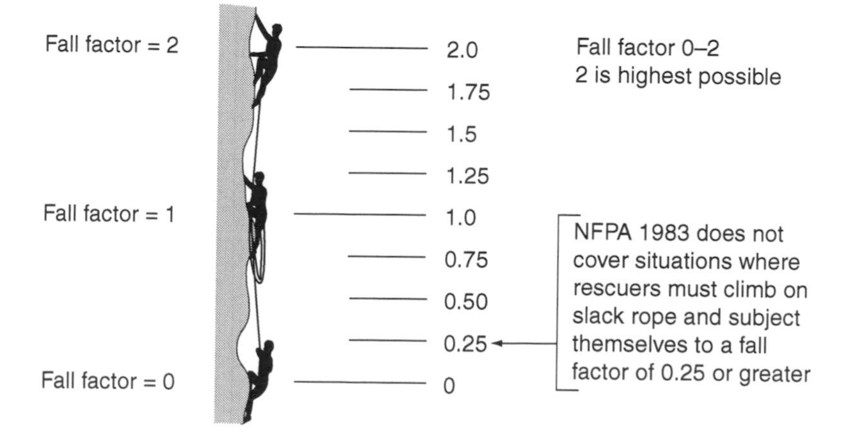

FIGURE 1–7

NFPA 1983 is not intended for situations where a rescuer must climb into a loose rope where the fall factor could exceed 0.25.

the most to the certification requirements, fearing a loss of business caused by passing along the cost of certification to the customer. For example, properly marked, NFPA-compliant ropes might cost three cents more per foot than noncompliant ropes. Most of the fears have since been eliminated, and NFPA 1983-compliant equipment is becoming more commonplace in catalogs across the country. The users must determine whether they really need NFPA 1983-compliant equipment. Rope equipment manufacturers in the United States make very high-quality, durable, strong products even without the urging of a national consensus standard. For legal reasons, however, it can be wise to have all NFPA-compliant equipment. The liability umbrella created by having *three* parties agree that the equipment is right for the application is indisputable.

NFPA 1983

Standard on Fire Service Life Safety Rope and System Components

Chapter 3—Product Labeling and Information. This chapter explains how equipment must be labeled to ensure compliance. It, as well as Chapters 4, 5, and 6, contains the following subsections:

1. rope
2. personal escape rope
3. harnesses
4. belts
5. auxiliary equipment

For example, 5–4 is Chapter 5—Performance Requirements, Section 4—performance requirements of belts.

The primary purpose of the stringent labeling requirements is to force manufacturers to get a third-party certification. Labeling also provides good, basic information and warning labels that urge the user to operate the equipment in a certain manner. For example, section 3–6.1.2 states, "The manufacturer shall provide information for the user to consider prior to reusing life safety rope, including that the rope be considered for reuse only if at least all of the following conditions are met:

a. Rope has not been visually damaged.
b. Rope has not been exposed to heat, direct flame impingement, or abrasion.
c. Rope has not been subjected to any impact load.

NOTE: There are several cross references regarding weights and measures used throughout rope rescue nomenclature. A common rope discussion may alternate between inches and millimeters all in the same sentence. It helps to understand both systems when discussing measurements (Figure 1–8).

FIGURE 1–8
Common rope size comparisons.

Diameter in fraction, in.	Diameter in decimal, in.	Diameter in decimal, mm	Equivalent circumference, in.
5/16	0.3125	7.937	0.9817
3/8	0.3759	9.525	1.1781
7/16	0.4375	11.11	1.3744
1/2	0.5000	12.7	1.5708
5/8	0.6250	15.87	1.9635

d. Rope has not been exposed to liquids, solids, gases, mists, or vapors of any chemical or other material that can deteriorate rope.

e. Rope passes inspection when inspected by a qualified person following the manufacturer's inspection procedures before and after use.

► **qualified person**

Defined by the Authority Having Jurisdiction (AHJ) generally as personnel employed by the local jurisdiction who are recognized as trained and proficient at performing a given task.

A **qualified person** is defined by the Authority Having Jurisdiction (AHJ), and this generally means personnel employed by the local jurisdiction who are recognized as trained and proficient at performing a given task.

► **Authority Having Jurisdiction (AHJ)**

The organization, office, or individual responsible for approving equipment, an installation, or a procedure.

Authority Having Jurisdiction (AHJ) is the organization, office, or individual responsible for approving equipment, an installation, or a procedure.

Another labeling addition for the 1995 edition is (section 3-5.2) where auxiliary equipment, mostly hardware, must be permanently stamped or marked with a G for general use or a P for personal use. This "at a glance" labeling is useful when unfamiliar equipment appears in a mutual aid call. Anything with a P marking is only for personal use such as ascending or emergency egress. Equipment with a G can be used for almost any heavy-duty rope rescue system application (see 5-5 for specific P and G applications). Unfortunately, some manufacturers are labeling products with P or G, thereby implying NFPA compliance when they are not compliant. To be NFPA compliant, the product must also have the NFPA and third-party certifiers' mark attached.

Chapter 4—Design and Construction Requirements. This chapter lists the minimum requirements for manufacturing:

life safety rope (4–1)

personal escape rope (4–2)

harnesses (4–3)

belts (4–4)

auxiliary system components (4–5).

► **life safety rope**

Generally designated as a one- or two-person rope with a factor of not less than 15.

In general **life safety rope** (4–1) must be designated as a one- or two-person rope with a safety factor of not less than 15. This requires a minimum breaking strength of 4,500 lbf (20 kN) for a one-person rope and 9,000 lbf (40 kN) for a two-person rope. It must also be of "block creel" construction with continuous filament fiber. Requirement 4–1 is flexible enough to allow the use of polyester, nylon, KEVLAR®, and even NOMEX® fibers but does not favor particular construction types like kernmantle, braid-on-braid, plaited, or hawser laid rope.

► **personal escape rope**

Used for emergency self-rescue and escape.

The personal escape rope (4–2) section is similar to the life safety rope section but defines a different safety factor. The **personal escape rope,** which is used for emergency self rescue and escape, allows for a 10:1 safety margin instead of 15:1. The reasoning was that a life safety escape rope would need to be very light in order to be

▶ **pounds**

A term used to measure an amount of weight or mass.

▶ **pounds force (lbf)**

The force created on an object by the gravity on the mass.

▶ **newton**

One newton is roughly 0.225 lbf, or about the weight of an apple.

Pounds, Pounds Force, and Newtons

Pounds is a term used to measure an amount of weight or mass. **Pounds force (lbf)** is a more accurate measurement than just measuring the weight of a static object. It is more accurate to measure the force created on an object by the gravity on the mass, or by forces created in a rope system by people, pulleys, angles, weather, and the equipment itself. **Newtons** are a measurement of force equivalent to lbf. One newton is roughly 0.225 lbf, or about the weight of an apple. A pound is equal to 0.453 kilogram or just a little less than half a kilogram. The prefix "kilo" means 1,000, so a kilonewton (kN) is 1,000 newtons or about 225 pounds force. For some people it is helpful to do the following:

1. Look at the first two digits of the pound measurements. For example, for 6,000 lbf, look at the 60.
2. Halve the 60 to 30.
3. Subtract 10% of 30 from 30, which is 30 – 3 = 27. The kilonewton equivalent of 6,000 lbf is 27 kilonewtons.
4. Or simply divide the pounds by 225 pounds force for 6,000 ÷ 225 = 26.67.

To find out how many lbf are in 40 kN, do the following:

1. Look at the first two digits of the kN, 40.
2. Double 40 to equal 80. Add two zeros, making it 8,000.
3. Add 10%, or 800, to get 8,800. The lbf equivalent of 40 kN is 9,000, which is very close to 8,800.
4. Or simply take 225 lbf times 40 kN for 225 × 40 = 9,000.

As a practical example, NFPA 1983 (5–5.5.2) says, "General-use descent control devices shall withstand a minimum test load of at least 6,000 lbf." 6,000 lbf equals 26.67 kN, which is close to the 27 we found in the above example.

The NFPA will require all documents to switch to metric units of measurement in the year 2000. Expect the 9,000 lbf carabiner to be listed as 40 kN in the NFPA 1983-2000 edition. Globally, kilonewtons (SI metric) is a more recognized measurement of force. Try to get used to it!

carried religiously on the fire ground. The warning label for a personal escape rope states, among other things, "Use this rope only for emergency self rescue/escape. This is a single-purpose, one-person, one time use rope. Rope must be previously unused, destroy after use!"

▶ **harness**

A webbing or rope configuration tied onto people to connect them to a safety tether or other component of a rope system.

NFPA 1983

Standard on Fire Service Life Safety Rope and System Components

The **harness** section (4–3) defines Class I, II, and III harnesses. Class I fastens around the waist and thighs or buttocks and is designed only to carry one-person loads or to be used for emergency egress. Class II fastens around the waist and thighs or buttocks and is designed to carry two-person loads as might be encountered when performing the "pick-off," described in Chapter 5. A Class III harness is defined as fastening around the waist and thighs or buttocks and over the shoulders. It is used wherever inverting may occur and is intended for two-person loads. Harnesses are now also permitted to be adjustable over a range of sizes, unlike in NFPA 1983–85 and 1990 editions. There are also details mandating synthetic and continuous filament fibers verses natural and spliced or knotted fibers.

The **belt** section is new to the 1995 edition of NFPA 1983. It allows for the use of a harness-like device, but without webbing, that fastens under the thighs or buttocks. It defines *ladder belt* as a positioning device for a person on a ladder, and an *escape belt* as intended for use only by the wearer as an emergency self-rescue device.

▶ **belt**

Harness-like device without webbing that fastens under the thighs or buttocks.

▶ **auxiliary equipment**

Any equipment that is not a rope, a harness, or a belt. For purposes of NFPA 1983, it refers to hardware and webbing.

Auxiliary equipment (4–5) is any equipment that is not a rope, a harness, or a belt and, for the purposes of the standard, refers to hardware and webbing. Again, personal-use and general-use labeling designations tell the end user whether the equipment is designed for use by a single person or two people. Two-person and rope "system" loads exceed the safe capabilities of equipment labeled P. The intent of the committee was to prevent rope equipment interfaces that could be self-destructive. For example, there were numerous stories of "hard" (mechanical) ascenders desheathing the rope when used as rope grabs. The ascenders were designed to help people *ascend*. Innovative rescuers were using them to grab the rope in haul systems, greatly exceeding their design capacity. The ascenders were convenient, quick, and easy to reset along the length of the rope, however, the conical toothed cams chewed into the rope so hard, usually at about 1,200 pounds force, that they completely desheathed the rope. Fortunately, attentive equipment manufacturers have taken steps to make mechanical devices that slide along the rope before desheathing it (see Chapter 3). Paragraph 4–5.10 categorizes rope grabs as being for personal use or for use as component parts of a rope system.

Chapter 5—Performance Requirements. Section 1 details life safety rope requirements, again establishing 4,500 lbf (20 kN) as the minimum breaking strength for a one-person rope, and 9,000 lbf (40 kN) as the minimum breaking strength for a two-person rope. Paragraph 5–3 signals the committee's continued efforts to prevent relatively high-stretch rope from being used for rescues (refer back to Paragraph 1–1.3) where the standard does not include equipment to be used when climbing above the anchor point, specifically where a fall factor of 0.25 or greater is possible. A life safety rope must stretch between 15% and 45%, at 75% of its rated breaking strength. For example, a 100-foot length of NFPA approved two-person rope tensioned to 6,750 lbf (30 kN) must stretch at least 15 feet and not more than 45 feet.

Paragraph 5–1.10 details the thermal requirements of rope to ensure that it does not melt at or below 400°F (204°C). The committee wanted to be sure that the rope would withstand the rigors of unusually high temperatures caused by rappeling and lowering operations that developed heat caused by friction.

Paragraph 5–5 details Auxiliary Equipment System Components, specifically, carabiners, ascending and descending devices, and webbing (Figure 1–9).

FIGURE 1–9

NFPA 1983 rope equipment performance comparisons.

	Carabiners							Descent control devices			Auxiliary equipment	
	Personal use			General use								
	Gate closed	Gate open	Min. axis	Gate closed	Gate open	Min. axis	Ascending device	Personal use	General use	Rope grabs	Personal use	General use
Minimum breaking strength	6,000	1,650	1,500	9,000	2,400	2,400						
Minimum test load without damage to device, rope, or other equipment							1,200	1,200	1,200	2,400	1,200	5,000
Minimum test load without failure								3,000	6,000		5,000	8,000
Paragraph no.	5-5.1.1	5-5.12	5-5.1.3	5-5.2.1	5-5.2.2	5-5.2.3	5-5.3	5-5.4.1 5-5.4-2	5-5.5.1 5-5.5-2	5-5.6	5-5.7.1 5-5.7-2	5-5.7.3 5.5.7.4

Chapter 6—Testing Requirements. The 1995 edition, for the first time, gave manufacturers and third-party certifiers definitive testing criteria. The testing requirements simply enforce what Chapters 3, 4, and 5 have stated, and an important phrase is introduced: the recurring "shall be tested in the manner of function." Previously, a device could be tested in a manner that assured a successful test. Additionally, before the 1995 edition there was considerable debate about the variables used to test equipment. For example, rope was sometimes being break-tested by slow-speed machines, and, in other cases, high-speed breaking techniques imparted shock loads that could produce wide fluctuations and render misleading results. Harnesses and belts now had to be tested assembled on a test torso weighing 300 pounds (136 kg), instead of simply testing individual components for breaking strength. The test torso was adopted from UIAA harness testing procedures as well as many of the testing criteria (Figure 1–10 and Figure 1–11).

Carabiner testing is now very specific and requires three pull tests: along the major axis, along the minor axis, and with the gate open along the major axis.

Referenced Publications This short section simply lists other publications that have been referred to throughout the document.

Appendix The appendix is included for informational purposes only, but, similarly to OSHA Codes' preambles, the appendix is loaded with important information. Much of what is in the appendix is a condensed version of committee debate that was never agreed upon and was not included in the standard. It also clarifies many of the dogmatic statements that are part of the document.

FIGURE 1-10

NFPA harness test matrix.

Test, paragraph number	Class I	Class II	Class III
6-3.2 (Upright)	Yes	Yes	Yes
6-3.3 (Head Down)	No	No	Yes
6-3.4 (Horizontal)	No	Yes	Yes
6-3.5 (Drop)	Yes	Yes	Yes

FIGURE 1-11

NFPA 1983–95 harness test dummy dimensions.

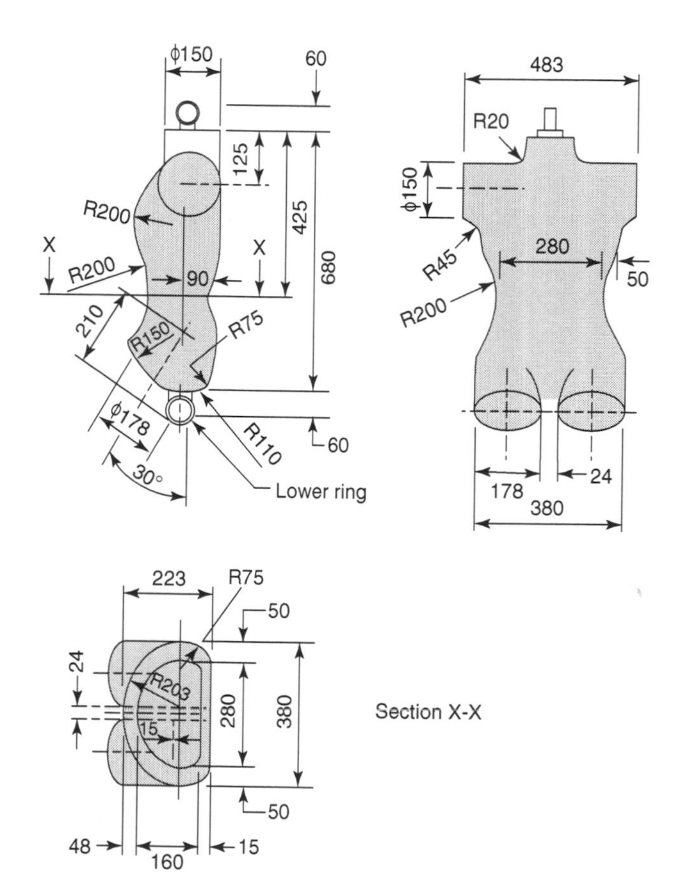

All linear dimensions are in millimeters, ±5 mm.
The dimensions are those of a dummy developed by the UIAA for testing harnesses
Note: Waist circumference at X-X is 850 mm.

NFPA 1983

Standard on Fire Service Life Safety Rope and System Components

Summary of NFPA 1983–2000 or 2001 (Proposed) Fortunately, the NFPA standards review process is a dynamic and progressive program that issues new editions every five years. As this book goes to print, it is impossible to say exactly what the NFPA 1983–2000 or 2001, or Fourth Edition, will look like. It is expected that NFPA 1983–2000 or 2001 will maintain the same basic chapter structure as the 1995 edition and will become effective sometime in late 2000 or early 2001. The following issues seem to be prevalent based on draft versions and committee member interviews. However, they should be used for informational purposes only until the final approved document is released.

Due to the failure in the United States of organizations other than the fire service to produce their own national-consensus rope standards, there was concern that other groups, for example wilderness rescuers, were being leveraged into using the fire service's NFPA 1983. In an effort to be sensitive to their concerns, the 2000 committee chose to highlight in writing in Chapter 1–1.2, "*This standard shall* **not** *apply to utility rope . . . mountain rescue, cave rescue, lead climbing operations . . . industrial fall protection . . . or recreational uses.*" It will remain to be seen if this softening effort on the part of the committee affects non-fire department teams in a positive manner.

NFPA 1983
Standard on Fire Service Life Safety Rope and System Components

A quick look at the definitions (Chapters 1–3) foreshadows much of the rest of the document. For example, "escape belt" is eliminated and the word "personal" is removed from all escape terminology—no more "personal escape rope," just "escape rope." "Throwline and portable anchor" has been added, and the terms "one-person" and "two-person" will probably be replaced by the terms "light duty" and "general duty." It would be unfortunate to see the helpful Ps and Gs that are stamped into NFPA 1983–95 compliant hardware to differentiate personal use and general use be eliminated.

Again, NFPA regulations will require metric units of measurement for NFPA 1983–2000 and all other documents as well. This means a one-person (or light duty) load will equal about 1.33 kN (300 ÷ 225 = 1.33), a two-person (or rescue) load will equal about 2.66 kN, and a 15:1 safety margin for a two-person load, if the 15:1 committee has not eliminated safety ratios, would be 2.66 kN × 15 = 39.9 kN, which is close enough to 40 kN to be the equivalent of 9,000 lbf.

Also, it appears as if the ultra-harsh testing criteria of the 1995 edition harness section will be softened somewhat to make it more affordable for manufacturers and more practical for rescuers who have to sit in the stiff, heavy, bulky monstrosities that could pass the tests.

It appears as if the life safety and escape rope elongation characteristics have been simplified from the 1995 edition to the 2000 or 2001 edition. Minimum elongation of less than 15% at 75% of the breaking strength will be changed to 1% at 10% of the breaking strength, and maximum elongation that shall not be more than 45% at 75% of the breaking strength will become 10% at 10% of the breaking strength. This also translates to allowing less stretchy ropes, an agreeable transition for properly trained rope rescue professionals. (Figure 1–12)

NFPA 1670 (Proposed) Technical Committee for Technical Rescue Operations and Training

NFPA 1670
(Proposed) Technical Committee for Technical Rescue Operations and Training

In February 1999 **NFPA 1670** was accepted by the Standards Council to be an American National Standard. NFPA 1670's details are paraphrased as:

FIGURE 1-12

Standards and laws that affect rope rescue operations. (*Courtesy Martin C. Grube*)

NFPA 471/472

Hazardous Materials

Following the example set by the NFPA Hazardous Materials Standard 471 and 472, proficiency levels provide the organizational structure for the document. In what is basically a level 1, 2, and 3 format, or beginner, intermediate, and advanced, both standards have adopted Awareness, Operations, and Technical levels.

Each chapter details the capability level that a team responding to technical rescues should maintain when performing technical rescues and when training. It provides a platform for fire departments and other rescue organizations to use to create their response categories. For example, every person in the organization could be trained to the awareness level in all of the disciplines. This would not be too difficult or even very expensive because most of the awareness training can be done in a classroom, in the station, or even in a lecture hall, by a knowledgeable instructor with a few audio-visual aids. Awareness basically means that a person has been trained to recognize a situation as a technical-rescue-oriented problem. This means that special equipment, special training, and special people are probably required to effect a successful rescue. At awareness level, the first responder should be able to do three things:

1. stop anyone else from becoming part of the problem
2. recognize and activate the next higher level of response (operations)
3. effect a noninvolved rescue, if possible.

A noninvolved rescue is a rescue in which the awareness level responders extricate patients from peril without placing themselves in peril. For example, confined-space patients may be winched out using pre-existing devices, or a fire service exhaust fan might be placed to blow fresh air onto the patients, thereby reviving them for self rescue.

At the operations level, some component of the organization would be detailed to perform actual, but relatively simple, technical rescue operations. Many fire departments use their truck (ladder) companies for this purpose. The "truckies" are given special equipment and special training to effect some rescues and are also capable of recognizing and supporting a more complex rescue scenario when working with technical-level providers.

The technical-level provider would be able to provide the highest level of response as ultimately defined by the AHJ, that is, his employer or the political entity for which he worked. It would be possible for a technical rescue provider to be multi-leveled. For example, technician level at rope and confined space rescue; operations level at structure collapse, water, and vehicle and machinery extrication; and awareness level at trench and excavation and wilderness rescue.

NOTE: For the purposes of this book, and in an attempt to synchronize terminology, most definitions used here are the same as those used in NFPA 1983 and NFPA 1670.

NFPA 1006

(Proposed) Standards for Rescue Technicians Professional Qualifications

NFPA 1670

(Proposed) Technical Committee for Technical Rescue Operations and Training

NFPA 1983

Standard on Fire Service Life Safety Rope and System Components

NFPA 1006–00

Standard for Rescue Technician Professional Qualifications

NFPA 1006 (Proposed) Standards for Rescue Technicians Professional Qualifications

The first edition of **NFPA 1006** will probably be approved by the NFPA Standards Council and made effective sometime in late 2000 or early 2001. In many ways, the document seems to be a repeat of NFPA 1670, but, on closer inspection of the draft editions, it is apparent that NFPA 1006 (Chapter 1–1 Scope) ". . . establishes the minimum job performance requirements necessary for fire service and other emergency response *personnel* (unlike NFPA 1983 who perform technical rescue operations." In other words, where NFPA 1670 focuses on the organization's obligations toward responding on technical rescue incidents, NFPA 1006 details the "requisite knowledge" and "requisite skills" that individual team members should have in order to function as a team in the technical rescue arena. The draft versions of NFPA 1006 are shaping up with the following foundation.

The document outline for NFPA 1006–00, *Standard for Rescue Technician Professional Qualifications,* 2000 Edition, follows.

Chapter 1 Administration
Chapter 2 General Requirements
Chapter 3 Structural Collapse
Chapter 4 Rope
Chapter 5 Confined Space
Chapter 6 Vehicle and Machinery
Chapter 7 Water
Chapter 8 Wilderness
Chapter 9 Trench and Excavation
Chapter 10 Referenced Publications
Appendix

Chapter 1—Administration This chapter covers scope, purpose, definitions, and a lengthy testimony about the AHJ's responsibility to carry out a technical rescue according to local protocol and procedure. The AHJ has a lot of flexibility, it would appear, in just how to tailor the standard to suit its own locality and service arena.

Chapter 2—Rescue Technician This reinforces the hazards of technical rescue and places even more responsibility on the AHJ and localities to determine the direction of their rescue members and response arenas. Entrance requirements are also included to make sure only qualified persons are involved in technical rescue.

Chapter 3—Job Performance Requirements This chapter suggests that the technically competent rescuer should have some management and assessment skills for establishing and controlling a technical rescue scene. Sad as it may seem, some departments still do not recognize when they have crossed the line from a routine or traditional rescue into the much higher hazard and competency levels of technical rescues. This chapter attempts to show those departments how to recognize that special fire and rescue operations require *special equipment, special training, and special people* (see the Hazard Curve Paradox, Chapter 2 of this text). The rest of Chapter 3 suggests that technical rescuers be competent in directing rope rescue operations and constructing various rope rescue systems, but there is no mention of the other disciplines at this time. Perhaps it is because a good rope rescue background greatly enhances a team's ability to perform all of the other disciplines, but the reverse is not necessarily true.

Chapter 4—Rope Rescue This chapter covers a plethora of rope rescue skills, from rappeling to building highlines, similar to the ones described in this book but without the details.

Chapter 4—Rope Rescue begins by putting the responsibility on the AHJ for determining where, when, and how their organization will handle rope rescue operations (4–1.1 and 4–1.2). Paragraph 4–1.3, Special Definitions, lists specific rope rescue definitions.

Awareness level functions for a team in rope rescue include (4–2.2) sizing up the problem, identifying the required resources, developing and implementing site control, and scene management. Operations level team capabilities specify a prerequisite of all awareness level skills, as well as basic knots and anchoring skills, rappelling, belay systems, patient packaging, and the engineering and use of rope systems in the low-angle environment. Technician-level team capabilities mandate a prerequisite of all awareness and operational level skills, and load distributing anchor systems, raises, lowers, and highlines in the high-angle environment, attending a patient and litter, and knot passing techniques.

Chapter 5—Surface Water Rescue

Chapter 6—Vehicle and Machinery Rescue

Chapter 7—Confined Space Rescue

Chapter 8—Structural Collapse

Chapter 9—Trench Rescue

Chapter 10—Referenced Publications

Appendix—Again, the appendix contains a great deal of information that is not part of the requirements of the document but is well worth the time spent reading it.

Urban Search and Rescue (USAR)

Search and rescue teams whose primary service area is in the urban semi-urban and industrial environments.

Urban Search and Rescue (USAR) was *developed by the Federal Emergency Management Agency (FEMA) in response to political pressure generated by several domestic disasters,* in part because of the perceived lethargic response by federal agencies to Hurricane Andrew in southern Florida. The Federal Response Plan (FRP) and Presidential Decision Directive #39 (PDD-39) dictate the responsibilities of federal assets in the event of domestic disasters. In a nutshell, the FBI is responsible for crisis management, including crime and terrorism prevention, detection, and legal aspects, and FEMA is responsible for consequence management, including disaster preparedness and response. The Emergency Support Functions (ESF) of the FRP include the breakdown of responsibilities according to function; for example, the Red Cross, the Army Corps of Engineers, and the United States Forestry Service, to name a few. ESF-9 is the USAR function.

Today there are twenty-eight certified teams maintaining a response cadre of sixty-three specialists in almost every emergency arena in five categories: rescue, medical, technical, search, and command. The typical equipment cache has a value of more than $1.5 million and is capable of being palletized by team personnel and loaded onto military aircraft for rapid deployment. The teams must maintain a wheels-up capability of less than six hours from the moment of activation by FEMA. The USAR teams carry a complement of rope rescue gear for use on deployments. Their rope rescue capabilities are considerable but vary somewhat from team to team. Primary rope rescue operations include high- or low-point remote patient access, confined-space entry and retrieval, USAR search dog transportation in the vertical realm, and rescuer and tool stabilization at height on damaged structures. After the terrorist bombing of the Alfred P. Murrah Federal Building in Oklahoma City, rescuers used numerous different rope systems to suspend themselves from the sides of the building to stabilize suspended and broken concrete slabs.

NFPA 1670

(Proposed) Technical Committee for Technical Rescue Operations and Training

Rescue technicians are certified in an intense training program based on NFPA 1670 that focuses on heavy lifting and moving, breaking and breeching, and stabilization and shoring. Additionally, each team has members that qualify at the technical level in each of the rescue disciplines, as outlined in NFPA 1670.

SUMMARY

There are literally dozens of organizations and regulations that affect rope rescue programs. The AHJ is the hub, bearing much of the control and the latitude that dictates the level of training and response that a team will maintain. With a good understanding of the standards, laws, and organizations affecting our business, it will be easier to progress to the basics of team development and the business of using ropes and associated equipment to rescue people from danger.

■ KEY TERMS

Gravity
Friction
Two in two out
NFPA 1983
Ropebotics
Fall factor
Qualified person
Authority Having Jurisdiction (AHJ)
Life safety rope
Personal escape rope
Pounds
Pounds force (lbf)
Newtons
Harness
Belt
Auxiliary equipment
NFPA 1670
Urban Search and Rescue (USAR)
NFPA 1006

■ REVIEW QUESTIONS

1. The three most important tools for use in rope rescue are ________.
   a. descent control devices, carabiners, and rope
   b. gravity, the human factors, and friction
   c. rope, eight-plates, and anchors
   d. maintenance cards, standard operating procedures, and national standards
2. Three important standards that have an influence on rope rescue equipment, training, and/or operations are ________.
   a. OSHA 29 CFR 1910.154, OSHA 29 CFR 1920.177, and NIOSH 1983
   b. NIOSH 1983, ANSI F-32, and NFPA 1910.134
   c. ASTM G-43, NFPA 32, and NIOSH 1983
   d. NFPA 1670, NFPA 1006, NFPA 1983
3. The good friction caused by the rope's contact with an anchored closed end rack is ________ equal to the downward pull of gravity on a rescue load.
   a. exactly
   b. less than
   c. more than
   d. approximately
4. A two-person or rescue load is generally assumed to be _____.
   a. 450 pounds
   b. 600 pounds
   c. 900 pounds
   d. 7 kilonewtons

5. In May 1993, NFPA Standard ________ effectively eliminated the "use once and destroy" portions of the Ropes, Harnesses, and Hardware Standard.
   a. 1500
   b. 1983
   c. 1006
   d. F-32

6. The severity of a fall that is expressed as a ratio calculated by dividing the distance fallen by the length of the rope used to arrest the fall is called the ________.
   a. gravity index
   b. descent modulator
   c. arresting value
   d. fall factor

7. In the Ropes, Harnesses, and Ancillary Equipment NFPA Standard, Chapter 2 Certification defines in detail the requirements of the equipment manufacturers to have their products tested by a third party—the first party being the NFPA, the second party being the manufacturer, the third party being ________.
   a. the end user of the equipment when requested in writing
   b. a certified testing organization
   c. any of a number of consumer groups
   d. the federal government

8. "General-use descent control devices shall withstand a minimum test load of at least 6,000 lbf." 6,000 lbf equals approximately ________ kilonewtons.
   a. 7
   b. 18
   c. 27
   d. 60

9. Thirty-five kilonewtons is approximately equivalent to ________ pounds force.
   a. 7,900
   b. 900
   c. 300
   d. 2,000

10. The Technical Committee for Technical Rescue Operations and Training Standard uses which of the following terms in rescue service capabilities leveling?
   a. beginners, intermediate, and advanced.
   b. I, II, and III.
   c. urban, wilderness, and rural.
   d. awareness, operations, and technical.

CHAPTER 2 Team Concepts

OBJECTIVES

Upon completion of this chapter, you should be able to:

- define the advantages of working as a team.
- explain the hazard curve paradox.
- define the team efficiency concept (TEC).
- list the three special requirements to effectively develop a technical rescue team.
- name five sources of rope rescue information.
- name a process to utilize structures for rope rescue training.
- explain the difference between traditional and nontraditional rescue services.
- discuss special rope rescue operations incident management, accountability, and safety management issues.

A DAY ON THE ROPES

The pace was calculated and fast—somewhere between a jog and a rolling canter. The team was working like a finely tuned machine. Not wasting movement, each team member was anticipating the moves of the other team members. Squad leaders were staying about two steps ahead of their respective squads, assuring that more work would be ready for them as they completed the present task. The squad leaders were also watching the overall effectiveness and safety of the group, thereby allowing individual team members to focus on their tasks. The operations officer was staying a strategic three steps ahead of the four squads, calculating the efficiency of the team and planning the next movements of the group to fine tune their efficiency. The safety officer was alert, monitoring the feverish activities of the team and always looking for suspected and unsuspected hazards.

Was this action part of a complicated technical rescue scene? No, it was the Tidewater Regional Technical Rescue Team (TRTRT) chipping in to help prepare for the Rescue '90 Conference in Montgomery County, Maryland. Time was running out. It was 2200 hours and the conference started at 0800 hours the next morning. The Montgomery County Fairgrounds trash cans were full to the brim—hundreds of them—and they had to be dumped before the start of the conference. It was a crummy job, but it had to be done, done now, and done in a professional and safe manner. It could have been approached with complaining and moaning: "I didn't come up here to dump garbage." But it was, instead, viewed as a vital and critical task, an opportunity to hone team skills in a different arena. It was not a grungy task, but instead an opportunity to make the team better.

Wearing helmets, gloves, and steel-toed boots, after the safety briefing, team members approached the task with a vengeance. Operations and planning officers laid out a strategic plan of attack based on the layout of the grounds and the location of the cans. Two sectors of four, five-person teams were staffed to work alongside two stake-body trucks. Each sector was assigned a geographic area containing about 125 trash cans. In each sector, advance teams scurried to pick up loose trash and place it in the cans. As the truck approached, a two-person team pulled out the trash in the plastic liner and handed it to members waiting in the back of the truck. Members on each side of the truck placed a new bag in the can, positioned the can properly on the ground, and picked up any remaining loose trash. If the can was faulty or otherwise in need of maintenance, the member placed a red tag on the can lid handle. Yet another team was set up with some basic tools, some duct-tape, nuts and bolts, and a rivet gun. They repaired or replaced the red tagged cans in true NASCAR pit crew style. The entire operation took twenty-five minutes to complete, or about ten cans per minute. A brief critique discussed ways to make the operation even more efficient in the future.

RESCUE TEAMS

▶ **team**
A group of people organized to accomplish a common goal.

A **team** is a group of people organized to accomplish a common goal (Figure 2–1). We use the word team almost every day. Sports activities immediately come to mind as teams that are organized to defeat other teams in competitive activities. Fire and rescue services are teams of a different type, since we, for the most part, do not compete with other fire and rescue services. Our common goal is problem solving and we aspire to be very good at it. It is obvious that we problem solve better as a team than as individuals. A team of ten people are infinitely more efficient at attacking a problem than are ten people working the problem individually.

FIGURE 2–1

A team is a group of people organized to accomplish a common goal. (*Heavy and Tactical Rescue School, Circa 1983, Courtesy Chase N. Sargent*)

Technical rescue managers know that teams work better than groups of individuals for five primarily reasons: specialization, sense of being, coordination and management, actions are not repeated, and networking concepts.

Teams allow individuals to *specialize,* or *concentrate on specific disciplines,* and develop a high level of expertise in that area. Despite what some individuals will tell you, it is completely impossible to know everything there is to know about technical rescue. However, with much self-motivation and self-discipline, a person can become very knowledgeable and even experienced in several specific areas. Teams facilitate the union of highly specialized people from differing technical backgrounds. Teams also offer a refuge and outlet for the human *need of being* part of a group. Humans feel emotionally satisfied when their actions contribute positively towards meeting the goals of a group, and when their endeavors are acknowledged by other group members. Teams are also highly effective tools in *coordinating and managing* resources. Span of control, and who is in charge are pre-established and do not have to be debated during the problem-solving process. Dovetailing the management efficiency of teams is that actions do not have to be repeated. Repetitious actions are expensive, extremely wasteful, and even dangerous on the rescue scene. And finally, team members can branch out in different directions when they are not problem solving to **network** and research technique or even to recruit new team members. This revitalizes and constantly improves the team's problem-solving capabilities. The team should have a clear understanding and acceptance of these basic team concepts.

► **network**

A group of people spanning a specific geographic area (local, state, national, or international) with common interests that work together to complement educational and practical experiences.

A rope rescue team may be either a separate entity or a section of a technical rescue team. The one difference between rope rescue and other technical rescue disciplines is that rope rescue uses rope as a tool. Rope has been used as a tool for almost as long as water has been used as a tool. All of the other disciplines have occasion to use rope to facilitate their rescues. For example, a car on its side can be safely stabilized using ropes. A victim in a structural collapse, a confined space, a roaring river, a cave, or even a trench accident may need to be extricated using rope and its associated equipment. The reverse is not true. A rope rescue seldom requires a hydraulic spreader, an airbag, a zodiac boat, an airshore, or a timber shore to effect a rescue. Rope skills are therefore most universal, once they are understood, accepted, practiced, and experienced. But, because of the overlapping effect of the rescue disciplines, the most efficient use of resources is a technical rescue team with specialists in every area.

One team organizational tool is to give team members major and minor classifications of study and response. For example, your major might be rope rescue, your minor confined space rescue, and you might have a particular interest in structural collapse. While you might be fairly well versed in the other areas of technical rescue, you cannot compete with those who are very good in their major where it is only your interest. Your major is where you know all the numbers and specifications of your equipment. You know the standards and laws affecting your discipline, and you maintain a working support network of people who have similar interests. Your minor is secondary to your major, but you are still a subject-matter expert on the topic. Your minor probably has many overlapping properties with your major. To draw a correlation with NFPA 1006, your knowledge base should far exceed the moderate requirements of the standard. Your particular interests may fall in the technician level of NFPA 1670, and you should have at least an operational level of expertise in every other discipline.

NFPA 1006

(Proposed) Standards for Rescue Technicians Professional Qualifications

NFPA 1670

(Proposed) Technical Committee for Technical Rescue Operations and Training

Remember that all team members must have at least operational level training in all of the disciplines your team is expected to handle. The concept here is that on a rescue call, the majors and minors handle the critical and pivotal strategies and determine the tactics and techniques of the operation. The rest of the team, who have an adequate but limited knowledge base, supports the majors and minors. This is a dynamic system that is flexible enough to vary with each call. In effect, it uses the best talent for each problem.

NOTE: A good technical rescue team maintains a balance of majors and minors that work together to problem solve rescue calls.

MANAGEMENT AND ACCOUNTABILITY

Special rescue operations incident management and accountability is not unlike the incident command systems that are in place for everyday emergencies. However, there are some differences worth highlighting. On all emergencies that require many people to perform multiple hazardous tasks, someone is designated to be in command, and a common goal is established and announced to everyone on the scene. In special operations incident command, the span of control may have to be narrowed due to the highly technical nature of the equipment and systems management. Typically the **span of control,** the number of people that can efficiently and safely be managed and continuously accounted for, is between five and seven. With fewer than that, something is likely to be overlooked or

span of control
The number of people that can efficiently and safely be managed and accounted for.

lost, someone will be overworked, or an accident will be more likely to occur. More than that and you may be wasting people, particularly managers, and the operation will not be as quick and efficient as it could be with the right span of control. Sometimes it becomes necessary to narrow the span of control to three or four people and a leader. In a rope rescue operation, for example, expecting one person to manage the people working the edge, the main line, the belay line, the attendants, and the ground recovery zone is overwhelming. It is more practicable to have one person assigned to oversee the operational (technical) aspects or the roof sector, with two smaller units (rig master and edge master) assigned to report directly to roof sector, instead of the rescue ops officer. In the sample incident management flowchart (Figure 2–2), the incident manager (IM) has divided the span of control into two divisions: rescue ops and medical ops. Additionally, the IM has some peripheral assistants who focus on safety, technical documentation, and public information. The assistants free the incident manager to project the rescue scenario several levels beyond what is currently happening and to have alternate plans and resources available as necessary. Safety officers, while independent of the technical operations, have the complete authority to alter operations that they consider inappropriately hazardous. The public information officer (PIO) keeps the press safely at bay, while allowing them as much

FIGURE 2–2

Example of an incident management flowchart for a rope rescue operation.

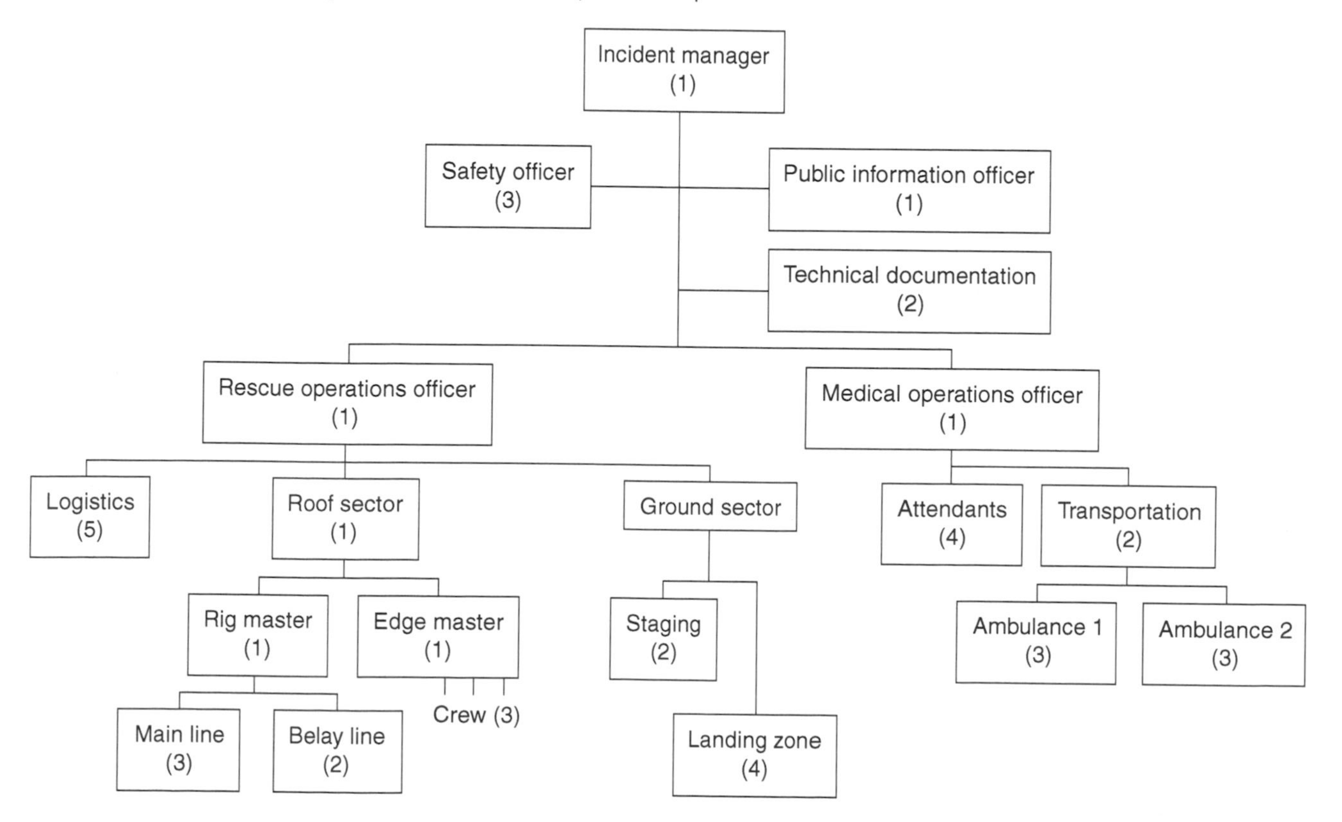

"B" roll[1] tape time as possible. The PIO also keeps them informed about the technical aspects of the operation in layperson's terms. And finally, one of the most important positions on the IM's staff, and the most commonly overlooked, is the technical documentation position. It is vitally important to document the flow of the incident, the time frames of activities, and critical landmarks during the rescue. There will be entry permits, health-exposure forms, post-incident analysis, critique comments, and many other clerical functions that should start at the onset of any incident requiring an incident management system. Also, when possible, the technical documentation staff should include a specially trained videographer to document actions and nonactions for later use in training, promotion, and even legal matters, if necessary.

A common problem can occur when some of the people in the incident management system are not well informed about the technical aspects of a special rescue operation. Many times, officers with a great deal of experience running working fires and other more typical emergencies can become a hindrance or be intimidated by the special rescue operation. The safety officer, for example, has the authority to shut down an operation that seems perfectly routine to you and your team, but may look awkward and unsafe to the uninformed. Likewise, less-informed IMs can sometimes direct rescue operations based on their fireground experience. For example, they might direct a tower ladder onto a soft surface to attempt an impossible reach that could be done very simply with a rope rescue maneuver. In addition, most traditional IMs want to position themselves where they can watch the action. However, many special operations calls are remote or in a hot zone, and IMs have to learn to be comfortable managing the incident strategy, not the operational tactics. It should become one of the highest priorities of every special operations team to deliver special operations incident management and technical awareness (Surviving Technical Rescue Operations) to every officer that could play a role in a technical rescue incident management system. Most technical rescue training organizations and agencies have introductory, "awareness" or "survival" classes for nonproviders and managers. They should be mandatory for all fire and rescue services managers but are often overlooked.

Unlike big fire scenes or jumbled automobile accidents, many special operations incidents do not present obvious hazards. For example, fire boiling out of every window on the second floor of a town house and people hanging out of windows on the third floor *looks* dangerous. The firefighters and rescuers are pumped with adrenaline and are on the lookout for everything and anything to go wrong. Likewise, an automobile accident with blood, groaning, and sharp shards of metal and glass protruding from smoking vehicles *looks* dangerous. People are on the alert and very cautious about their movements and actions. However, special operations envi-

---

[1] *"B" roll* is a term for topic-specific videotape footage used as filler or as background for news broadcasts and video documentaries.

SAFETY *Most confined spaces look safe, but the dangers within them are many, and confined spaces can be very hard to get out of once in them.*

► **Personnel Accountability Report (PAR)**

A roll call.

NOTE: One of the best accountability tools available to the IM is a simple sketched map of the rescue area complete with the location (or suspected location) of the patient(s), and the location of each assigned unit and the associated hazards.

ronments do not always present themselves as being particularly worrisome. It is sometimes easy to get lulled into a false sense of security (Figure 2–3).

A simple rope rescue lowering operation can instantly turn sour if the elevated anchor attachment breaks, the tripod or bipod suddenly collapses or tips, or a rope gets cut on a sharp edge and recoils dangerously toward the rope handlers. Good teams stay alert for the worst-case scenario and always practice with a special level of caution.

Accountability on the special operations scene should be a product of whatever accountability system or policy is currently in place. Usually this requires *at least* tracking every person assigned to the call, and their assignments. A roll call, or **Personnel Accountability Report (PAR),** should be called for whenever a landmark event happens or a task or assignment is completed. They may also be timed as predetermined by policy or by the on-scene IM, possibly every 15 minutes during the incident. And, of course, an official PAR should be taken upon termination of the event to make sure no one is left behind.

The sketch should show the location and boundaries of the hot or high-hazard zone and there should be a single point of entry/exit control where *everyone* entering or exiting the danger zone must pass. The hazard zone boundaries could be, for example, the top floor of a highrise, and the point of entry/exit control the base of the ladder or step access to the roof. In an industrial facility or on board a ship, the boundaries might not be so clear cut and fire line tape (yellow hazard tape) might need to be used to create a boundary with a single point of entry/exit control. Assign someone to secure the

FIGURE 2–3

Rope rescue used to complement a confined-space operation.

boundaries and place someone near the point of entry/exit to track everyone coming and going. Make sure the number coming out is equal to the number that went in, plus any patients that may have been rescued. Train the team in good IM practices and team discipline. Prevent freelancing that leads to a less-efficient team effort and always blurs the accountability and tracking of people. Make sure everyone understands that geographic boundaries will be established on every call, particularly special operations calls; that they must be respected; and that entry and exit must only take place at the designated point of entry/exit control.

Another team developmental concern is determining whether your team will be held within a single department or organization, whether it will be within the confines of your local jurisdiction, or whether it will be a regional concept. There are merits and demerits to all three options, including legal implications, physical boundaries, funding sources, and potential target rescue sources. One of the most important concepts to understand when developing a team and determining organizational structure is the hazard curve paradox.

THE HAZARD CURVE PARADOX

Fortunately, enlightened fire and rescue service managers realize the importance of the "Big Three" factors in increased technical rescue survivability. They are *special people, special training, and special equipment* (Figure 2–4). Eliminate or even compromise any of these elements and successful special rescue operations become prohibitively perilous. Understanding the hazard curve paradox and concentrating scarce resources on the proficiency training of special rescue teams will greatly increase the chances of performing a successful technical rescue.

The **hazard curve paradox** compares funding and other allocated resources with the hazard profile of a given operation. Specifically, technical rescue operations, such as rope, confined space, trench, swift water, cave, helicopter operations, structural collapse, etc., have poor statistical support. That is, compared with fire- and EMS-related calls, technical rescue represents only a small fraction of the calls. To the uninformed, this lack of statistical bolstering

► **hazard curve paradox**
Compares funding and other allocated resources with the hazard profile of a given operation.

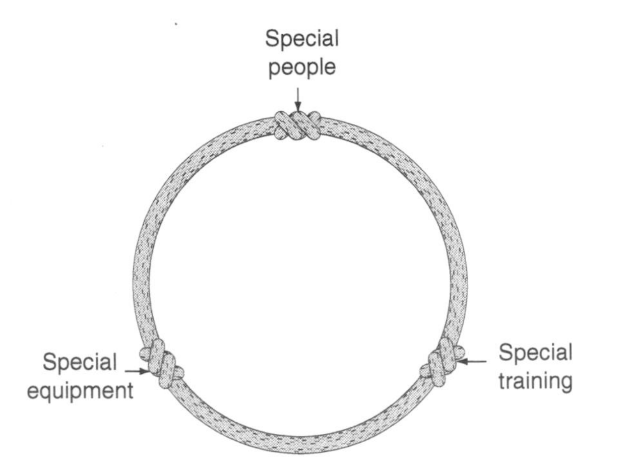

FIGURE 2–4
The Big Three required to successfully perform special rescue operations: special people, special training, and special equipment. Technical rescue operational failures can always be traced to a loss of, or compromise of, any one of the Big Three.

represents an opportunity to underfund or completely ignore funding for technical rescue operations. The hazard curve is thus created, because technical rescue calls by comparison are much more hazardous to perform. The paradox therefore is that the most dangerous calls to rescuers are the least funded and supported because of weak statistical support (Figure 2–5).

Special People

Technical rescue operations require people who are particularly well suited for working in an extraordinarily stressful environment. While traditional fire and EMS services are very stressful by nature, so many calls are run on a daily basis that the adjective *typical* can be used accurately, to describe them. This is not so with technical rescue calls. By their very nature, and as defined in the hazard curve paradox, technical rescue calls suffer from a lack of statistical support. Typical, or high statistical support, allows us to become proficient at handling certain types of emergencies with almost complete confidence, such as structure fires, car accidents, EMS calls, and so forth.

The special rescue technician requires a level of enthusiasm and self-motivation that is unparalleled. Once again, the hazard curve paradox rears its ugly head. True proficiency in the technical rescue disciplines requires almost constant training and continuous research. Unfortunately, technical rescue remains a peripheral service to traditional fire and EMS activities like pre-fire planning, inspections, apparatus maintenance, and training. This leaves little time to concentrate on learning new tactics and perfecting special-rescue strategies. This level of unanticipated or forced underpreparation again compounds the hazard curve.

Special people who are required to perform especially dangerous and complex tasks require an extraordinary support network to stay efficient and to prevent burnout. The most effective

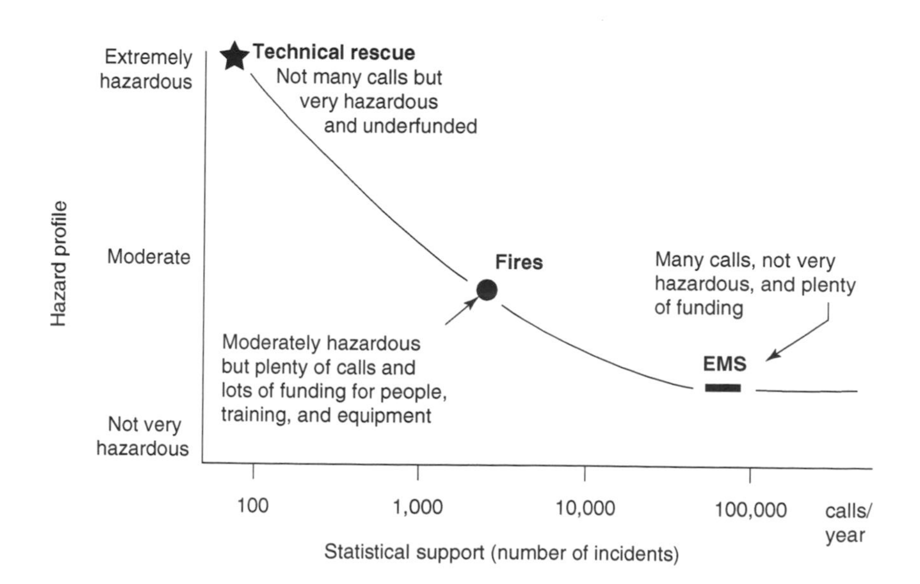

FIGURE 2–5 Hazard curve paradox: technical rescue calls have low statistical support (relatively low numbers of calls compared with fire and EMS calls) and therefore receive less funding. Paradoxically, technical rescue calls are markedly more hazardous, and the practitioners, because of relatively low call volume, are therefore usually less experienced than regular fire and EMS providers and should thus receive increased funding to help compensate for the differences. Many fire and rescue services managers do not understand this dangerous discrepancy.

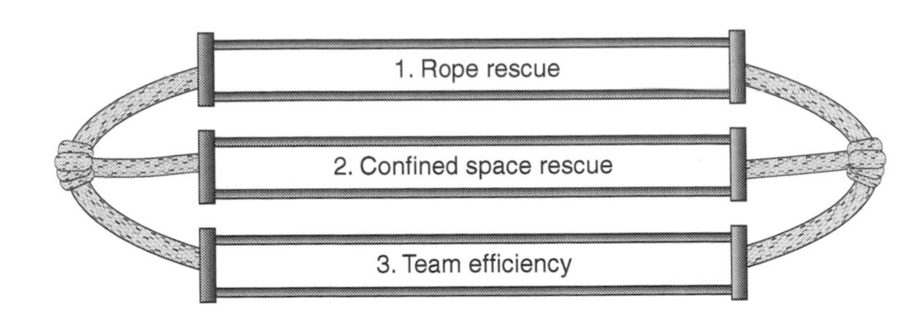

FIGURE 2–6

A typical class system with team efficiency concept (TEC) being used. The "hidden" training module should always be team efficiency. For example, when a rope rescue and a confined space rescue class are being taught in conjunction, team efficiency training makes up the third module and should be announced and made an integral part of the training.

▶ **team efficiency concept (TEC)**

Refining team member interaction to be productive without waste in every team endeavor.

way to develop and manage that support is through a clearly stated **team efficiency concept (TEC),** which suggests that ultimately safety and efficiency are synonymous. Any targets (victims or patients) of the fire and rescue services, and all personnel associated with the rescue benefit because of a vastly increased survival profile. When practiced specifically to fine-tune team proficiency, developmental team weaknesses will surface. The idea is that no one person is skilled in every area. Learning each other's strengths and weaknesses allows teams to compensate and adapt to changing rescue situations.

Once the team efficiency concept is stated as a formal team goal, it must be rehearsed at every opportunity. It could be a parallel third module during a two-module training program. For example, a quarterly in-service on rope rescue and confined-space rescue could have an emphasis on the third module, team efficiency (Figure 2–6). Likewise, take every opportunity to practice the team efficiency concept, for example, moving chairs before a class or a meeting, dumping garbage at a banquet or symposium, or setting up static displays at a conference or public event. Once it has been clearly stated that the team efficiency concept is an organizational goal, do not pass up any chance to hone the skills.

NOTE: Consider establishing a major and a minor for specialization control and team balance.

The overall skill characterizations of a given team vary extensively. There simply is too much information available for any single rescue technician to be an expert in every area. However, a person may specialize in one or two areas and actually develop expert proficiency in those areas. Team managers, after providing adequate base knowledge for all members, can expect extremely high overall levels of proficiency using specialization.

Ultimately, the well-rounded team will be able to call on the proper formula of majors and minors to most effectively carry out a rescue.

Special Training

Once the target rescue sites and customers have been identified for potential technical rescue calls, and team personnel selection has been completed, it is critical to recognize the importance of *special* training. It helps to visualize the technical rescue training cycle.

The example shown in Figure 2–7 indicates a cycle that is never ending. It may even be visualized as a spiral where the third dimension (vertical) reflects the research phase to maintain a cutting-edge philosophy toward new equipment and techniques.

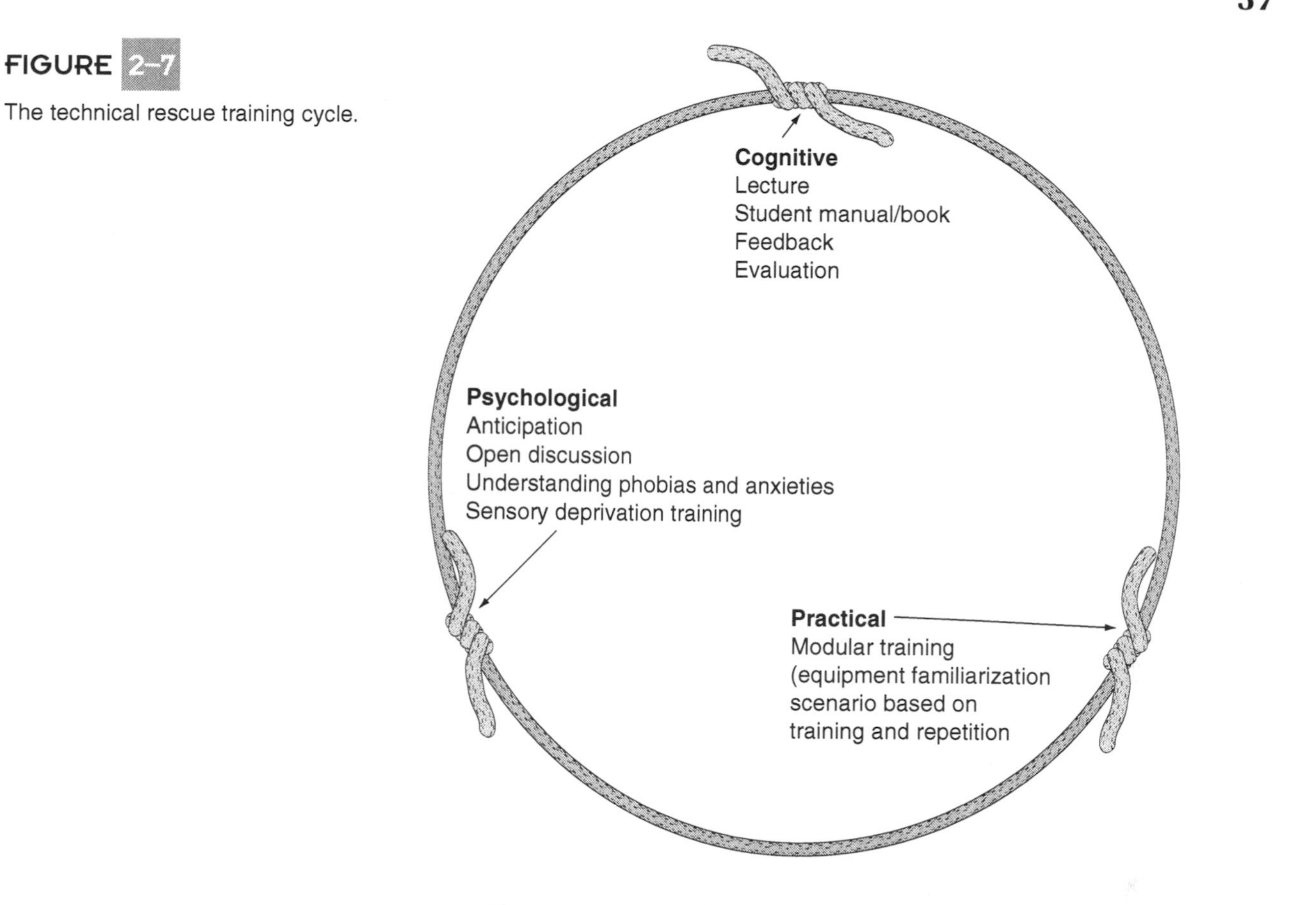

FIGURE 2–7
The technical rescue training cycle.

There are many phases of training (Figure 2–8). Most special rescue training programs begin with cognitive training. This traditional classroom approach is necessary to lay down ground rules for the program. Rescue workers and serious fire/rescue services recognize the importance of identifying hazards, equipment familiarization, standards and legal requirements, and accepted tactics and strategies.

The fire service, in particular, has long recognized the importance of practical training to reinforce classroom objectives. This is even more important with special rescue operations due to the myriad technical parts and pieces of task-specific equipment. For example, there are dozens of new rope rescue equipment components being designed and marketed annually that help to speed high-angle rescues and make them safer. There are intricate seismic listening devices so sensitive that they make it possible to literally hear a pin drop in a collapsed structure from yards away, exothermic cutting devices and surgical explosive breaching techniques that open paths through almost any barrier, emerging confined-space communications devices, and supplied-air breathing apparatus that are dramatically increasing rescuer safety in the most hazardous environments. The examples are almost endless, but they can all be dangerous without practical training to develop proficiency.

The traditional training model suggests cognitive and practical training for traditional fire and rescue services, however technical rescue instructors do not have the luxury of stopping there. Some level of psychological training is also mandatory. The primary goal

FIGURE 2-8

The training spiral indicates that training is never ending and that one should seek to achieve higher goals.

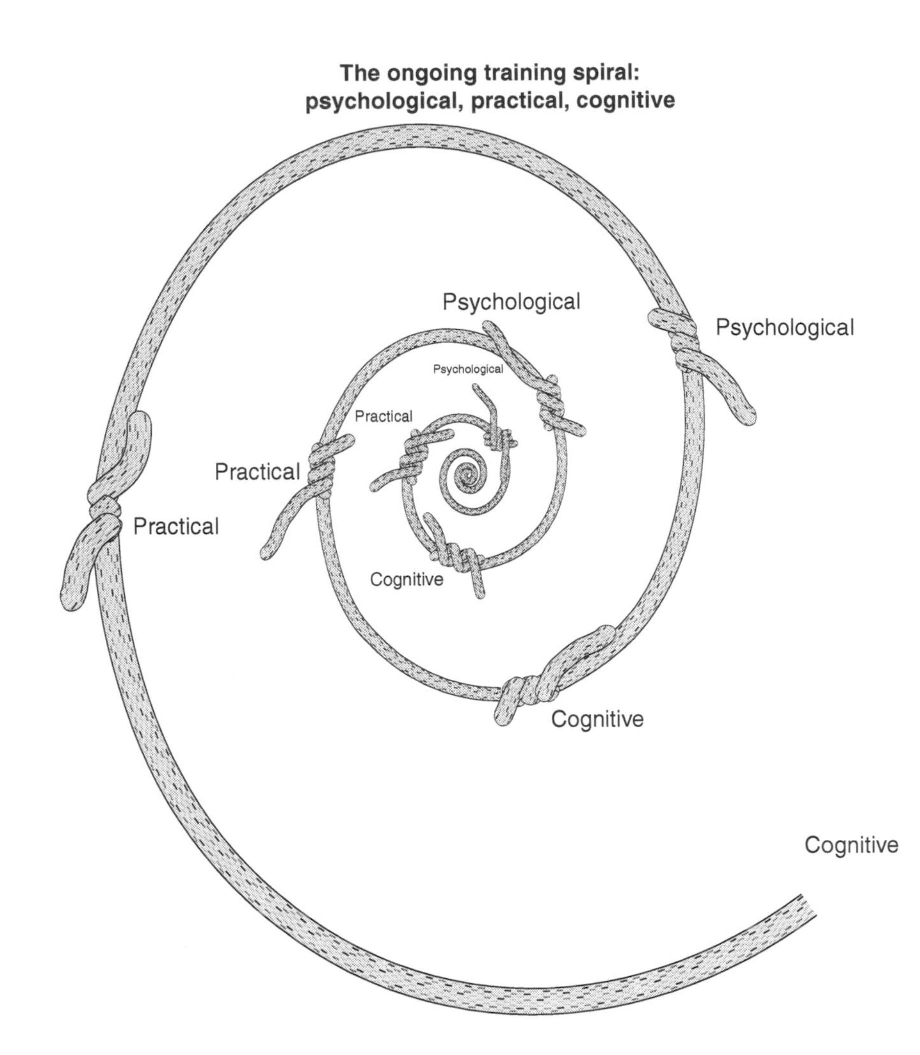

SAFETY
Rescuers who are trained in recognizing and understanding normal, self-preserving psychological responses have a vastly increased survival profile on a call.

► **contextual training**
Involves developing students' ability to handle anxieties related to the practical applications of a given discipline, and entering the environment in a progressive sequence to enhance skill levels.

of psychological training is to reduce or eliminate the feelings of anxiety that can create confusion during complex and stressful technical situations and, if possible, turn feelings of anxiety into performance-enhancing tools.

Comprehending concepts like avoidance reactions, subconscious fears, and the phobic wave, help tremendously toward objectifying individual feelings. This helps rescuers to control their psychological responses instead of letting the psychological responses control them. The latter inhibits team safety and efficiency.

From a programmatic viewpoint, there is a significant and recognized method for combining the three major areas of the training cycle. **Contextual training** is recognized as a substantive method for traditional fire and rescue training. Contextual therapy is likewise a commonly accepted method for treating anxiety-related disorders. Their marriage in technical rescue training has proven to be a tremendous asset in accelerating the proficiency of rescue technicians.

Basically, contextual rescue training strategies involve combining skill builder modules and real-time scenarios in a gradient manner, from simple to more complex, from easy to hard, and from low to high. For example, in confined-space training under controlled circumstances, rescuers enter progressively more confined and complex spaces. Because the environment is controlled and therefore relatively safe, some other means of stress must be introduced to the students.

There are many ways of introducing artificial stress to developing and advanced rescuers. Using time as a tool is most beneficial. Timed skill builders might include a five-person team building a 4:1 compound mechanical advantage system for speed and accuracy in fifteen seconds, assembling a breathing apparatus, or shoring a door jamb with a pneumatic shore. This technique can be pushed with remarkable results in terms of speed and efficiency. It is very important to push these teams to new levels of *in*efficiency. That is the point where they are rushing so fast to accomplish the task that they make mistakes in terms of time and efficiency. It is also very important to recognize and identify that moment of inefficiency. Technicians need to recognize when they are approaching this point and click it back a notch to prevent dangerous levels of anxiety.

Caution
Teams that consistently practice and operate at one rate of speed often fail to recognize when they might be approaching a dangerous level of inefficiency *in a true emergency.*

Other artificial stressors are primarily reserved for the more advanced skill levels. Following the contextual training model, they involve making the training situation progressively more complex and difficult. The amount of stress that a person develops, or does not develop, is directly proportional to the amount of sensory impact on the nervous system. **Sensory deprivation** techniques, advanced training techniques where one or more of the senses is removed or inhibited, lend themselves well enhancing contextual training. Simply stated, sensory deprivation is used once the individual team members have become reasonably comfortable with a training environment, such as confined spaces, a high angle, or cave. Through repeated exposure, their stress level is reduced and their operating efficiency and safety is increased. To re-elevate the stress level, some primary senses are removed. This could include impairing vision with a blindfold, hearing with a running ventilator fan, or feeling with gloves or mittens. These are all simple deprivation techniques that increase technical rescue preparedness for individuals and teams.

► **sensory deprivation**
An advanced training technique where one or more senses is removed or inhibited.

Training without the benefit of all the senses increases stress that can be used to simulate real-time emergencies. For example, experienced rescuers are blindfolded, effectively removing their ability to see. They then must don their harness, reeve their descent control device, and rappel down a wall (with obvious safety spotters and belay).

NOTE: Sensory deprivation training requires special considerations to assure safe practice including but not limited to personal spotters, special belay, and communication techniques.

Using rudimentary techniques of sensory deprivation for technical rescue training involves careful monitoring of individual responses to practical exercises and skill builders. The goal is to elevate individual stress levels under *controlled* circumstances, so the student learns to function with a level of fear and to appreciate the accomplishment. The ultimate result is rescuers who are familiar with elevated feelings of anxiety and who can use them to their benefit during a true emergency.

Failure to closely monitor and control emotional responses can have a detrimental effect on an individual's progress. Never train with sensory deprivation techniques without providing the student with relief valves:

1. Begin cognitive training by explaining the techniques you will be using, such as time factors, narrow crawlways or tunnels, low/no light conditions, noisy confines, extreme heights, squeeze boxes, etc. There should be no surprises!
2. Always be sure that the environment is controlled and does not have the potential to become immediately dangerous to life and health (IDLH).
3. Inform the students before any training begins that *they* may terminate any evolution at any time by saying the word "stop!" They will be encouraged, because of learned avoidance reactions, but not coerced to complete any module or skill builder. They will not be belittled for failing to complete any activity.
4. Remind students of psychological control techniques.

Special Equipment

Another common misconception that many emergency service managers have is that typical fire and rescue service equipment is suited for technical rescue operations. Or, as it is sometimes called, the "close enough to work in a pinch" syndrome. Once again, the hazard curve paradox can lead unprepared teams into a false sense of security. For example, standard self-contained breathing apparatus (SCBA) for structural firefighting tasks is very poorly suited for most confined space entries due to its limited supply of air and increased physical profile. Yet, many fire service managers continue to rely on SCBAs to protect firefighters during that infrequent confined space entry. Structural firefighting equipment and personal protective clothing should not be used during confined-space entry any more than it should be used for swift water or underwater rescues, yet many departments remain underprepared or, worse, falsely prepared.

The training cycle can once again illustrate how the hazard curve paradox hinders special rescue operations dealing with technical rescue equipment. Highly technical and task-specific equipment requires a tremendous amount of time to master, and the technical rescue disciplines are evolving so fast that technicians must be allowed time to research and master their specialties. Some departments have gone far in reducing the traditional workloads of rescue technicians.

In other words, technical rescue has finally come of age as a bonafide, specialized branch of modern fire and rescue services. As a result of new OSHA laws, NFPA and ANSI standards, and an ever-increasing need, technical rescue stands to expand and solidify into a major branch of fire and rescue services. The hazard curve paradox and a focus on the big three helps us to understand some of the obstacles and the solutions associated with this transformation from traditional to specialized operations.

Jumping Training Hurdles

Once you have accepted that rope rescue education is a never-ending process, you will discover that training opportunities are everywhere. If you expect all of your training to be sponsored by your agency, you will do yourself a serious disservice for two reasons. First, your organization probably cannot afford to train you constantly. Training is costly and time consuming, and you probably have many other job responsibilities. Second, if you work for an organization that *only* trains in rope rescue, or any other single discipline, your scope and your perception will be limited. In addition, single-source rescuers can develop a dangerous "this is the only way to accomplish this task" attitude. Sadly, some rescue instructors have adopted a "do it my way and my way only" mentality ("ropebotics"), sighting safety as the reason for their attitude. Some have even fashioned some shade-tree test results that support their single-minded techniques. Unfortunately, this type of teaching scares many potentially excellent rescuers from using the techniques they have learned. I have heard many students, after attending one of these classes, say, "The one thing I learned in that class was to *never* attempt a rope rescue. It's too exacting, and I could never design that exact system. It would be too unsafe."

Consider a more open-minded approach. There are many ways to accomplish the same task safely. The systems discussed in this book are safe, proven, and redundant rope rescue transportation tools, but there are many other systems that are just as safe, proven, and redundant. Attend as many classes by knowledgeable instructors as possible. Take the good techniques you learn and remember them for later use. Filter the techniques that you do not fully understand or that seem illogical to you. You can learn a lot about what *not* to do by attending some schools, and that, too, can be educational. You can learn much about instructors' rope techniques by closely observing their professional working demeanor and self-discipline while teaching. Are they collected or hyper? Do they wear the same personal protective equipment that they require the students performing similar tasks to wear? Can they discuss alternative solutions to rigging problems, or does it have to be done their way with no discussion allowed? Ultimately, the more classes you attend, the more tools you will have for your strategic tool box.

Read everything you can find (Figure 2–9). Catalogues are full of information. The performance characteristics of equipment are constantly changing and you should strive to stay abreast of all your equipment specifications. Finally, there are many good fire- and rescue-oriented magazines and periodicals that include rope rescue stories and techniques. A favorite is called *Nylon Highway,*[2] published twice a year by the Vertical Section of the National Speleological Society. While it is primarily devoted to vertical cavers, it has excellent articles of interest to anyone learning rope rescue.

Develop your own experience history. Do not wait for classes to come to you. Make them yourself. After getting a base knowledge from a reputable source, find a simple, low-level site and practice. If you do not have a training tower, a big tree or a fire or rescue station can be a great training site. Have a local structural engineer inspect the overhead ceiling trusses, and if they are found to be strong

---

[2] *Nylon Highway* can be obtained from the National Speleological Society, Vertical Section, P.O. Box 401, Marietta, GA 30061.

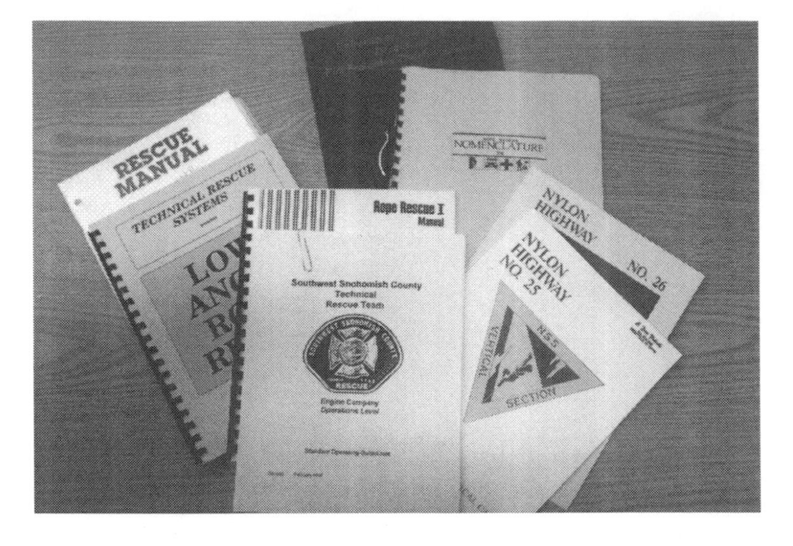

FIGURE 2–9

Information can be found from hundreds of sources, and every effort should be made to accumulate and research as many as possible.

enough, consider using them to practice ascending and building pulley systems. Engineer raise and lower systems until they become automatic. In addition, most local building owners, industrial facilities, and some landowners will allow you to use their structures or sites if you follow some simple rules (Figure 2–10):

1. Do your liability homework before you even ask. Your organization's attorney can draw up a simple liability release that will satisfy most building and landowners. They need to know that you will pay for any damage you might accidentally do to their building, and they need to be held harmless in the event of an accident. If you want to train at a wilderness-oriented site, always obtain permission from the landowner. Ask if there are any special considerations that the owner would like from your team. Also, ask for permission in writing from the owner to prevent any problems with the local police.
2. Take whatever steps are necessary to prevent making marks on the building. Boot scuff marks, rope indentations, or friction burns in railings will make it hard to get the building back again. Mind your ropes, and do not stomp on the landscaping or drop equipment in the petunias. Police your site before leaving and leave it cleaner than when you arrived.
3. Make sure you and your team are on your best behavior. Watch your language and show respect to the facility inhabitants. Open doors for them, do not crowd their elevators, and use "yes, ma'am" and "no, sir" at every opportunity. Take time to answer their questions thoroughly and patiently. Consider making professional-looking flyers to post during the days before the drill so occupants will not be intimidated when your team shows up, banging and clanging around the building.
4. In the early phases of discussions with the owners, it is a good idea to explain your training scenarios and some of the techniques that may have to be used on their building should

FIGURE 2–10

Most building owners will allow the use of their buildings for life safety rescue training if they are approached well in advance. Liability releases, courtesy, and a "leave no mark" policy assures everyone is comfortable with the training opportunity.

traditional methods prove not to be viable. Most tenants who live outside the reach of modern aerial ladders will actually appreciate having you use their building. Industrial facilities may want to have your practiced expertise at their site. Consider also the public relations value of the training but make sure you authorize any publicity with the building owners first. After working for more than a year on permission to build a highline between two bank buildings, stockholders of one of the banks stopped the whole process just weeks before the drill because "they did not want to be connected in any way with that other bank!"

5. When the training evolution is over, let your contact person know you have finished and secure the building as

FIGURE 2–11

Good teams work hard to coordinate, train, and respond to emergencies together in a professional and uniform fashion. (*Courtesy H. Dean Paderick*)

requested. If you damaged something, tell the contact person and make arrangements to fix or replace the object.

6. Finally, send the building contact person, main office, or landowner a letter of appreciation and a nicely framed picture of their building or site taken during the drill. You might even want to send a picture of your team (Figure 2–11).

■ SUMMARY

Technical rescue, and specifically rope rescue, is serious business. A team that trains, works, and researches together is much safer and more efficient than a group of individuals. *Good* technical rescue teams look beyond simple training classes and practice to become more efficient with every class and with every opportunity. Efficiency training will not be automatic. It must become a clearly stated goal, and all members must be encouraged to participate in the process. *Great* teams study the effects of the hazard curve paradox on team efficiency and develop strategies to isolate and enhance special people, special training, and special equipment.

■ KEY TERMS

Team
Network
Span of control
Personnel Accountability Report (PAR)
Hazard curve paradox
Team efficiency concept (TEC)
Contextual training
Sensory deprivation

■ REVIEW QUESTIONS

1. Technical rescue managers know that teams work better than groups of individuals for five primary reasons: specialization, sense of being, coordination and management, _____, and networking concepts.
   a. emotional enrichment
   b. actions are not repeated
   c. increased survival profile
   d. economic feasibility
2. In special operations incident command, the span of control may have to be narrowed due to _____.
   a. the highly technical nature of the equipment and systems management
   b. the narrowing of the golden hour
   c. the hazard curve paradox
   d. relatively poor survival profiles
3. A roll call, or Personnel Accountability Report (PAR) should be called for liberally, whenever a landmark event happens, _____, or timed as pre-determined by policy or by the on-scene IM, possibly every half hour on the hour during the incident.
   a. new crews arrive for assignment
   b. patients are safely packaged
   c. just after shift change
   d. a task or assignment is completed
4. The hazard curve paradox is a comparison of _____, with the hazard profile of a given operation.
   a. the personality profile of victims
   b. team member experience
   c. the funding and other resources allocated
   d. the way individual teams manage a given rescue
5. The traditional training model suggests cognitive and practical training for traditional fire and rescue services, but technical rescue instructors do not have the luxury of stopping there. Some level of _____ training is mandatory.
   a. psychological
   b. physical
   c. sensory deprivation
   d. sensitivity
6. Successful technical rescue teams have to make a conscious effort to break from traditional thinking by emphasizing the big three. They are _____.
   a. special awareness, operations, and technical training
   b. technical accountability, incident management, and hazard profiling
   c. technical assets, motivation, and customer service
   d. special people, equipment, and training

7. The rescue team's common goal is _____.
   a. problem solving
   b. to research and buy the safest equipment
   c. to practice until it hurts
   d. accountability
8. The span of control is an incident management term meaning _____.
   a. the amount of psychological composure a rescuer has during a stressful incident
   b. the number of people that can efficiently and safely be managed and continuously accounted for
   c. the budgetary (fiscal) restraint placed on specialty teams with little statistical support
   d. the manager's (chief or other leader's) control over the events that occur during an emergency incident
9. The amount of stress that a person develops or does not develop is directly proportional to the amount of _____ the nervous system.
   a. adrenaline in
   b. injury to
   c. sensory impact on
   d. emotional trauma to
10. Your organization's attorney can draw up a simple _____ that will satisfy most building and landowners.
    a. maintenance agreement
    b. list of rules
    c. code of conduct
    d. liability release

CHAPTER

3 Equipment and System Components

OBJECTIVES

Upon completion of this chapter, you should be able to:

- identify the characteristics of a rescue quality rope.
- describe five uses for the prusik loop.
- correctly measure, cut, and tie a set of prusik loops.
- describe the friction interface correlation between prusik diameter and the host rope.
- discuss mechanical rope grabs and ascenders and their pros and cons.
- identify the characteristics of a rescue quality harness.
- identify the characteristics of rescue quality hardware.
- identify the characteristics of rescue quality personal protective equipment.
- compare the qualities of aluminum versus steel hardware.
- define the major aspects of a carabiner.
- define the advantages and disadvantages of the three primary types of descent control devices.
- describe material and construction types of rescue quality rope and its proper care, maintenance, and use documentation.
- correctly tie, don, and test an improvised webbing harness.
- correctly tie, don, and test an emergency egress improvised webbing harness.

INTRODUCTION

It is easy to classify all of our rope rescue gear as either hardware or software. Hardware is made from some kind of metal, and software from some sort of fiber. Chances are, if it is hardware, it is steel, aluminum, or maybe even titanium, and, if it is software, it is probably nylon, polyester, or plastic. While this represents an extraordinary oversimplification of the materials involved, for practical (not scientific) purposes, it has proven helpful for rig masters to break down the components of a rope system into these simple, understandable categories. In this chapter, we will study the major components used to engineer a rope rescue system. Less commonly used components will be discussed in specific-use chapters.

HARDWARE

► **hardware**
According to NFPA 1983, "a type of auxiliary equipment that includes but is not limited to, ascent devices, carabiners, descent control devices, pulleys, rings, and snap links."

Hardware, according to NFPA 1983, is "A type of auxiliary equipment that includes but is not limited to, ascent devices, carabiners, descent control devices, pulleys, rings, and snap links." Swivels, buckles, rigging plates, O-rings, D-rings, litters, edge rollers, helmets, and steel-toed boots can also be categorized as hardware.

NFPA 1983
Standard on Fire Service Life Safety Rope and System Components

The materials used in hardware construction have evolved in thousands of ways in the last fifty years. **Metallurgy,** the study of the science and technology of metals, has increased the safety profile of rope rescue in the past thirty years enough to make it a viable, and even survivable, tool for modern fire and rescue services. A brief discussion of the metals used in our rope rescue tools is in order. Other than those used for exotic electronic and military applications, none of the metals are 100% pure. Adding other elements, usually more valuable metals, to the base metal greatly improves its utility and workability. The term **alloy** is used to describe the combination of two or more elements, where the majority element is called the *base* element, and the other elements are called the *alloying* elements. Metal alloying elements are primarily used to improve a metal's hardness, tensile strength, conductivity, shock resistance, and/or machinability.

► **metallurgy**
The study of the science and technology of metals.

► **alloy**
Combination of two or more elements where the majority is the *base* element and the other are the *alloying* elements.

Aluminum

► **aluminum**
The most common metal used today in the construction of rope rescue tools.

Aluminum is the most common metal used today in the construction of rope rescue tools. It is plentiful in the United States and therefore relatively inexpensive. It is very lightweight compared to most other metals and, since it contains no iron (nonferrous), it will not rust, and it is nonmagnetic. Aluminum can be worked very easily into shapes needed for rope rescue tools, primarily DCDs and connectors in the form of carabiners.

Of course, not all aluminum is alike and very little is 100% aluminum—for good reason. Pure aluminum is so soft that you can press your fingernail into its surface. Therefore, all the aluminum used in rope rescue is aluminum alloy or combinations of exact percentages of other metals. Alloys combine metals with needed characteristics to make a new metal that is much more useful than any of the individual parts. It works something like a metallurgical team.

Mined from bauxite, aluminum was first discovered in 1727, but it was retained only in small amounts until it was electrolytically reduced into usable portions in 1885. Aluminum has a much greater strength-to-weight ratio than steel, and it is generally easier to work into useable configurations. Aluminum's yield strength is reduced drastically above about 375°F, so it cannot replace steel in large structures where there is a possibility of fire or locally applied heat. However, the introduction of even minute amounts of other elements into aluminum makes possible an infinite variety of alloys. The Aluminum Association of America uses a numbering system to detail the following alloy and aluminum contents:

Series No.	Alloying agent(s)
1000	99% pure aluminum
2000	Copper
3000	Manganese
4000	Silicon
5000	Magnesium
6000	Magnesium *and* silica
7000	Zinc
8000	Other elements
9000	Unused to date
X000	Experimental

The most commonly used aluminum alloys in the manufacture of rope rescue hardware are 2024, 6061, and 7075. The 2000 series, aluminum-copper alloys, is the oldest and most widely used. Copper content ranges from about 2% to 6%, and until the development of the 7000 series, these aluminum alloys were the highest strength alloys available. The 2000 series is more susceptible to corrosion and is almost impossible to weld safely. The first high-strength lightweight aluminum alloy was 2024. The 6000 series (aluminum-magnesium-silica) alloys are not as strong as the 2000 and 7000 series, but they have a greater corrosion resistance, are more ductile (flexible), and are commonly used in bridge railings, moderate strength structural components, and furniture. The 7000 series has the highest strength rating of any aluminum-alloy group. It is obtained by adding 1% to 7.5% zinc and 2.5% to 3.3% magnesium. Chromium and copper also provide added strength but lower the weldability and corrosion resistance.

By far, the most common aluminum alloy used today is 6061. Most rod or angle stock, tubing, aluminum carabiners, rigging plates, swivels, pulleys, fire service ladders, vehicle body and structural components, and ladder truck outrigger pads are made of 6061. It is composed of 97.9% aluminum, 0.25% copper, 0.6% silicone, 1.0% magnesium, and 0.25% chrome. It has high electrical and heat conductivity qualities and is soft enough (ductile) to be worked into usable shapes without breaking. The other commonly used aluminum alloys in rope rescue equipment are 2024 and 7075. Alloy 2024 is 93.4% aluminum, 4.5% copper, 0.6% manganese, and 1.5% magnesium. It is harder than 6061 and has a higher tensile strength of 27,000 lbf compared to 6061's 18,000 lbf tensile strength. A modern "aircraft" aluminum alloy, 7075 has very high tensile strength properties. It is 90% aluminum, 1.6% copper, 2.5% magnesium, 0.3% chrome, and 5.6% zinc, and it has a remarkable tensile strength of 33,000 lbf. It has lower electrical and heat conductivity qualities and is much harder and less ductile (more brittle) than either 6061 or 2024. Alloy 7075 is relatively difficult to work due to its hardness and more subject to tearing or breaking under extreme shock loads than are 6061 or 2024. (See Figures 3–1A through 3–1G.)

Alloys are only the beginning of the aluminum materials story. Much of a metal's strength is a result of the alignment of its molecules. Molten aluminum ingots are pressure rolled, either hot or cold, to align the molecules and strengthen the metal. The pressure heats the aluminum, giving it plastic-like qualities that, again, help to create molecular alignment. Another common method aligning molecules is to extrude the metal under pressure, literally forcing the aluminum through small holes. Forging aluminum is a process taken from blacksmiths, where the aluminum is hydraulically pressurized (formerly hammered) into shapes or into plates from which shapes can be cut. Forging is an excellent method of aligning molecules and is the process by which much rope rescue equipment is manufactured. Casting is a process in which molten aluminum is poured into molds and cured. Casting metal tends to scatter the molecules into nonlinear disarray. Early aluminum Jumar ascenders were cast into molds and were known to break under even minor shock loads. There are some military specification (mil-spec) cast ratings, often called Jewelers Cast, that are very strong and are sometimes applied to rope rescue equipment.

Heat treatment plays an important role in the effectiveness of aluminum alloys. The Aluminum Association's rating system lists a heat treatment rating after the composition rating. The most common heat treatment method used for aluminum is T-6. The rating is written as 6061 T-6 or 7075 T-6. Like the grain in a hardwood baseball bat, a greater alignment of molecules results in greater tensile strength. Heat treatment methods again plasticize the aluminum molecules, loosening them so they can become oriented in a more linear fashion. Solution heat treatment is a science in and of itself. Metal is brought to a certain temperature and kept there for a certain length of time, each of which is manipulated to create maximum strength. The amount of time in a quenching solution, usually oil or water, is varied for maximum effective tensile increase. The heating allows further molecular alignment, and the sudden quenching "freezes" it in place. The T rating is an abbreviation of the *treatment* used. Heat-treating aluminum doubles or sometimes triples its tensile strength. The annealed strength of 6061 is 18K lbf (thousands of pounds force), but treated with T-6 it is 45K lbf. Likewise, 2024's annealed strength is 27K lbf and 2024 T-6's is 69K lbf. The annealed, but not heat treated, tensile strength of 7075 T-0 is 33K lbf, whereas 7075 T-6 is 88K lbf.

Aluminum does have negative characteristics that a rig master should be aware of, such as its capacity to oxidize when left exposed to oxygen. A shiny new aluminum carabiner begins to oxidize almost instantly after manufacture. Oxidation can create a white powdery coating and, like rust, it can undermine the performance of equipment either by hindering the movement of its

FIGURE 3–1

Metals worked into rope rescue tools: (A) Metal rod stock ready for shaping and hardening. (B) Rod stock cut to length for making carabiners. (C) Carabiner blanks (in box) and rappel racks (hanging in back) ready for additional machining and hardening. (D) Rod stock stainless steel being shaped into an open-end rappel rack on a shaping jig. *Note:* "Eye" has been welded and short leg has been threaded to receive brake bar retaining nut. (E) Parts inside of a giant tumbler. Water and river rock tumbler smooths burrs and rough angles. (F) Carabiner blanks back fresh from heat treatment processing. Note remnants of oil quenching bath on blanks. (G) Final stage of carabiner blank machining. The next step is tumbling and then assembly of gate lock and spring components. (*Courtesy Seattle Manufacturing Corporation, SMC*)

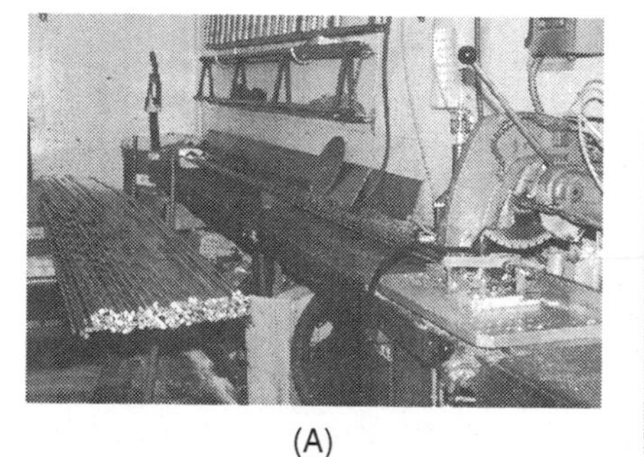

(A)

(B)

(C)

(D)

(E)

(F)

(G)

parts and/or by removing enough metal to weaken an aluminum component. Oxidation can also cause a very high-temperature spark that can ignite some flammable gases and vapors. This would not be good in a highly flammable rescue environment, like the point of entry on a gasoline tanker. To a lesser degree, shiny aluminum is a visual detriment to tactical teams working at night who do not want to be seen.

Because aluminum is a very soft metal, frictional surfaces of noncoated aluminum wear very quickly. Shiny new aluminum eight-plates may only last for a couple hundred rappels when used with sandy or muddy ropes. The heat of a long lower on the bars of a brake bar rack soften the aluminum even more and wear is accelerated.

Manufacturers have addressed the problem of aluminum's softness and tendency to oxidize by coating it with various materials, commonly referred to as hardcoat and softcoat anodization.

► **anodization**

A process whereby aluminum is subjected to the electrolytic action of the coating material in an energized solution, causing a strong molecular bond.

Anodization is a process whereby the aluminum is subjected to the electrolytic action of the coating material in an energized solution, causing a strong molecular bond. Simply put, anodized solutions stick better and last longer than paint. Softcoat applications are only marginally tougher than paint and act only to keep reflectivity down and to prevent oxidation from the time it leaves the factory through the first couple of rappels. Softcoat is usually base black or bright red and relatively inexpensive. Hardcoat is usually gray in color and much harder than softcoat. It is, in fact, much harder than the aluminum to which it is applied. Hardcoat extends the life of the frictional surfaces of aluminum equipment dramatically. Thousands of rappels or lowers on clean ropes and hundreds of rappels or lowers on muddy and sandy ropes are the normal life span of descent control devices (DCDs) with hardcoat anodization. Like all good things, hardcoat has its disadvantages. It is relatively brittle, and dropping hardcoated devices on rocks or concrete can chip the hardcoat easily. Once the hardcoat has worn away on even a tiny spot of a friction surface, the aluminum underneath wears away rapidly. This can create an incredibly efficient software cutting device (Figure 3–2). An eight-plate, for example, that is start-

FIGURE 3–2

Magnified view of hardcoat anodized surface worn away from aluminum eight-plate creating extremely sharp anodized edges that can rapidly abrade rope.

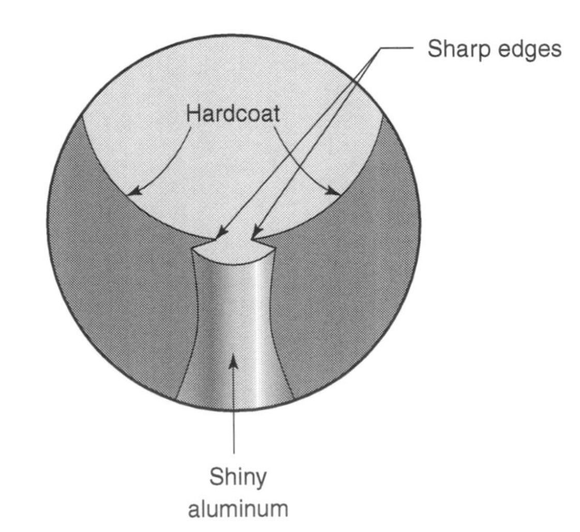

ing to show shiny aluminum through the hardcoat is an almost perfect tool for rapidly destroying a good rope. The cut-away edges of the hardcoat are filed razor sharp by the friction of the moving rope. The rope will have the characteristic spiral pattern inherent with eight-plate use, but with seriously frayed fibers highlighting the spiral. Several rappels on a worn hardcoat eight-plate will render the rope very fuzzy and of questionable life safety value. The bottom line is, destroy any hardcoat aluminum friction devices when the hardcoat has worn away and shiny aluminum is showing through.

Steel

▶ **steel**

Metal primarily composed of iron that has been refined from naturally occurring iron ore.

Steel is popular because of its tremendous strength and inexpensive cost relative to other metals. **Steel** is primarily composed of iron that has been refined from naturally occurring iron ore. It is graded according to the amount of carbon that has been mixed with the iron to produce desired effects. Carbon is usually added in percentages ranging from 0.20% (low-carbon steel) to a maximum of 2% (ultra high-carbon steel) to make steel harder. However, carbon contents of greater than 1.2% are unusually brittle and difficult to work into useful components. Iron with contents of carbon between 2% and 6% is classified as *cast iron* and has very little application in the production of rope rescue equipment. Because steel is composed primarily of iron, it is magnetic (*ferromagnetic*), and when it is exposed to oxygen, it forms a coating of ferric oxide (FeO), or rust. Other common alloying agents are sulfur, manganese, and sometimes phosphorus, used to increase machinability, to retard magnetic aging and to reduce corrosion, respectively.

▶ **stainless steel**

A type of steel that is very strong, has a relatively low iron content, and is alloyed to be almost rust free and nonmagnetic.

Stainless steel First produced in the United States in 1914, **stainless steel** today is becoming increasingly more common as a material in the manufacture of rope rescue equipment. Carabiners, eight-plates, brake bar racks, pulley side plates, axles, ball bearings, and rope grab shells are but a few of the devices being made from stainless steel because of its strength and unparalleled corrosion resistance. Stainless steel's base element is iron; however, a large percentage of chromium, 12% to 30%, is added to increase corrosion resistance. Chromium binds with ferric particles to produce a clear surface chromium oxide film that forms in the presence of oxygen. This film essentially enables stainless steel to be nonporous, insoluble, and self healing, meaning that through repeated machining and everyday use, it stays almost completely corrosion free.

Stainless steel's other common alloying agents are nickel, copper, manganese, and silicon that, in various amounts, increase annealed strength, electrical conductivity, and machinability. Some stainless steel, such as the American Iron and Steel Institute's (AISI) standard Type 400 series stainless steel used in the manufacture of industrial chemical pipes and containers, while extremely corrosion resistant, *is* magnetic. Most stainless steel used in rope rescue equipment is of the AISI Type 300—nonmagnetic, tough, and ductile stainless steel.

NOTE: In the following discussions on DCDs, there is mention of how to use them for rappeling, or for lowering operations. This is intended for reference only, as detailed instructions on rappeling are included in Chapter 5, and instructions for using DCDs in lowering operations are included in Chapter 7.

The manufacturers' responsibility is to research the myriad materials options and develop a product with the perfect blend of alloys, with the proper heat treatment technique, and covered if necessary with the most advantageous coating. Finally, its shape and interacting components must serve its intended purpose without causing self-destructive action.

Descent Control Devices

▶ **descent control device (DCD)**
Metal configurations (friction producers) that rub the rope and slow its movement.

Descent control devices (DCDs) are simply friction producers, metal configurations that rub the rope and slow its movement. By anchoring a DCD, we are effectively causing the anchor to carry a majority of the load, so the operator need only control a small amount of the remaining force, just enough to allow the load to move or stop when necessary. It is up to us as operators to control the interaction of the rope with the device, adjust the friction, and therefore control the descent of the load. Again, as the massive gravity of the earth attempts to pull objects toward its center, the rope gives us substance on which to attach the DCD. Either the rope or the DCD must be securely anchored, allowing the other to move freely. If the rope is anchored, the system is set up to rappel; if the DCD is anchored, the system is set up for a lowering operation. The movement causes friction, and both the rope and the DCD develop heat, which is dissipated into the air. The descent energy is altered. Instead of suddenly slamming into the ground, creating enough energy to explode a human body, the same amount of energy is just spread out over time and distance.

Almost anything can be used as a DCD if it is strong enough to withstand the rope rubbing across its surface. Wrapping the rope several times around an anchored object produces a wonderful DCD. Trees, standpipes, steel railings, a big rock, a mainsail mast, or even the smooth edge of a parapet wall will cause friction with the rope and, therefore, some or even complete control.

▶ **Dulfersitz**
Body rappel, a method of descending in which the body becomes the DCD and the rope is anchored at the top.

Early rope transportation methods relied on throwing a couple of round turns around the anchors, tying the ropes around the people's waists and lowering them to the target. Later, people learned to **Dulfersitz,** or body rappel. In this method of descending, the body becomes the DCD, and the rope is anchored at the top. The rope is wrapped around the arms, waist, back, and legs, in any number of configurations to gain enough friction to control descent. Dulfering in the vertical environment is very dangerous, and next to impossible for all but the strongest and most agile humans. It can be used to aid in the descent of a scree slope (Figure 3–3) or low-angle embankment, but all Dulfers expose the rappeler to rope burns and the possibility of release and fall from the rope.

Attempts at rappeling, rather than being lowered on the rope, increased rapidly during World War I. European alpinists started using "swami" belts—leather, cotton, or wool webbing wrapped around one's waist, to support that person's weight. Alpinists also were successful at running several round turns around a steel O-ring for friction control (Figure 3–4). This technique made it hard to load anywhere but at the end of the rope due to the threading of

Person performing Dulfersitz (body) rappel on scree slope.

Laid rope wrapped around ring for descent control.

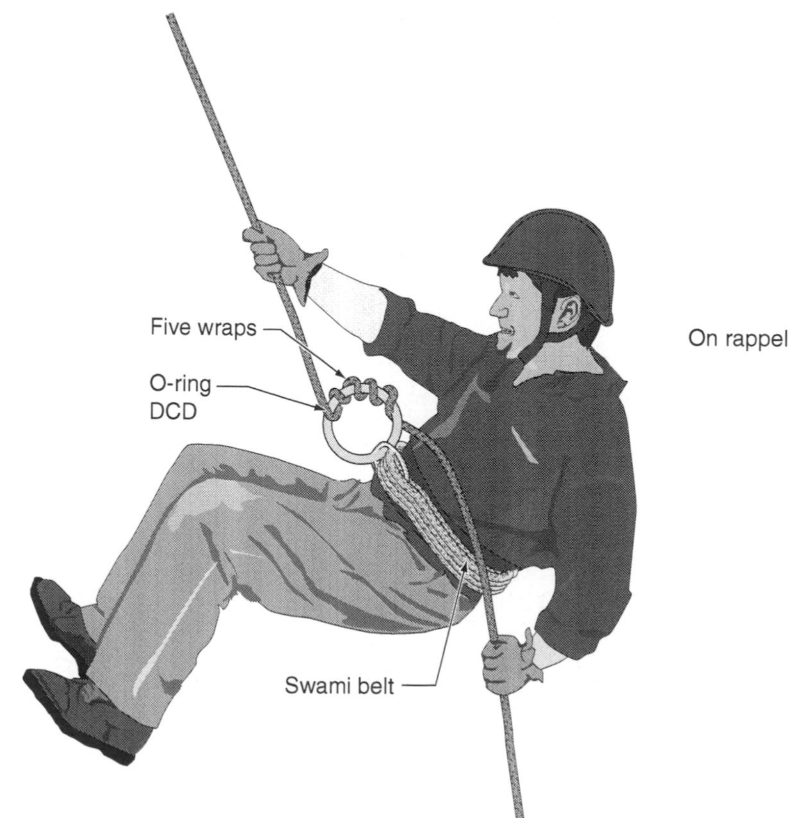

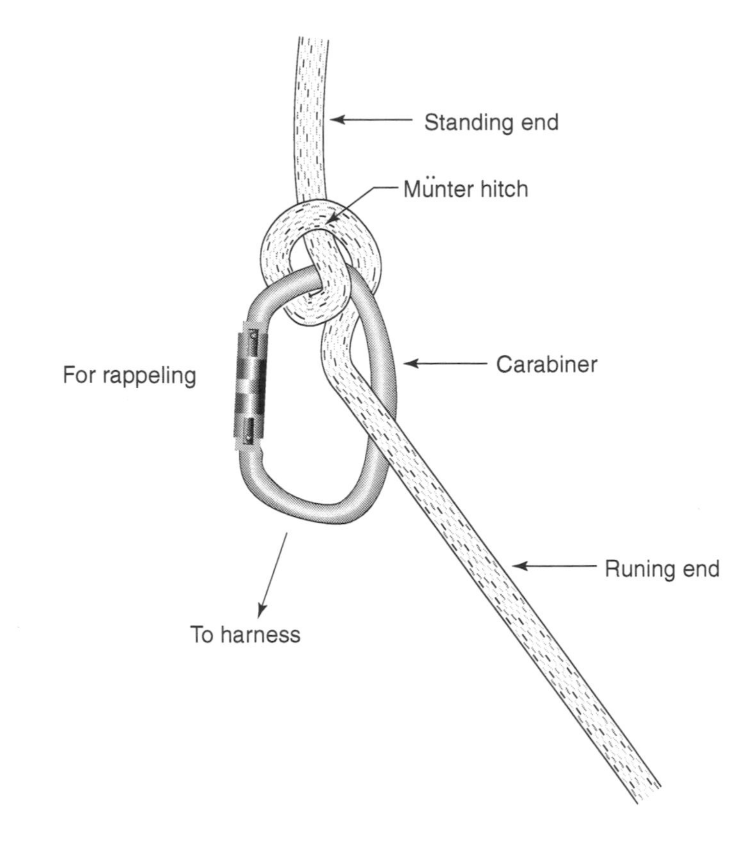

FIGURE 3–5
A münter hitch attached to a carabiner and a rappeling person's harness as a descent control device.

the O-ring. Also, stopping to work hands free was slow and physically demanding.

Firefighters were rappeling on their ladder belt hooks as early as the late 1800s. Steel snap-link devices used for cargo handling and sailing-ship rigging were gaining popularity as moving DCDs on natural fiber ropes as early as the 1920s. Military applications became frequent in the 1950s with the advent of modern steel oval carabiner designs. The problems with early carabiner and big hook rappels, ignoring inherent problems with natural fiber ropes, was determining the proper number of wraps and the presence of a gate by which the rope could inadvertently be forced while on rappel. Too few wraps resulted in too fast a rappel, and too many wraps made sliding the rope very difficult. There are dozens of stories of firefighters and military people falling out of their gated devices because they were wrapped improperly. The downward braking force necessary to slow descent also opened the gate, unwrapping the rope. Even automatically locking gates could be wound open by improperly wrapping the device. The carabiner and big hook rappel can be successfully performed with plenty of low-level practice as long as the rope is wrapped away from the gate.

The münter hitch can be used as a DCD when attached to a carabiner (Figure 3–5) that is secured to a harness. It requires plenty of low-level practice and a comfortable understanding of most open (less friction) and most closed (most friction) (Figure 3–6). It is not as easy to control as a dedicated DCD. However, in a last-ditch bailout situation, it may be the only available alternative if all that is available is a carabiner, some rope, and a last-chance belt.

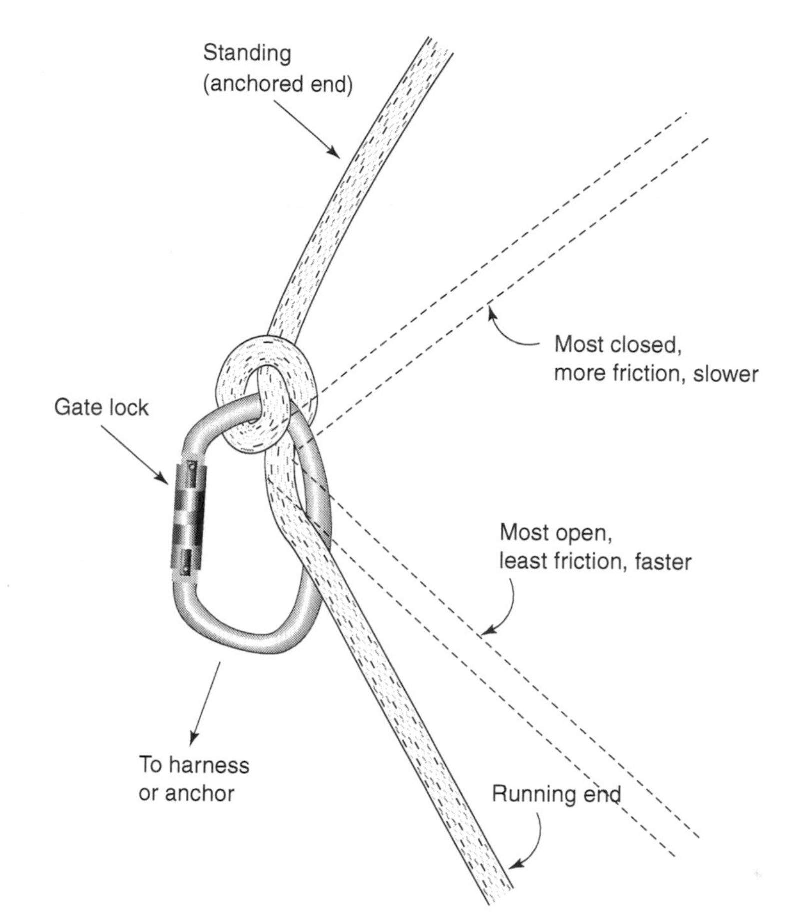

FIGURE 3–6
The münter hitch in the most closed position for maximum friction and in the most open position for less friction.

Literally thousands of devices have been crafted to be used as DCDs for rope movement control. DCDs can be placed in one of four categories: eight-plates, brake bar racks, tube type, or miscellaneous (Figure 3–7).

► **eight-plates**
Relatively inexpensive and the most common of descent control devices.

Eight-plates (Figure 3–8A) are easily the most abundant of the DCDs for a number of reasons. Eight-plates are fairly inexpensive. At print, aluminum ones, depending on size, cost from $20 to $35 and steel ones run from $35 to $60. Eight-plates do not have any moving parts and are therefore somewhat easier to use than other DCDs. Thousands of people are trained only in the basic levels of rope work and only learn to rappel on eight-plates, many more progress through the advanced levels to learn to use and then buy other DCDs. Also, there are thousands of people who do not require the versatility of the other DCDs. Rock climbers, short-drop cavers, and recreational rappelers are comfortable with the utility, weight, and compactness of the eight-plate descender.

An eight-plate is reeved for rappeling by passing a bight through the large top opening and laying it over and behind the bottom neck (Figure 3–8B). The eight-plate is then attached to the harness via a carabiner. The friction can be doubled for rappeling with a patient in tow simply by double passing the rope through the large top opening. Instead of stopping with the first wrap around the back of the neck, pass the bight back through the opening and then over the bottom neck.

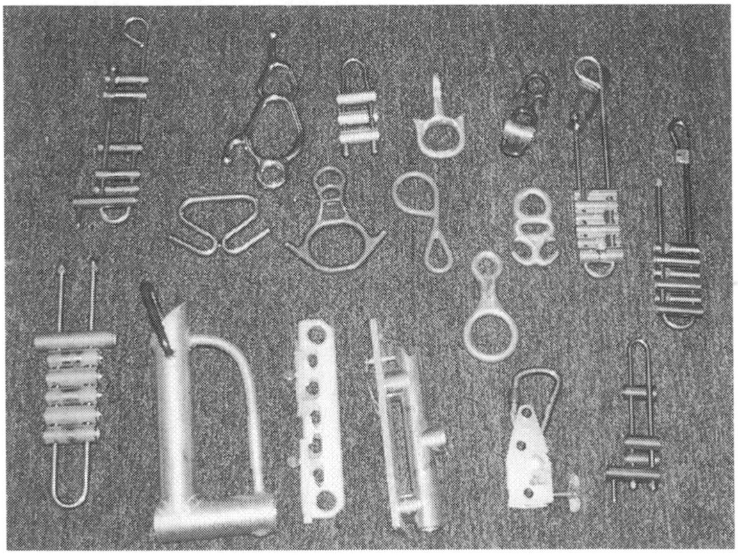

FIGURE 3-7

Various descent control devices (DCDs). (*Courtesy Martin Grube*)

Wrapping an eight-plate's large upper ring with five round turns (Figure 3–8C) is a great emergency escape technique that mimics the carabiner wrap's advantages (see Chapter 6). It does not have the disadvantage of the gate inadvertently popping open. Stashed in a personal escape bag, the five wraps cannot come off the rope like they could if the eight-plate were wrapped in the traditional manner.

► **brake bar rack**

A versatile descent control device made popular by caver John Cole in the mid-1960s.

The **brake bar rack** (Figure 3–9A), made popular by caver John Cole in the mid-1960s, is an enormously versatile DCD that is very useful to rope rescuers. Because the rack has some moving parts, however, it is shunned by many people who have not had the opportunity to experience its simplicity and marvel at its advantages. Reeving the rope through interlocked carabiners was the early, and much weaker, predecessor to the rack. Machined-steel rod stock or cut-aluminum rod stock bars were later added to carabiners, perpendicular to the long axis, as friction surfaces. However, the brake

FIGURE 3–8

(A) Various eight-plate designs. (B) Reeving the rope onto an eight-plate, traditional method. (C) One method of reeving an eight-plate for personal escape. (*Courtesy Martin Grube*)

(A)

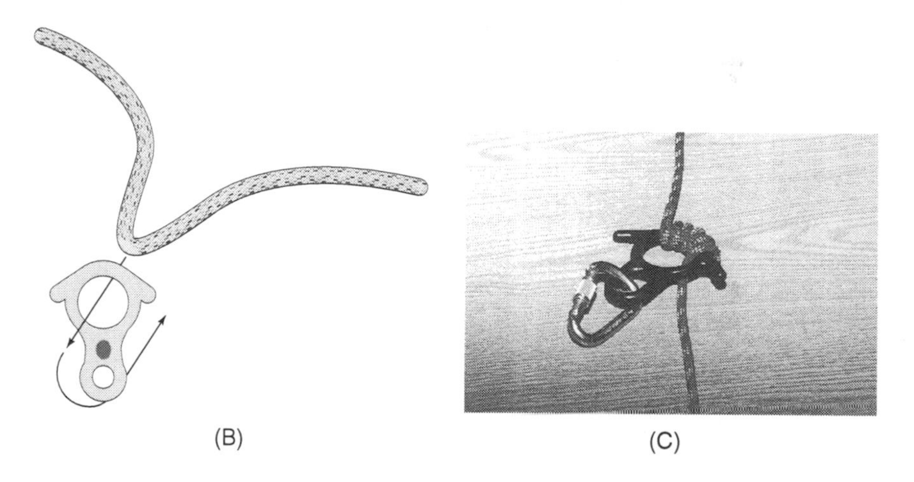

(B) (C)

bar carabiners placed considerable force on the gate of the carabiner, causing some failures and several fatalities. The control and flow of the rope on brake bar carabiners was considerable, but it only took the invention of a strong U-shaped, stainless steel rack frame to eliminate the carabiner gate weakness problems.

Essentially, the open-end rack best suited for rappeling is a 28-inch piece of rod stock, type 304 stainless steel, that has been bent into a U shape (Figure 3–9B). One end is curled into an "eye," welded, and dye-penetrant inspected. Some manufacturers prefer to coil the eye to eliminate potential problems with rare inferior eye welds. The eye is used to attach to the rappeler's harness or anchored to use in lowering systems and can be oriented parallel to or perpendicular to the plane of the rack. This 90-degree rotation is done by the manufacturer. The preference is entirely personal and of almost no real technical consequence, as any rack will freely rotate 90 degrees at its carabiner attachment for ease of tying off. Racks are oriented for either right-handed or left-handed rappelers. The open side of the rack, the nut side, should be oriented toward the preferred hand. This keeps the open-end accessible for easy control and quick tie-offs. Six brake bars are then added to the rack

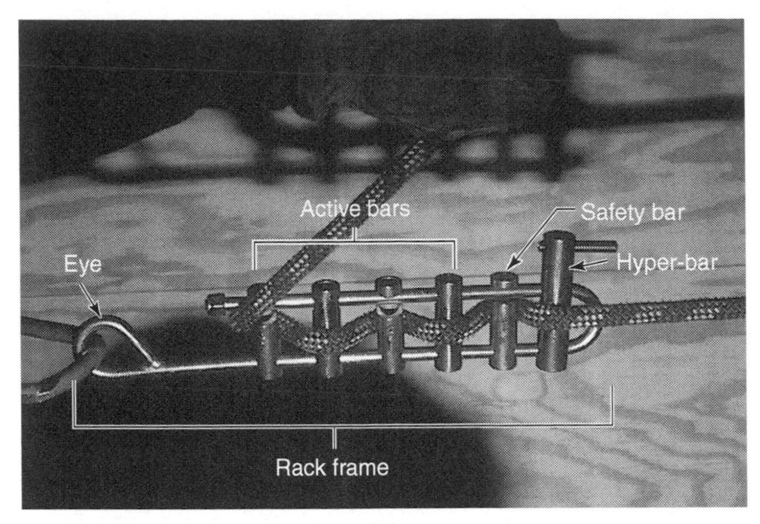

(A)

(B)

FIGURE 3–9

(A) The open-end brake bar rappel rack. *Note:* The second bar from the top has a straight slot as a safety reminder so that the rack cannot be inadvertently reeved backward. The rest of the bars have slanted slots that allow them to stay on the rack until needed. (B) After welding the eye and threading the short leg bar retaining nut, the rack is hand shaped on the jig into the familiar rappel rack configuration. (*A, Courtesy Martin C. Grube; B, Courtesy Seattle Manufacturing Corporation, SMC*)

in a specific order to ensure proper rope loading. It is helpful to use a top bar manufactured with a training groove to train the rope towards the center of the rack during use. A **hyper-bar** (Figure 3–9B) is a bar that is longer than normal bars, extending an inch or so outside of the rack frame. Hyper-bars usually have a shear pin driven perpendicular to the bar near its end to facilitate a rapid lock off. The second bar should always be arranged with a straight slot to prevent the rope from accidentally being reeved onto the rack incorrectly. As opposed to a slanted slot bar, the straight slot bar cannot bear any weight if the rack is reeved backwards.

► **hyper-bar**

A longer than normal top bar, extending an inch or so outside of the rack frame.

The remaining four slanted slot bars are then added to the rack. The slanted slots in the bottom four bars help to keep them in place on the rack. This way, when some friction is needed, the bottom active bar is in place and is not freewheeling around near the eye of the rack. The other end of the rack frame is threaded and a self-locking nut attached to prevent the brake bars from coming off the rack when it is not being used.

NOTE: There are horror stories of rappelers accidentally reeving a rack backwards. With all slanted slot bars, a rappeler could get a little tension on the rope, enough sometimes to even load the rope and start rappeling. When the tension on the rack exceeded the capacity of the slots in the bars to hold onto the rack, the bars would come flying off and the rack would be detached from the rope—not a good thing.

Brake bar material is usually either 6061 T-6 aluminum or 301 stainless steel. Aluminum bars are nonanodized and provide a great deal of friction and control. They do, however, wear out quickly, and they leave aluminum oxide on the rope. At best, it discolors the ropes and keeps gloves dirty, and at worst, it is a mild deteriorant to nylon in the presence of moisture. Stainless steel is a vastly superior material for brake bars. It is much harder than aluminum and, therefore, it will last much longer. Its hardness, however, gives it a lower friction coefficient that is easily overcome by adding a bar or two when additional friction is required. Tubular stainless steel stock is the preferred choice for bars and was the only choice for years. Aluminum was cheaper, so most of the racks produced during the 1980s came with aluminum bars. Few people put hundreds of miles on racks, and aluminum bars fit the bill nicely. One manufacturer replaced tube stock bars with stamped and bent stainless plate U-shaped pieces that were very sloppy. There was no question, however, about which way to reeve the rack with the open-back stamped bars. The opening in the back prevented attaching the rope to the rack backwards. Tubular stainless bars were virtually impossible to get from about 1985 through 1995. In 1995, due to the demand for real brake bars, stainless tubular top bars could be found at rope shows periodically throughout the country. Several manufacturers have hinted that tubular stainless bars are going to make a comeback. The ideal rack would have top hyper-bars with protruding pins that allow for fast and easy tie-offs.

▶ closed-end rack (or closed frame rack)

Modification of the John Cole brake bar rack.

The **closed-end rack (or closed frame rack)** (Figure 3–10A) is a modification of the John Cole brake bar rack. It (Figure 3–10B) is made from stainless steel rod stock with double-threaded nuts at the end of each leg. Most manufacturers use bars made from rope-grabbing 6061 T-6 aluminum for massive amounts of friction and to keep the price down. The aluminum wears out relatively fast, though, with sandy or muddy ropes and leaves a smear of gray aluminum dust on every rope that goes through the rack. Bars made from stainless tube stock are greatly preferred over aluminum ones for multiple rope rescue operations.

The closed-end rack design improves on the open-end rack by balancing and, therefore, sharing the load on both legs of the rack. This prevents neck bending that can occur when an open-end rack is overloaded, shock loaded, or loaded sideways over an edge. The bars are positioned on the rack so that every even bar (2, 4, and 6) can pivot away from the rack, and every odd bar (1, 3, and 5) is prevented from pivoting (Figure 3–10C). In this manner a bight of rope is passed through the plane of the rack frame to add a bar. The closed-end rack is relatively awkward to use for rappeling, because adding bars requires passing bights of the rappel rope through the plane of the rack frame while hanging on rappel and popping in another bar precisely where the waiting bight should be. It is, however, unquestionably the best device available for engineering rescue load lowering systems. When properly anchored, the closed-end rack is infinitely and easily adjustable by one person for loads exceeding 1,000 lbf. Friction is controlled *first* by closing the angle of the rope

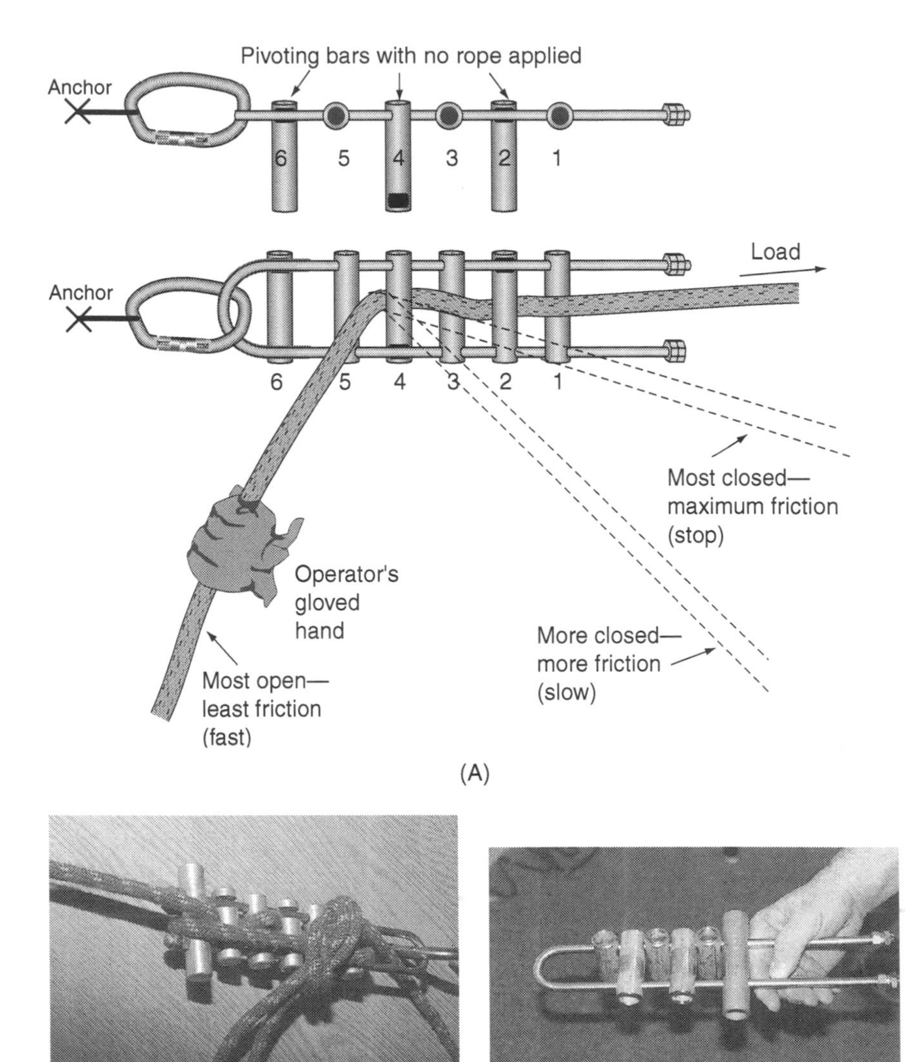

FIGURE 3-10

(A) The closed-end rescue (or rigging) rack. Much stronger for rescue lowering than the open-end rack, this rack is balanced and has the ability to apply almost infinite friction to a rope or ropes. Fine-tuning a lowering operation is performed by changing the angle of approach the rope has to the rack. Closing the rope-to-rack angle increases friction by crushing the bars together and increasing surface area on the bars, slowing or stopping the lowering. Opening the rope-to-rack angle decreases friction by allowing the bars to spread and decreasing surface area on the bars, which allows the rope to move through the rack. (B) The closed-end rescue rack with all bars activated for maximum friction and the rack tied off to hold the load steady. (C) A closed-end rescue rack. *Note:* Every other bar pivots away from the rack frame. Bights of rope must be passed through the plane of the rack and the bars swung into position. The more bars are added, the more friction is available to control the lowering. (*B and C, Courtesy Martin Grube*)

NOTE: The anchor side of the rack is the U bend in the rack frame, and the load side is toward the doubled machine nuts at the end of each leg. This looks backwards from the open-end rappeling rack.

on the anchor side of the rack with respect to the rope on the load side of the rack.

Closing the rope angle compresses the bars, increases the surface area of the rope in contact with the bars, and dramatically increases the friction. In a shock load situation, any active bars are automatically compressed, forcing the load to stop. Friction is controlled *second* by passing a bight through the plane of the rack and adding one of the next pivoting bars. Two ropes can be reeved onto the rack when necessary to perform two line lowers; however, it is preferable to use separate racks for finer control and system redundancy. Additionally, the closed-end rack causes kernmantle rope to align its fibers nicely to prevent kinking and coiling, which are common problems with eight-plates and tube style descenders.

▶ **tube type DCD**

Anchored pipe-like mechanisms with rope spiraled around them like stripes on a barber pole.

Tube type DCDs come in various forms and are basically anchored pipes with rope spiraled around them like stripes on a barber pole. They provide excellent control for lowering systems, and friction can be added or subtracted by adding or deleting wraps on

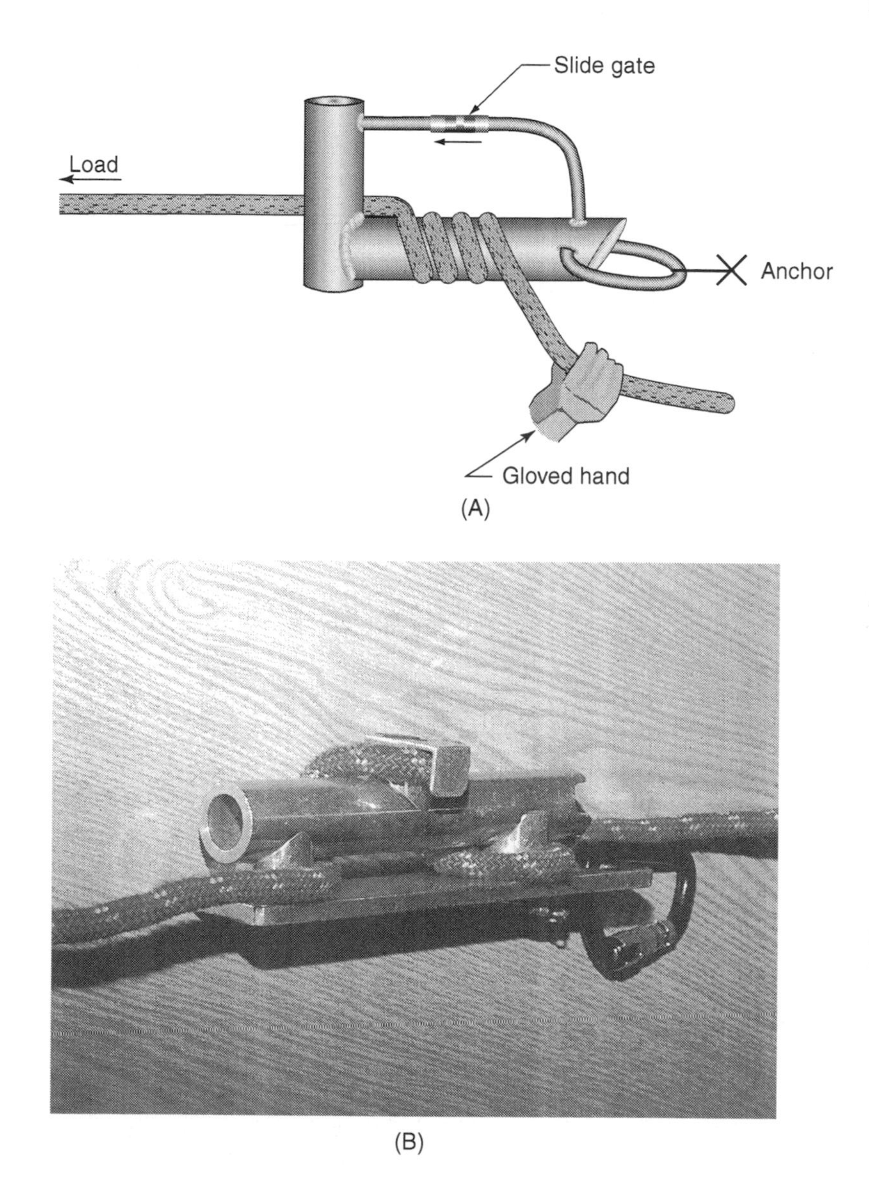

FIGURE 3–11

(A) A single tube style DCD develops friction because of the large amount of rope to device surface area winding around the tube. It is essentially a high strength tensionless and adjustable friction anchor, but it does cause the rope to develop twists as the moving rope spirals around the tube during a lowering. (B) Some tube style DCDs reduce rope twist by counteracting one twist with another (one tube rotates the rope clockwise, the other tube counterclockwise), effectively balancing the twists. This arrangement is superior to the single tube style DCD. (*B, Courtesy Martin Grube*)

the tube or on other surfaces of the device (Figure 3–11A). In some shock load situations, it can be difficult to counter the increased force rapidly enough to arrest a fall. Most tube type DCDs are made from tubular or machined 6061 or 2024 T-6 stock. Each tube has a hole machined in one end for an anchor attachment and a retaining mechanism that prevents the rope from coming off the tube during use. Some tubular devices cause the rope to twist, similar to an eight-plate. To correct this problem, some manufacturers have channeled the rope partially around the tube, over a post midway that reverses the direction of the spiral, and back onto the tube (Figure 3–11B). This effectively counteracts most of the spiral, thereby eliminating rope twist.

► **miscellaneous DCD**

Myriad devices sometimes used for special applications or for recreational use.

Miscellaneous DCDs (Figure 3–12) include the myriad devices that are sometimes used for special applications or for recreational uses. As stated before, almost any sturdy device that will accept a rope and a carabiner can be used as a DCD. Bobbin type DCDs (Figure 3–13) direct the rope around a series of two or three bobbins that

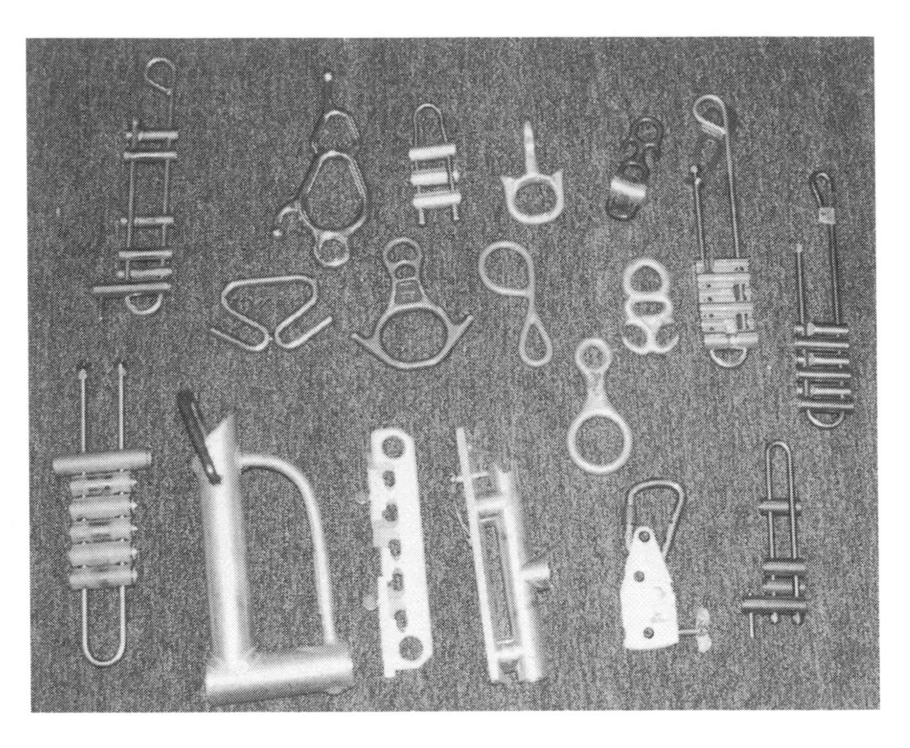

FIGURE 3-12

DCDs come in all sizes and shapes, producing all different advantages and disadvantages. The common denominator is that they develop friction when a moving rope passes over their surfaces. (*Courtesy Martin Grube*)

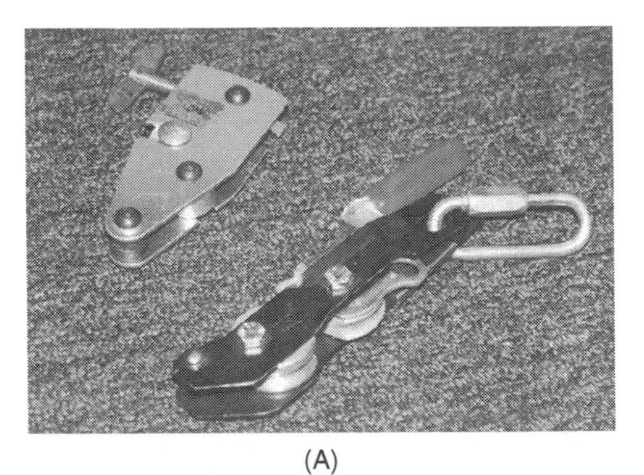

(A)

(B)

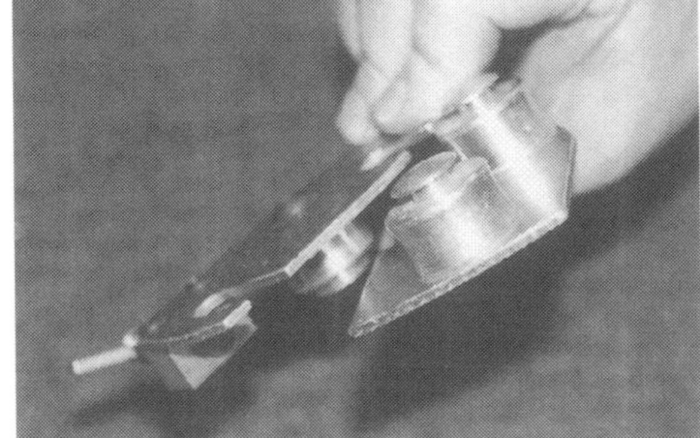

(C)

FIGURE 3-13

(A) and (B) Bobbin style DCDs gain friction by having the rope pass over a series of nonrolling metal bobbins. Friction is adjusted using the handle for leverage or by adjusting the space between bobbins with a screw-like mechanism. (C) Bobbins are exposed by shifting the spring-activated side plates. (*Courtesy Martin Grube*)

provide the friction. A handle, or screw-type device, controls the interface between the bollards for variable friction. Advantages are that the rope tracks straight through the device and prevents spiraling. Disadvantages are that preset bobbin interfaces do not allow for changes in rope texture. As a rappel or a lower is being performed, the load can move at different rates for used sections of the rope than for new sections of the rope.

Spindle type descenders coil the rope around a spindle, and a case keeps the rope from coming off the spindle. It is possible in shock-loading situations that some spindle type devices can cause the case to be displaced and the rope to separate from the device while under load.

Marking Equipment for Identification

Part of the maintenance program for your personal equipment and your team's equipment should be to mark it for identification (Figure 3–14). Often when you set up rope systems, gear from many sources gets mixed together. Color-coded equipment is quick and easy to identify during the breakdown and inventory process. Each of the fire stations in many cities has its own color code, so any time a piece of equipment turns up out of place, it is a simple matter to return it. Years ago, I color coded my gear fire-engine red. After several years, so many people across the state had their own gear that red was being reused. Eventually it led to some confusion, so I started putting a black stripe around the red, or a black dot on a red field. I am constantly surprised when I find that a missing carabiner turned up in a school somewhere hundreds of miles from home. I've even had people who recognized my color code mail me gear that was accidentally stowed in someone else's bag. Some manufacturers of paint make paint pens like you might use on model airplanes. They are great for marking your gear quickly and cleanly. When painting equipment, make sure not to paint any of the interacting parts, as the paint will gum them up. I like to paint carabiners on the gate lock. It wears off somewhat during use, but enough usually stays down in the grooves for it to remain visible. Also, do not paint DCDs where the rope is likely to rub. It will just wear off.

Software can be marked easily with a heavy permanent marking pen. There was some discussion years ago that the chemicals in some of the markers could deteriorate webbing, harnesses, and rope. Most of those concerns have since been dispelled, but it is wise to consult the manufacturer of the equipment before marking software with any type of paint or marker.

Stamping or engraving hardware (Figure 3–15) is another excellent method of marking your gear. Stamped gear may be a little harder to recognize at a distance and, for the less than scrupulous, harder to remove than paint. Make sure to check with the manufacturer before choosing a metal marking device. Most quality hardware manufacturers have no problem with either stamping or engraving, but some recommend engraving only. Stamping subjects equipment to a fairly severe shock, while engraving actually removes minute amounts of the metal. Neither method really adversely affects the equipment, but always check with the manufacturer before choosing.

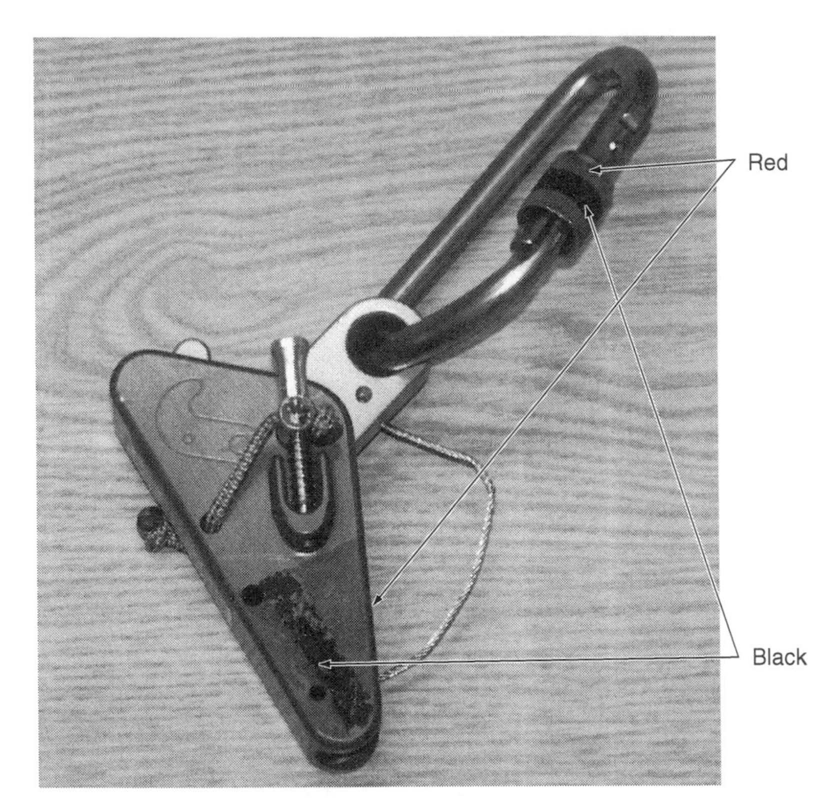

FIGURE 3–14

Color coding makes it easy to identify, track, and inventory equipment. (*Courtesy Martin Grube*)

FIGURE 3–15

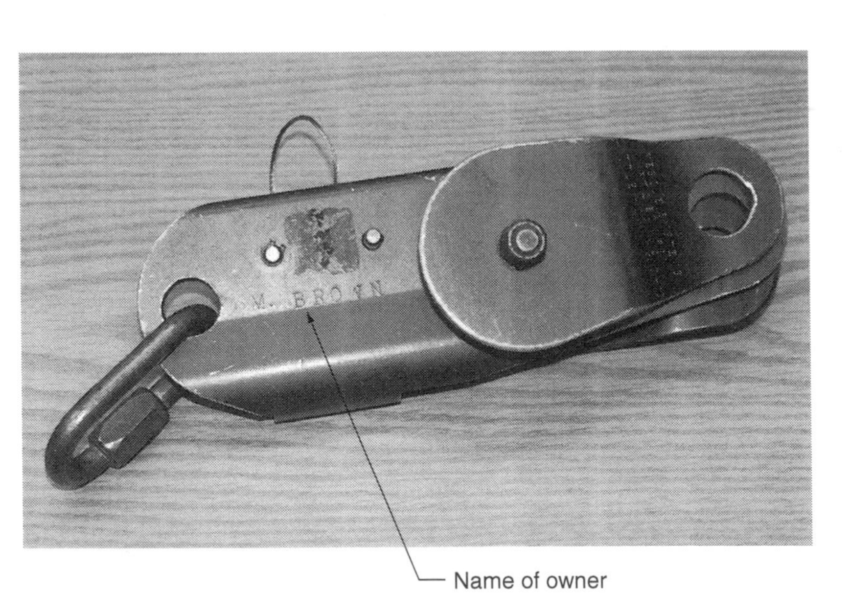

With the manufacturer's approval, some equipment can be stamped to provide specific identification. (*Courtesy Martin Grube*)

Carabiners

▶ **carabiners**

Devices that cause rope system components to become interrelated and useful.

Carabiners are the connectors that bring it all together (Figure 3–16). Like the nuts and bolts of a machine, **carabiners** are the devices that cause rope system components to become interrelated and useful. A good rescue quality (R/Q) carabiner is easy to use, quick to apply to other components, and strong enough to hold components together under the rigors of rescue work.

Although many excellent R/Q carabiners are manufactured from aluminum and some from titanium, most R/Q carabiners are made from chromium molybdenum, carbon alloy rod stock, or

FIGURE 3–16

Carabiners are used to connect pieces of equipment together and come in hundreds of different shapes and sizes. (*Courtesy Martin Grube*)

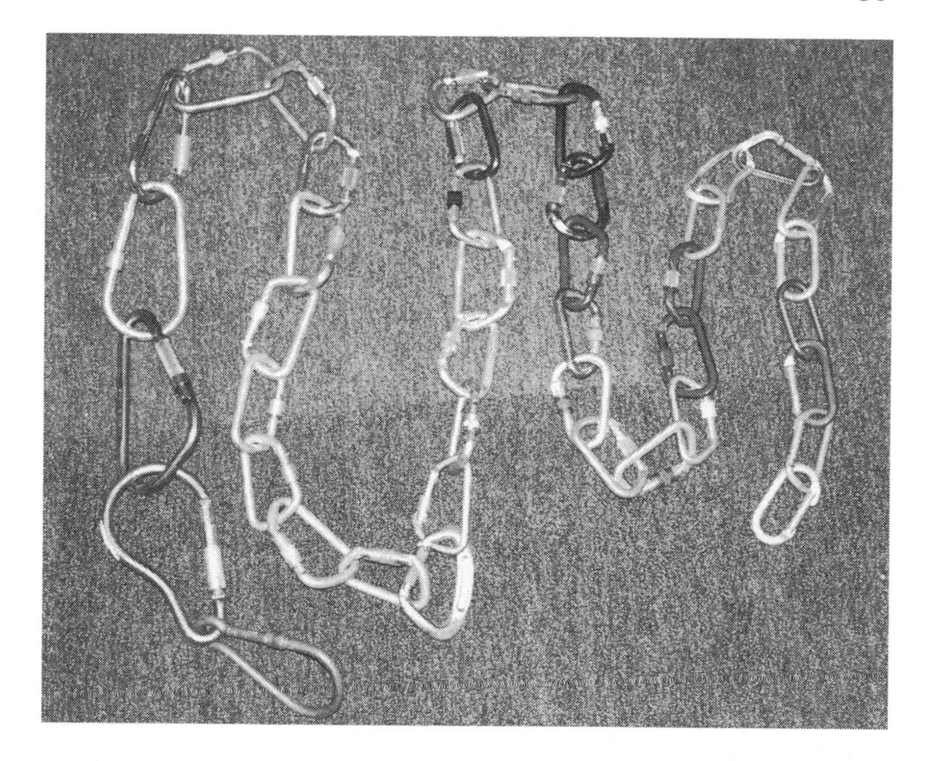

forged steel. Steel rod is cut into lengths and bent into rough C shapes, called the frame, and then machined into its final shape. Forged frame blanks are hydraulically pressed into molds at very high pressures, maintaining the original molecular orientation. The frames on both rod stock and forged steel are then usually heat treated, oil-bath quenched, and tempered. Gates are cut from the same rod stock or forge blank stock and machined to tolerance, threaded, have a spring mechanism inserted, and are attached to the frame with a pin.

With only two or three moving parts, one might assume that the carabiner is as simple as a lock washer and has as few parts. However, carabiner science is as complex as the value that is placed on carabiners. For the purposes of this book, its parts will be defined as *frame, gate, gate lock, nose, spine, latch, latch pin, gate lock pin stop, and slot* (Figure 3–17). It is important also to recognize a carabiner by the way it can be loaded. The long axis is the manner in which all carabiners are intended to be loaded. Short axis loading places dangerous forces on the gate of the carabiner and should be avoided.

The shape of a carabiner contributes greatly to its versatility, utility, and durability. Original frame designs were oval, which allowed loads to place equal forces on the spine side and gate side of the carabiner. Newer designs are increasingly D, or delta, shaped to force interacting components toward the stronger spine side of the carabiner. Webbing and rigging slings tend to spread the load out along the ends of the carabiner and transmit unwanted forces toward the gate, latch, and lock. This can be reduced by lapping wider webbing over itself in the corners, using manufactured rig straps with steel D rings, or connecting wide rigging straps to the carabiner with steel O rings.

The aspects of a carabiner.

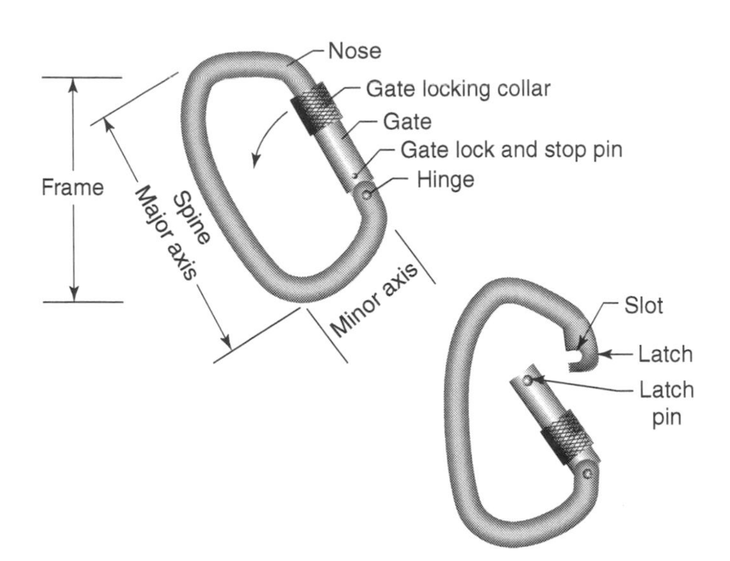

Forged steel auto-locking carabiner. (*Courtesy Martin Grube*)

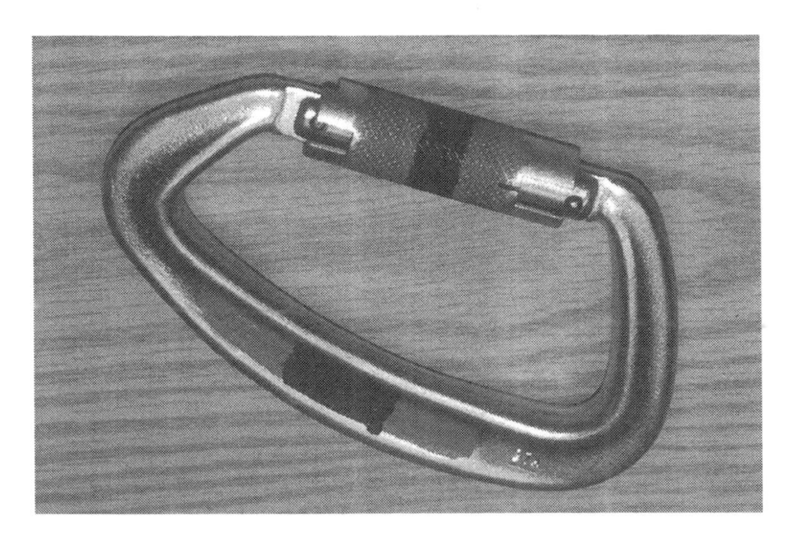

Some excellent forged carabiners have been designed with a curved spine (Figure 3–18) to allow for some shock absorption via the minute uncurving of the spine when loaded. Regardless, every effort should be made to use the tremendous tensile properties of the metal on the spine side of the carabiner by loading it along its major axis and keeping as much of the load as possible away from the gated side.

The gate of a carabiner can be either locking or nonlocking. For rescue work, locking carabiners are preferred in almost every situation. Some locking gates are made using a threaded spinning collar that travels the threads machined into the gate. The gate lock travels over the nose and latch of the frame to prevent the gate from opening. There are other, more elaborate gate lock designs including auto-lockers, tandem movement lockers, and even combination lock lockers, for specialized applications.

The latching mechanism of the gate should be designed so that the gate stays closed if the gate lock is accidentally or inadvertently unlocked. This is called a positive latch. A common design uses a pin and slot (Figure 3–19) technique, in which the pin in the gate

FIGURE 3–19

Pin and latch design helps keep the carabiner closed even if one is accidentally left unlocked. (*Courtesy Martin Grube*)

FIGURE 3–20

Machined gate carabiner latch design does not have a positive latching action, and the carabiner opens under relatively light loads if accidentally left unlocked. (*Courtesy Martin Grube*)

fits into the slot on the nose and, under loading situations, prevents the gate from opening. In other positive latching mechanisms, gate tips of various shapes interface with the nose of the carabiner. Some carabiner gate and latch designs are not positive latching (Figure 3–20) and, if the gate lock is left off, the carabiner frame opens under load and can cause catastrophic failure. Such machined gate latching mechanisms should be avoided. Most positive latch carabiners maintain approximately 90% of their capacity unlocked. Some nonpositive latching machine gate carabiners maintain only about 10% of their ultimate breaking strength when unlocked and should be avoided where human error could occur.

Gravity can have an influence on carabiner locks just as it affects all of the components in a rope rescue system. The pics on the sheath of a rope or the rattling from a helicopter operation, ventilation fans, or other machinery can transmit vibrations to the carabiner

NOTE: Locking a spinning collar gate lock is as simple as giving it a spin with the thumb. A properly maintained and lubricated gate lock will spin to the hilt easily. Then, finger tighten the lock only. It is not necessary to crank down on the locking collar, as the carabiner may be extremely difficult to unlock after being so loaded. This can be very dangerous when you need to quickly add another system component to the carabiner. The best way to unlock a stuck carabiner gate is to use cordage wrapped around the collar with two round turns. One person holds the carabiner by the spine while another person pulls on the cordage wrap, causing the lock to move in a counterclockwise direction, in effect loosening the collar like an oil filter wrench. It is always wise to carry a multi-purpose utility-tool-and-pliers combination on one's belt. The pliers can be used to nudge the gate collar open if it becomes really jammed. Sometimes the hole in the device the carabiner is attached to, such as one of the small holes in an eight-plate, can be leveraged against the gate collar and used to ratchet the lock open.

► dynamic roll out

Rare but dangerous condition that occurs when there is torque in a system and the components twist against themselves.

NFPA 1983

Standard on Fire Service Life Safety Rope and System Components

gate. When the carabiner is riding in the vertical position or is exposed to engine and tool harmonics, the gate locking collar can vibrate open, toward the hinge, and obvious problems can occur. The carabiner is weaker unlocked than locked, and other equipment could accidentally get caught in the carabiner. Always try to ensure that vertically positioned carabiners have a gate locking collar that vibrates locked. Duct tape will work in a pinch if all else fails to keep the collar locked, but it will also severely inhibit opening the gate should it become necessary.

An additional consideration is commonly referred to as **dynamic roll out.** This rare but dangerous condition can occur anytime there is torque in a system and the components twist against themselves. An item rolling against the unlocked gate of a carabiner can open the gate and roll completely out, a potentially catastrophic result. To minimize this possibility, rig systems torque-free through experience and good planning. Consider adding a swivel, webbing, or a rope component between pieces of hardware where dynamic roll out is possible.

Most newer carabiners have some sort of manufacturer's marking on their spine sides. It has finally become acceptable to imprint the NFPA 1983 recommended markings, per Chapter 3-5.1.2: "*The following statement and information also shall be legibly printed on the product label(s). All letters shall be at least 1/16 in. (1.6 mm) high.*"

"Meets NFPA 1983 [200X ed.]."

Certification organizations label, symbol, or identifying mark

Name of manufacturer or trademark

Manufacturer's lot number

Minimum rated breaking strength (This figure shall be prefaced by the letters MBS.)

And in Chapter 3-5.2: "*Auxiliary equipment also shall be stamped or otherwise permanently marked according to intended use and load ranges with a 'G' for general use*" . . . (9,000 lbf MBS), *or a 'P'* " . . . (6,000 lbf MBS) "*for personal use as designated in accordance with 4-5.2.*"

While this seems like a lot of information to put on a carabiner, it is possible and is being done, and it is fairly easy to read with the naked eye. It should be noted that this labeling is only required when a manufacturer wants to market an NFPA-approved carabiner. The cost of the third-party certifier, such as Factory Mutual, Underwriters Laboratory, etc., is passed along to the buyer. Reputable manufacturers are accustomed to the NFPA process. In fact, their non-NFPA carabiners are often the same as their NFPA compliant carabiners. They just have not been tested by the third-party certifier, so they are a little less expensive. Beware, however, of less scrupulous manufacturers who include only a small portion of the information to make it look like the carabiner might be NFPA compliant. NFPA markings are very strict, and the third-party certifier must ensure that the markings are compliant, along with the carabiner or other auxiliary equipment. The decision to purchase

NFPA-compliant carabiners lies completely with rescuers or their organization as the AHJ.

Some carabiner manufacturers use the Euronorm or European Committee for Standardization (CE) standards, previously known as the Union of International Alpine Associations (UIAA). CE markings include the minimum breaking strength stamped onto the frame of the carabiner with a little picture showing the manner of testing along its major axis with the gate closed, along its minor axis, and along its major axis with the gate open.

► **mechanical rope grabs and ascenders**

Devices constructed to grip the rope for raising or holding loads or for climbing.

Mechanical rope grabs and ascenders (Figure 3–21) are devices constructed to grip the rope for raising or holding loads or for climbing. They are made of aluminum, steel, or even stainless steel. Many variations have been used over the years, but all have essentially two primary parts, a shell and a cam. The shell encompasses the rope and gives the cam something to leverage against. Some shells have handles to assist in climbing the rope. The cam can be the attachment point, and force is applied to the rope in proportion to the amount of leverage applied to and by the levering action of the cam. Other cams are spring loaded to hold the cam on the rope, so increased tension on the handle or shell increases load on the cam.

Much discussion has been generated about the detrimental effects the cam interface has upon the host rope. All cams have some type of interface arrangement that bites into the rope when load is applied. The cam face may be ribbed, or it may have little spikes that bite into the rope. The trouble with the more aggressive cam face designs (Figure 3–22) is that they can tear the sheath off the rope if a certain amount and kind of force is applied. Less aggressive cam faces do not bite the rope adequately, especially if the rope is wet, icy, or muddy.

The crux of the mechanical rope grab matter is that some of them are excellent for personal use, mainly ascending. They bite the rope well without slipping, even on slippery ropes, and advancing up the rope is much quicker than with any prusik or similar friction-type hitch. But excellent ascenders usually make poor

FIGURE 3–21

Mechanical rope grabs come in many variations. Like all equipment, they have associated advantages and disadvantages. (*Courtesy Martin Grube*)

FIGURE 3–22
Two different types of cam/rope interface designs. The one on the left has an aggressive conical tooth element that bites the rope very hard. This design can damage the rope if used incorrectly. The cam interface design on the right is not as aggressive and can in most cases support more of a load before damaging the rope. (*Courtesy Martin Grube*)

rope grabs for heavy-duty rescue system raises because of the heavy loads and potential shock loads encountered. They can desheath the rope, sometimes at a force as low as 800 lbf, which is common in rescue work. Other rope grabs can be used for ascending and, under close observation, are very good tools for raising rescue loads. For this reason, mechanical rope grabs are designated by the job they most safely and efficiently perform, that is, ascenders for individuals to climb the rope, and rope grabs for heartier, less sheath-destroying units. *More discussion is included in the prusik section near the end of this chapter.*

Miscellaneous Hardware

There are probably hundreds of combinations of metal shapes that represent tools for the rope rescue trade. However, there are several pieces of hardware that deserve special mention here because of the enhancing effect they have on the engineered systems. **Rigging plates** (Figure 3–23A) are simply flat pieces of aluminum or steel that have been manufactured with holes in them for the attachment of other pieces of equipment. Simple by design, they are wonderful housekeeping devices (Figure 3–23B) that can be used as a sort of manifold from the main anchor. Rarely are anchors in a convenient or safe place to work. By running ropes from the anchor to the place where you prefer to work and attaching the ropes to a rigging plate, you can effectively anchor and focus the rest of the system components to the rigging plate in an orderly fashion. They also work well as a housekeeper for a Stokes litter bridle by organizing the legs of the bridle, the attendant lines, the barf line, and so on.

► **rigging plates**
Flat pieces of aluminum or steel that have been manufactured with holes in them for the attachment of other pieces of equipment.

Swivels (Figure 3–24) **and snap links** are devices designed as attachment points for other system components that pivot on an axis to eliminate unwanted torque and twisting and are also known as **anti-torque devices (ATDs).** Rope systems sometimes develop twists that hinder the operation, even to the point where it becomes a safety problem. Good rig masters keep some swivels in their pockets to place into the system to relieve torque build-up. One caution

► **swivels and snap links**
Devices designed as attachment points for other system components that pivot on an axis to eliminate unwanted torque and twisting.

► **anti-torque devices (ATDs)**
Another name for swivels and snap links.

(A)

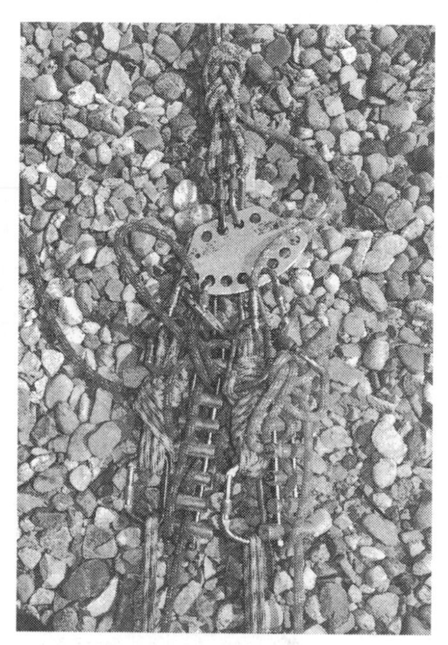

(B)

FIGURE 3-23

(A) Rigging plates are the perfect housekeeping device. They help keep ropes and equipment from becoming entangled, and they can be used to multiply anchor site attachment points. (B) This rigging plate acts as the focus for this lowering operation. Tandem 1/2-inch ropes go to separate anchors. Care must be taken to assure that the rigging plate is not the only link in the system. A back-up rope completely bypasses the rigging plate in the highly unlikely event of plate failure. (*Courtesy Martin Grube*)

FIGURE 3-24

Swivels act as ATDs (anti-torque devices), cleaning up twisted ropes and equipment. The swivel on the right has a snap hook release component that can be used to release loaded equipment under certain circumstances. (*Courtesy Martin Grube*)

with swivels: they are addictive! Snap links are great for releasing loaded system components and are particularly well suited for releasing falling loads for belay practice.

Pulleys are incredibly important to rope rescue systems and are discussed in detail in Chapter 8.

SOFTWARE

Ropes, cordage, webbing, harnesses and seats, gloves, eye protection, rig straps, and edge padding are all considered software.

Rope and webbing are the system components that add flexibility, elasticity, and flow to rope rescue systems. W. H. Carothers, a chemical engineer, working for E. I. duPont deNemours and Company, patented the synthetic polymers that changed rope rescue from extremely hazardous to relatively routine. Synthetic ropes do not degrade or rot the way natural fibers do, so the invention of nylon in 1938 created untold thousands of new applications and occupations, one of which was rescuing people using ropes.

Manila, hemp, cotton, sisal, jute, silk, hair, and grapevines (Figure 3–25), formerly the fibers of choice, were all living things before they were wound or woven into linear configurations. Therefore, they started dying, degrading, and getting weaker from the day they were made. Prior to 1938, the only option, when rope was needed to accomplish a task, was to use natural fiber.

FIGURE 3–25

(A) Three-strand hawser laid rope. Each fiber is exposed to the effects of abrasion and sunlight at some point along the rope's length. Rope strength is gained by the torquing action of the rope under tension, increasing inter-fiber friction.
(B) Natural fiber (manila) rope is still found on some fire and rescue apparatus and is best left for non-life support applications. (*A, Courtesy Delmar; B, Courtesy Chase N. Sargent*)

(A)

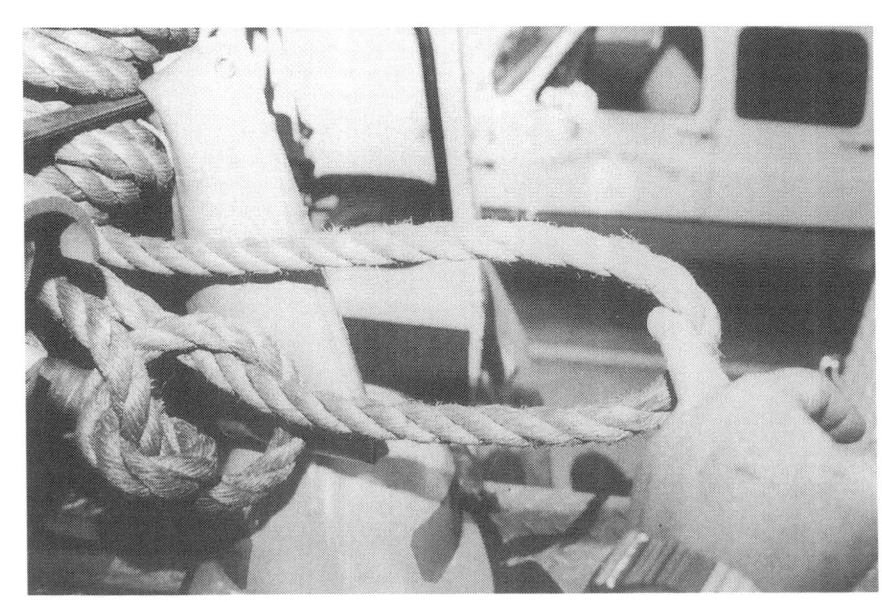

(B)

Nylon

In the 1930s, a vast majority of the world was at war and with that war came the need for ropes for myriad uses. Synthetic ropes were invented more out of a need to replace the dwindling supply of natural fiber ropes than a desire to develop stronger ones.

Dupont introduced a rope in 1940 that looked like the hawser laid manila of the time but was made from nylon fibers (called Perlon in Europe), each thinner than a human hair. Collectively, the twisted and laid fibers made a remarkably light and strong rope with about a 3,000-lbf breaking strength in the 7/16-inch version. Even more striking was its ability to resist degradation.

There are many manufacturers of synthetic fibers, and today's ropes are distant cousins of that first nylon rope. Basically, the nylon polyamide is a hydrocarbon molecule that can be extruded from its solution form into very strong filaments. The filaments are gathered into yarns and then bundles. Most of today's R/Q nylon is either type 6 or type 6.6. Most American-made rope is type 6.6, which has a slightly higher melting temperature (450°F or 230°C) than type 6 (380°F or 200°C). Nylon is inherently stretchy, 15% to 25%, and can therefore absorb shock loads very well. During normal use, it can be bent, knotted, and loaded repeatedly with no appreciable damage to the fibers.

Nylon's disadvantages are few, but they are important to understand when using it as your primary rescue tool. Nylon deteriorates in the presence of ultraviolet (UV) radiation of which the sun and fluorescent lights are the most common sources. UV radiation has a drying and bleaching effect on the fibers, causing them to become brittle and break. Rope bags (Figure 3–26) will offer some protection from the sunlight, certainly more than a rope coil will. Unused portions of rope should always be put back in the bag to keep sunlight off. Also, check your department's resource warehouse to make sure spare ropes are kept away from fluorescent lighting.

FIGURE 3–26

Rope storage bags help to keep the harmful effects of chemicals, sunlight, and abrasion off of life safety ropes. (*Courtesy Delmar*)

FIGURE 3–27
Commercial tube type rope washers are an excellent tool for keeping ropes clean and ready for rescue work. (*Courtesy Delmar*)

Also, most acids and strong alkalis have a devastating effect on nylon. Acid splashed on a nylon rope will cause it to melt away in seconds. A common mistake is storing nylon rope in the trunk of a car or the back of a truck. If a battery or jumper cables are transported in the same compartment, the hint of acid attacks the sheath of the rope. Bag ropes and jumper cables separately and keep them in separate compartments of the vehicle.

Nylon loses a considerable amount of strength when saturated with water. The hydrogen in the water weakens some of the molecular attraction of the hydrogen bonds, causing it to lose up to 15% of its tensile strength. Additionally, nylon becomes very heavy when saturated, an important consideration when rigging towards the edge of equipment and anchor-strength windows. Finally, water will cause a nylon rope to shrink a bit and to lose some of its hand, or soft flexible feel.

Eventually it will become necessary to wash the rope (Figure 3–27 and Figure 3–28) to remove dirt and metal oxides. The most important rule in the maintenance of ropes is to *always follow the manufacturers' recommendations.* They understand their rope more than you do and will be happy to supply you with thorough instructions on maintaining them. Keep a rope log (Figure 3–29) on all your ropes. If you are responsible for a large cache, a three-ring binder with plastic pocket sheets works well for documenting and tracking individual ropes.

In the days of natural fiber ropes, whipping the end of a rope to prevent unraveling was a true art. There are several whipping techniques (Figure 3–30) using small-diameter (1- or 2-mm) twine of various colors that work very well if you have extra time and can find any manila rope. Nylon and polyester ropes, however, are very easy to cut and whip to prevent unraveling. Simply heat the ends

FIGURE 3-28 Ropes should be inspected closely during the washing procedure. (*Courtesy Delmar*)

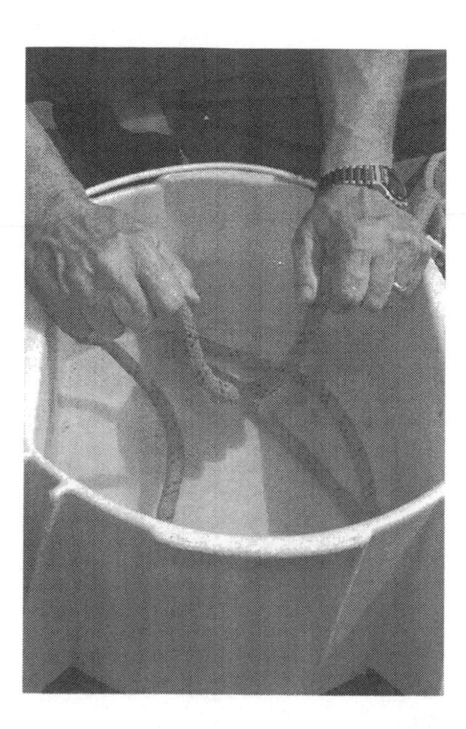

FIGURE 3-29 Sample rope inspection and history log.

Mayberry Fire and Rescue Department Life Safety Rope Inspection and History Log						
Date	Location	Type of use	Sheath fray, %	Other damage	Additional comments	Inspected by

of the rope to their melting point at 450°F and the fibers will fuse together like glue. Let it cool off and you have whipped the ends for the life of the rope. In the field, if you find you need to cut a rope and have no heat producing tools, simply tape the rope where you want to cut it and cut in the middle of the tape wrap. Athletic tape or 100 mph tape works best. This method is only temporary, as the tape always pulls off and the rope begins to unravel. There are also several good rope-cutting tools on the market. They are adaptations of a soldering iron where the heating elements are connected with a steel blade. Turn it on and when it reaches the right temperature, push the blade through the rope (Figure 3–31A, Figure 3–31B, and Figure 3–31C). It is cut and whipped in an instant. If you do not want to buy a bona fide, premanufactured rope cutter, use a lighter or small propane torch and a sharp knife. Gently heat the part of the rope where you want to cut it. Melt, or glaze, an area around the sheath of the rope to fuse the outer fibers. Cut the rope in the

FIGURE 3-30

Rope ends can be whipped clean using whipping thread or twine. (*Courtesy Delmar*)

FIGURE 3-31

Synthetic rope is easily whipped using a hot knife or commercial rope cutter. (A) Bring the cutter blade to full heat. (B) Rotate the rope through the blade. (C) Seal the ends against the cutter's blade. *Note:* Always use gloves and make sure your work area is well ventilated. (*Courtesy Martin Grube*)

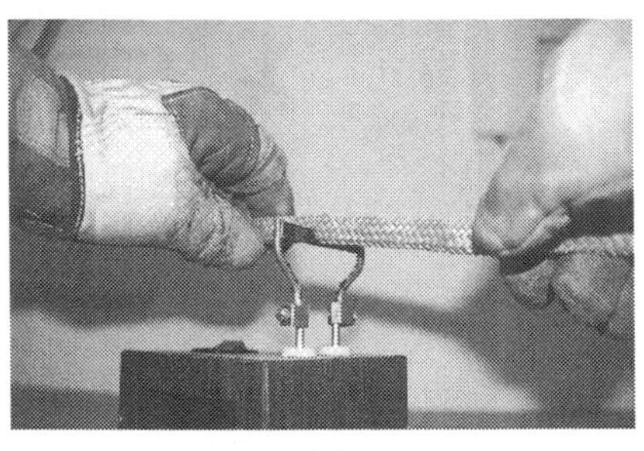

(A)

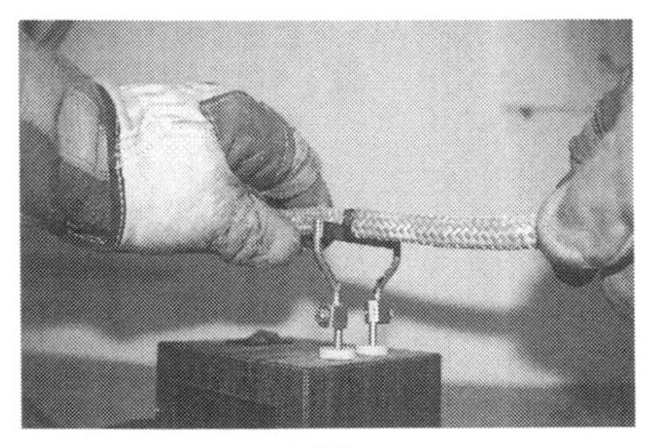

(B)

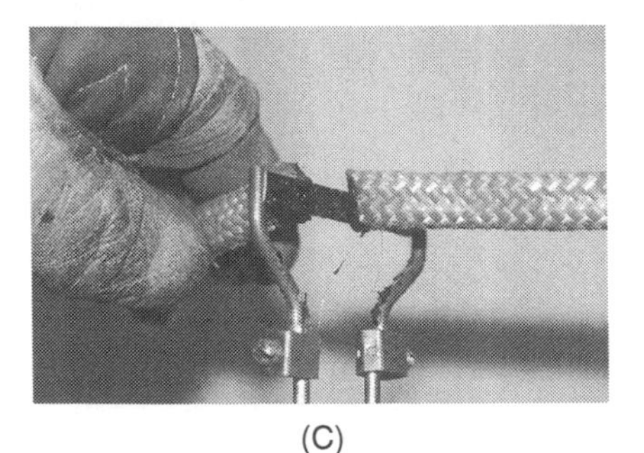

(C)

middle of the glazing and then heat the loose inner fibers until they also are fused. It is not necessary to get the rope so hot that it continues to burn on its own. Heat the fibers till they just start to turn black and then stop. With a little practice you can customize the ends of your ropes and cords to reflect your attention to rope system detail.

FIGURE 

Clear heat-shrink heat tubing holding rope identifier on the end of the rope.

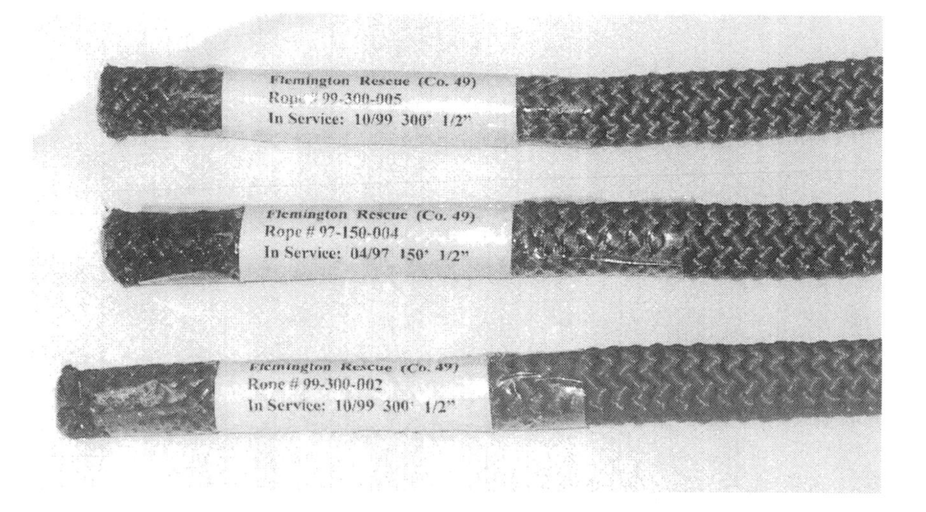

FIGURE 

Heavy plastic tubing acts as a rope protector over sharp edges.

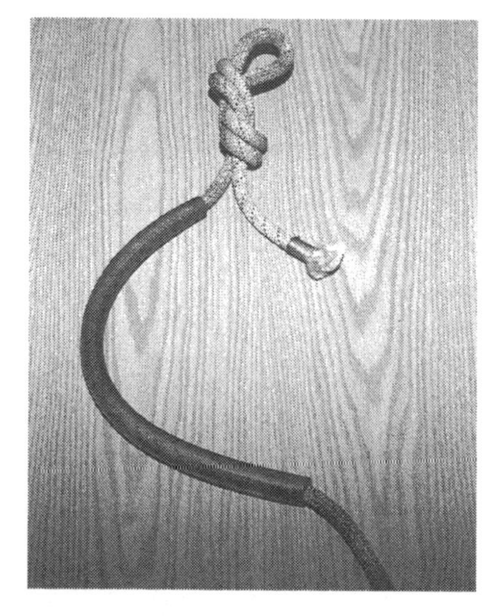

CAUTION
Never heat any synthetic fiber in such a manner that you are breathing the smoke. The fumes are toxic and may cause cancer. At the least, you will get a nasty headache and red, watery eyes. Always heat the fibers in a well-ventilated area or wear a breathing apparatus if you are going to cut a lot of rope.

You can also identify your ropes by color coding them with a commercial chemical whipping product. This glue-like substance comes in a variety of colors and works well if you have to color identify a lot of ropes. Chemical dips can be very messy, are usually very toxic, and will eventually dry and crack off the rope. A better method of marking your rope is to buy commercial, clear, heat-shrink tubing (Figure 3–32). Make a label showing the rope information and a reference number and glue it to the end of the rope. Then apply the tubing and heat it until it is snug over the label. Marking and labeling both ends of the rope will prevent someone from cutting a chunk off the unmarked end.

Another common cause of damage to nylon ropes is mechanical in nature. It is estimated that 90% of rope failures result from inadequate edge protection. There are a number of ways to pad your loaded rope, some of them as simple as using carpet samples or cut-up fire hose between the rope and any sharp edges. There are many good manufactured edge protectors available as well (Figure 3–33 and Figure 3–34). Remember that edge protection is as important as any

FIGURE 3-34

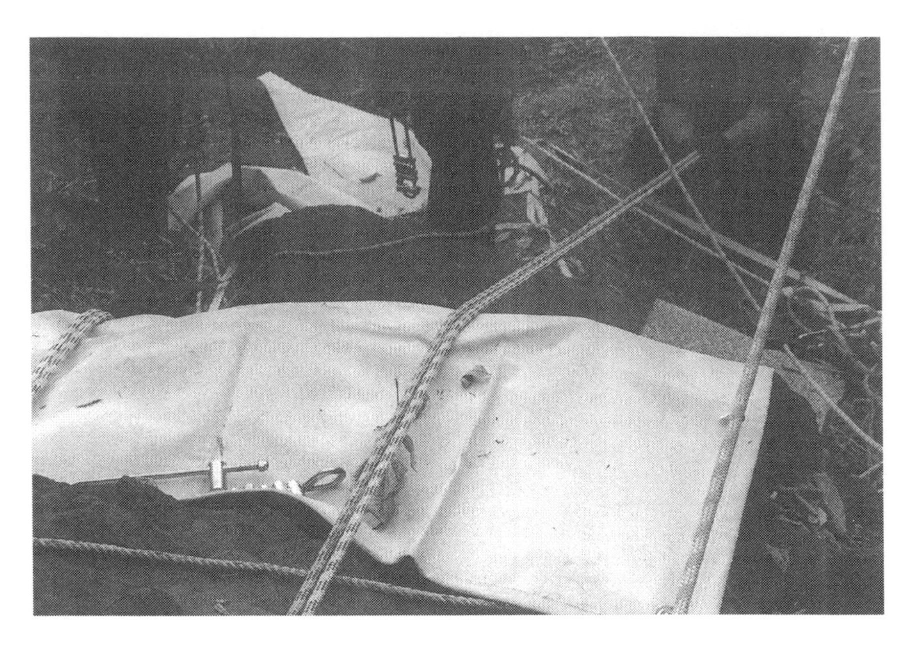

Canvas covers can provide excellent rough edge protection. (*Courtesy Martin Grube*)

other piece of equipment in a system. Obviously, sharp and rough edges can cut or abrade ropes, but there are many other mechanical causes of rope damage. Stepping on the rope forces microscopic particles into the fibers, and every time it is run through a DCD or knotted, the fibers suffer some damage. Rope is a lifeline. Get the team into the habit of protecting all rope surfaces and never walking on *any* equipment, such as extension cords, airlines, and the like. Do not drag the rope across the ground, slam it in doors, or let it get near moving machinery.

Polyester

More and more rope manufacturers are working with polyester rope. Polyester, with brand names of Dacron, Kodel, and Fortrel, is slightly stronger than nylon, has smaller elongation numbers (9% to 13%), and is much less prone to damage from UV radiation or chemicals. This is one reason the seatbelts in most cars are made from polyester. It is always wise to use good housekeeping rules around any ropes, but polyester handles sun, water, and certain chemicals better than nylon. Additionally, it is almost impossible to tell nylon from polyester with the naked eye. It is more slippery and there is a noticeable olfactory difference; polyester smells a little sweeter, and when it is burned to melt an end or to cut it, it smells like plastic. Also, a quick rappel will show that nylon is bouncier, at 4% stretch with 600 lbf applied versus 2.5% or 3% for polyester depending on construction type. Polyester loses almost none of its tensile strength when wet and maintains about the same melting characteristics.

While it is possible that polyester will replace nylon as the rescue rope of choice in the not-too-distant future, there are still some concerns that must be dealt with (see Chapter 9). Many teams have grown accustomed to the inherent stretch that has become nylon's calling card. Nylon is very forgiving due to elongation factors in the high load 20% and 30% range. Because polyester has noticeably less stretch than nylon, common rope grab techniques, such as prusiks and gentle cam and polyethylene-coated designs, do not work the same on polyester as they do on nylon. Extra attention is in order when using polyester to make tensioned rope systems or any other system where stretch is considered part of the rigging safety formula. More testing needs to be done on the interaction between polyester rope and the common adjuncts to rescue by rope, but the future of polyester looks extremely bright.

Other Rope Materials

At this point in technology, there is no perfect fiber for every purpose. Currently, nylon and polyester are all that is available. There have been some interesting developments, however, that make the next generation of rope fibers seem very exciting.

KEVLAR® is the strongest common synthetic. It is much stronger than steel but with very little elasticity (3% at break). Its ability to withstand shock loads is not good at all and knots weaken it considerably. Knots with sharp bends can weaken KEVLAR® rope by as much as 60%, so that repeated bending of KEVLAR® causes the fibers to break like coat hanger wire. Some construction methods, like loose braid patterns, can introduce flex, like the overlapping sections of bullet-proof vests. KEVLAR® escape belts seem well suited for firefighters in heated environments, and KEVLAR® sheathed escape ropes show great promise as emergency life safety tools. It is hoped that scientists will be able to introduce more stretch into the remarkable strength of KEVLAR® type fibers.

NOMEX® is most commonly seen in firefighters' turnout gear. Its advantage is that it does not burn but merely decomposes at temperatures exceeding 750°F, or 400°C. NOMEX® has about half the tensile strength of nylon, however, and its ability to withstand repeated bendings and loaded knots is relatively poor. At print time, some ropes are being manufactured as escape lines in the 8-mm size range with NOMEX® sheaths and KEVLAR® cores.

Polypropylene and polyethylene are molecularly similar materials that float on water and are very resistant to chemical damage but possess less than half the strength of nylon and polyester. They are sometimes used in water rescue because they absorb less water than glass and therefore float handily on the surface. They have limited rope rescue properties, however, due to their poor shock-absorbing and tensile characteristics.

Construction Methods

Construction methods are the many combinations in which a manufacturer can twist, bend, tension, and otherwise bind interacting fibers to create a rope for a given purpose. Natural fiber rope, like manila rope, contains many relatively short fibers, from 3 feet to 15 feet long, depending on the grade of rope. The fibers are wrapped into bundles and laid into a collective twisting pattern, forming the familiar hawser-laid ropes. The twist pattern, right-hand S, left-hand Z, or balanced S&Z is the mechanism that binds the fibers to give manila its moderate tensile properties. When tensioned, the

FIGURE 3-35

(A) Synthetic fibers beam from fiber manufacturer going through twister to make S and Z core fiber yarns. (B) Core fiber bundles spools ready for braiding machine. (*Courtesy Sterling Rope Company*)

(A)

(B)

twisting action of the rope tightens the fiber bundles together, effectively increasing the friction between the fibers. The hawser-laid pattern *does* make for good gripability in the fingerhold like grooves between the braids. Unfortunately, each fiber rotates to the surface every several inches exposing it to abrasion. Usually a natural fiber rope is just a twisted bundle of broken fibers held together by the torque action of the rope construction method.

NFPA 1983 (Chapter 4-1.4) states, "Life safety rope shall be of *block creel* construction; load-bearing elements shall be constructed of continuous filament fiber." **Block creel** construction is defined as "rope constructed without knots, or splices in the yarns, ply yarns, strands or braids, or rope. Unavoidable knots might be present in individual fibers as received from the fiber producer." The block creel method eliminates the need to use lumpy hawser-laid construction that relies on twisting fibers together for frictional strength. Additionally, fiber-weakening knots and splices are eliminated from the beginning, and a manufacturer is able to effectively bury a vast ma-

► **block creel**

As defined in NFPA 1983, "rope constructed without knots, or splices in the yarns, ply yarns, strands or braids, or rope. Unavoidable knots might be present in individual fibers as received from the fiber producer."

FIGURE 3–36

(A) Colored fiber yarns are bundled for sheath braiding. (B) Core and sheath fibers coming together for the first time in the braiding machine. Capstan drum pulls sheath yarns out of complex braiding rotation on drum at bottom. (C) Braid point forming die, monitors yarn tension and funnel shed (rope diameter) capacity. (D) Cordage braider joining core and sheath fiber yarns. (*Courtesy Sterling Rope Company*)

(A)

(B)

(C)

(D)

jority of the fibers inside a sheath, where they are protected from abrasion, ultraviolet radiation, and minor chemical contact.

The most common R/Q ropes are manufactured (Figure 3–35A and B, Figure 3–36A through D, Figure 3–37A and B) using a technique called kernmantle, which translated from German means core/sheath. Essentially, parallel fibers carrying approximately 85% to 90% of the tensile strength of the rope are gathered for the core of the rope (see kernmantle rope in Figure 3–38). The core is then covered with a braided sheath. This core/sheath arrangement

FIGURE 3–37

(A) Inspection and custom measuring process. (B) Finished rope spools. (*Courtesy Sterling Rope Company*)

(A)

(B)

gives an almost perfect combination of strength and durability. Different manufacturers tweak the tensioner on their braiders or add some fibers or bundles here and there to create performance characteristics that differ slightly from those of other ropes. Most often, this involves varying sheath thicknesses and braided tension. For example, to create a very abrasion-resistant rope, some manufacturers simply tighten the tension in the sheath braiders. This makes a very stiff hand that is, in fact, more resistant to abrasion from rocks and edges but makes the rope hard to knot and manage

FIGURE 3-38
Kernmantle (core/sheath) rope diagram.

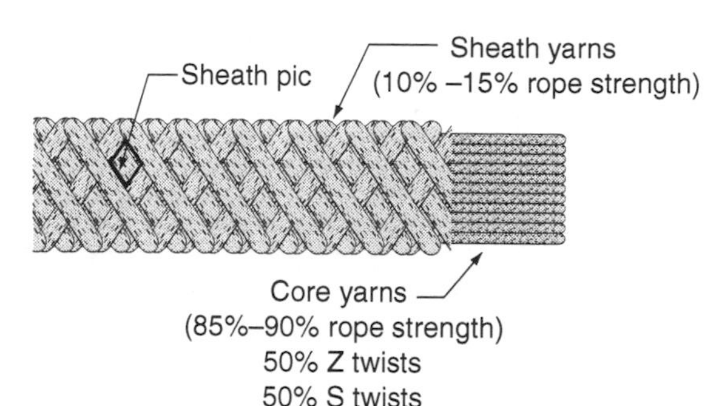

through hardware. Some manufacturers have resorted to high-tension sheaths for durability but have increased the number of sheath braiders to thirty-two or forty-eight, instead of the United States' standard sixteen. This gives the sheath pics more points over which to flex, efficiently countering the high tension in the braids and giving characteristics of good hand *and* abrasion resistance.

How a manufacturer tweaks the core fibers also contributes to a rope's performance characteristics. The most common R/Q ropes are low-stretch kernmantles. The parallel core fibers do not allow for any mechanical stretch characteristics. The elongation of the rope is therefore dependent on only the inherent stretchiness of the nylon fibers. There is also a **high-stretch, or dynamic, kernmantle.** By spiraling fiber bundles in the core or by spiraling the core fibers as a unit, manufacturers can introduce a certain amount of mechanical stretch into the rope that can be very useful, or not, depending on the application. In the event of a shock load on a rope, such as a climber falling off a wall or a worker falling off a scaffold, the force of the fall is counteracted by the unwinding of the spiraled fibers, like an elongated spiral spring.

▶ **high-stretch, or dynamic, kernmantle**

Rope with high elongation characteristics such as stretch of between 30% and 75%.

In general, rescuers choose to use high-stretch ropes *only* when they have to climb above an anchor point and a fall is possible. They, therefore, go out of their way to rig *above* a target and lower or rappel rescuers on low-stretch ropes. Also, tensioning techniques, such as pulley systems, raises, highlines, and the like would become very unpredictable using high-stretch ropes. There are times when it is necessary to climb above an anchor point, as in accessing a disabled target on a power tower. It is wise, therefore, to keep a couple 50-meter climbing ropes in the cache. Also, practice climbing, placing protection, and working belay techniques.

It seems logical to somehow try to integrate the advantages of nylon and polyester into a kind of super rope, minus the disadvantages. Such a rope would combine polyester's better abrasion, water, UV radiation and chemical resistance, and lower elongation characteristics with the steel-like breaking strength and elastic tensile properties of nylon. Recently some manufacturers have started marketing nylon core and polyester sheath kernmantle ropes. Because braided rope naturally stretches more than parallel fibers, like a set of Chinese finger cuffs, the braided polyester stretches to a point and then grips the nylon core material adding to the rope's overall strength. The trick is to make the nonstretchy polyester elongate via the mechanical flexing of the sheath braid exactly the same amount as the parallel core fibers of the more stretchy nylon. This would help prevent sheath creep and perhaps result in the nearly perfect combination rope.

Some manufacturers use a process called plaiting to construct their rope. Plaiting is a braid structure with more than three strands. It generally produces a very strong rope, but every fiber eventually surfaces and, therefore, is subjected to abrasion and separation. Plaited ropes, like laid ropes, have bumpy surfaces that, in tactical situations, cause vibrations that can make attached equipment rattle. The noise could prove hazardous.

Braid-on-braid ropes have been around since the early 1970s. A small tubular braided rope is surrounded by a larger tubular braided rope. These ropes have a soft, if not sloppy, hand, making them easy to tie and rig but relatively hard to untie once loaded. Due to the all-braid construction, braid-on-braid ropes have greater elongation factors than low-stretch kernmantles.

Webbing

Sometimes round rope is just not as useful as flat rope, or webbing (Figure 3–39). Webbing is used primarily in harnesses, in patient packaging, and as anchor attachments. In harnesses, webbing spreads the load over a greater surface area, making it more comfortable than a harness made from rope. In anchor attachments, webbing fits around and over obstacles easier than rope and, because it is flat, it can be doubled to increase the strength of the attachment with no appreciable increase in bulk. Coincidentally, tubular webbing makes great outer-edge protection for rope, supplied air respirator airlines, and intercom lines, because the lines can be pulled through the interior of the webbing to create a safety shell called the umbilicus.

FIGURE 3–39

Tubular nylon webbing tied in a water bend. (*Courtesy Delmar*)

Webbing comes in two forms, flat or tubular. Flat webbing (like a seat belt) is similar to a piece of rope that has been repeatedly run over by a giant steamroller. It can be found in almost any width from 1 mm up to 24 inches. In rope rescue work, flat webbing is usually 1 to 2 inches (25 to 50 mm) wide and, depending on the grade, carries a tensile strength between 3,000 and 8,000 lbf.

Tubular webbing is constructed using either shuttle loom (spiral) or needle loom (chain-type) construction. Shuttle loom is preferred as it is somewhat stronger than needle loom and cannot be unwound by cutting the edge seam, as can needle loom webbing. You can tell the difference by rolling the webbing between your fingers and looking for a seam. The seam indicates needle loom webbing that can be unwound like the stitches on a baseball. Shuttle loom is a continuous spiral of nylon with no seam stitch.

With webbing, the load bearing elements are always exposed to abrasion or to whatever else harms software. Webbing can be cut with a hot knife, which will prevent fraying just like rope. Again, always follow the manufacturers' recommendations on the care and maintenance of your webbing.

Webbing is connected most securely using a water bend (see Chapter 4) and may be connected quickly using a square bend. Both configurations are relatively bulky and require special dressing and pretension attention, as webbing is particularly susceptible to slippage until properly tensioned. Sewing is the most secure method for connecting two ends of webbing. Sewn webbing is much smaller than a knot, and it helps stabilize the floppy ends of the webbing. Sewn-webbing devices can be made locally or at home with a heavy-duty sewing machine using heavy nylon or polyester thread. There are many factors, including thread diameter and strength, bobbin tension, thread pattern and length, etc., that affect sewn-webbing performance characteristics. Unless you have a particular expertise in sewing heavy material, it is best to buy premanufactured webbing equipment.

Harnesses

Harnesses are classified in four categories: belt, and Class I, Class II, and Class III. A belt is used for support and stabilization, such as on an aerial ladder, or as a last-chance emergency bailout belt; a Class I harness supports the thighs, waist, and buttocks and is designed for a one-person load; a Class II harness supports the thighs, waist, and buttocks and is designed for two-person loads, as in performing a pick-off; and a Class III harness is designed to prevent a person from inverting if he or she becomes unconscious.

Improvised harnesses are made from webbing, usually 2-inch tubular webbing, and are tied onto a person to be lowered. They can be used when many people have to be trained to rappel, or be lowered or raised, and no manufactured harnesses are available, usually due to cost. Essentially, there will always be a harness as long as there is some webbing.

The **emergency egress improvised harness** (Figure 3–40) is made from an 18- to 25-foot length of 2-inch tubular webbing. It is

► **improvised harnesses**

Harnesses made from webbing, usually 2-inch tubular webbing, which are tied onto a person to be lowered.

► **emergency egress improvised harness**

A harness tied onto a person in an emergency using rope or webbing to allow escape from and untenable environment.

FIGURE 3-40 Emergency Egress Improvised Harness

Nylon webbing can be used to quickly tie an improvised emergency egress harness.

A This harness is tied by finding the middle of the length of webbing, which should be marked at mid-length for quick identification.

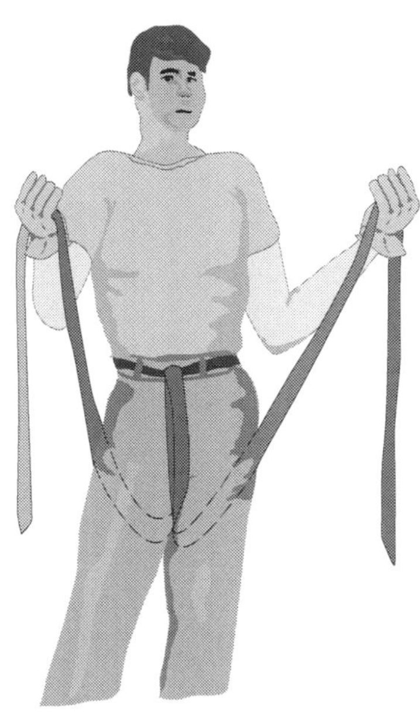

B A bight is formed in the middle and brought up between the legs.

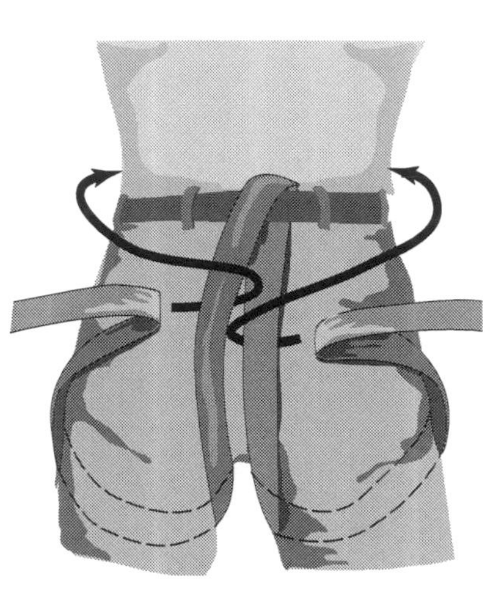

C The two long tails separate behind the buttocks, surround the thighs, and are passed through the center bight.

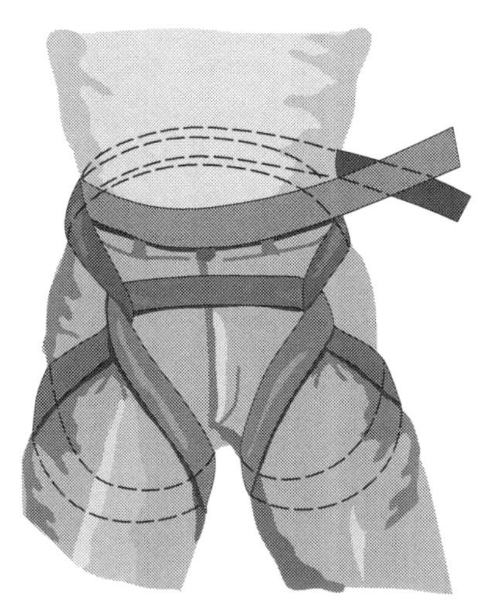

D Each tail is then passed toward the back of the hips, crossed, and brought to one side for joining with a square bend.

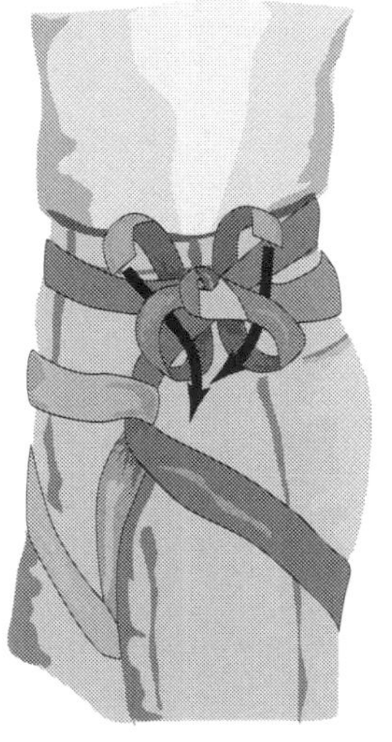

E Half hitches safe the square bend and hold it in position.

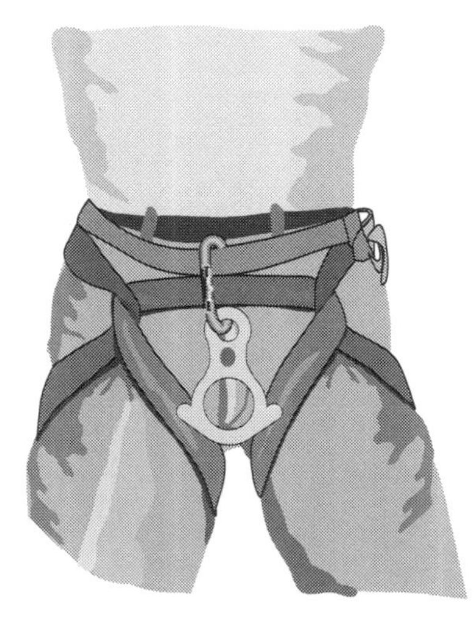

F The webbing should always be flat and tight against the body, and the harness should be applied to the pelvic girdle with tension on the hip points and not the abdomen.

FIGURE 3–41 Emergency Egress Improvised Harness

Modified Swiss seat is an improvised harness tied using synthetic webbing and has more redundancy than the emergency egress seat.

A Find a spot near the middle of the 25-feet webbing. (One side will need to be about 2 feet longer.)

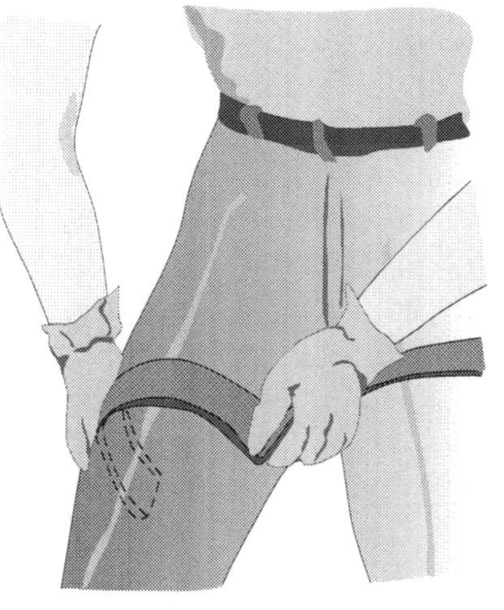

B Measure and then tie leg loops using an overhand knot.

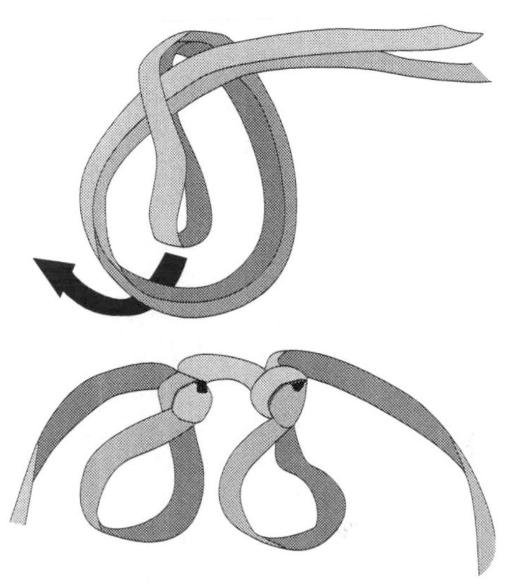

C Space between the leg loops should be about 3 or 4 inches only.

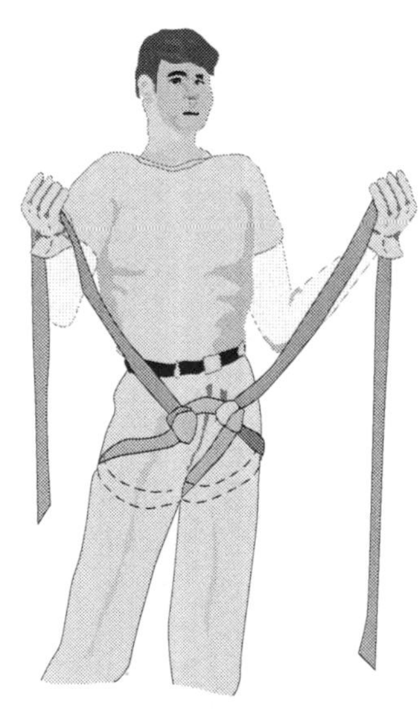

D Wrap the running ends around the waist. At the first pass, make a square bend.

Square bend

E Finish by wrapping excess webbing (if any) multiple times around hips (swami style).

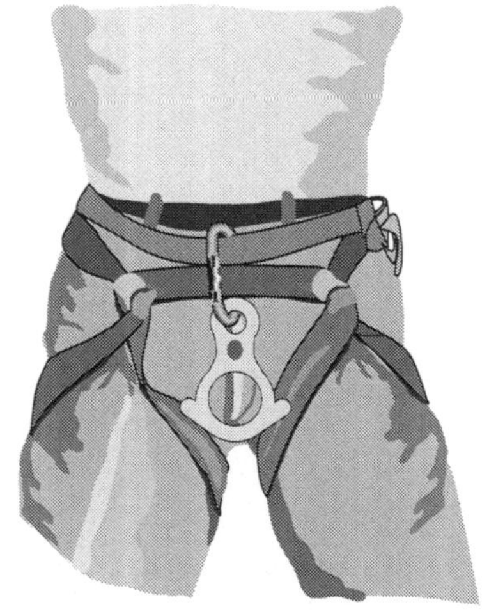

F Conclude with a square bend and overhand knot safeties on both sides of the bend.

► **modified Swiss seat**

A safer harness than the emergency egress harness but takes more time to tie.

specially designed to be put on in a hurry, less than one minute with practice, and can be carried rolled or daisy-chained in a coat pocket for quick access. The steps for securing the harness are shown in Figure 3–40.

The **modified Swiss seat** (Figure 3–41) improvised harness is a safer webbing harness than the emergency egress harness but takes longer to tie, about 3 to 5 minutes. In general, where pre-sewn harnesses are not readily available, the modified Swiss seat is the best option. Because of the thigh loops and separate square bends, one part of the harness could be cut, but another part of the harness would survive intact to keep the rescuer from falling. Using 20 to 25 feet of 2-inch tubular webbing, this harness has some redundancy that eliminates many of the worries of the emergency egress seat. To tie the modified Swiss seat, start with two loops custom tied with overhand knots for the thighs to fit into snugly. The distance between the thigh loops should be only about 3 or 4 inches. The remaining lengths of webbing are brought up and wrapped around the waist. When they meet, a square bend is tied at the hip and the long tails advance to the other hip where they are again connected with a square bend. The final tails are safety knotted with *overhand knots* on each side of the square bend. It takes some patience to get the thigh loops sized and the long tails positioned so that the square bends and safeties match up. With practice, a rescuer should be able to tie the modified Swiss seat in less than three minutes.

SAFETY *All improvised harnesses are tied around the body and secured with a square bend. The square bend allows a degree of tightening unavailable with the water bend. Because the square bend is inherently loose and, therefore, requires backup knots, and because knots in webbing are hard to identify as being correct, it is possible to tie the harness incorrectly and lose contact with the anchor. ALWAYS PRE-LOAD ALL IMPROVISED HARNESSES IN A SAFE LOCATION BEFORE COMMITTING HUMAN WEIGHT AND LIFE TO THEM. ALWAYS HAVE SOMEONE ELSE CHECK THE HARNESS BEFORE PUTTING IT IN A LOADED SITUATION. After a prolonged break and before every loading situation, recheck and have someone else recheck the improvised harness.*

The **hasty hitch harness** (Figure 3–42) is an improvised webbing harness used when the victim has no reasonable harness, is conscious and in good medical condition, and time is of the essence. The hasty hitch is not particularly comfortable for the victim, and the rescuer must make sure that the victim's arms stay down because the victim can slip out if unconscious.

► **hasty hitch harness**
An improvised webbing harness used when the victim has no reasonable harness, is conscious and in good medical condition, and time is of the essence.

Presewn harnesses (Figure 3–43) are manufactured harnesses that are designed for quick donning and eliminate worries about knots and such. There are light rock-climbing harnesses and heavier padded rescue harnesses to suit every taste and application. Probably the most useful harness designed today is a combination Class II (Figure 3–44) and Class III (Figure 3–45). The seat portion of the harness should be comfortably padded with a D-ring attachment front and back and adjustable waist and leg straps. The upper, or chest, portion of the harness can be attached to the seat harness when necessary via an interconnecting and adjustable webbing and buckle system. Better Class III harnesses have a load-bearing D-ring attachment point located between the shoulder blades. Class III harnesses should always be used in confined-space vertical entries where unconsciousness is a possibility and to prevent inverting. The Class III harness is also considered preferable for patient litter attendants.

► **presewn harnesses**
Manufactured harnesses that are designed for quick donning, eliminating worries about knots and connections.

A good rescue harness is invaluable for a rescue professional. It should be comfortable to wear when walking around and should be comfortable to sit in while on a rope for long periods of time. Because the harness will probably become one of your most important investments, it is worth the time and money to invest in a top-quality harness. Spend some time researching your options. Features like quick-release leg and waist connectors, color-coded webbing and buckles, chest sections that do not ride up under the arms, and multiple attachment points are valuable features to look for when shopping for a harness. Ask for opinions from people who wear specific harnesses. Buying an NFPA 1983, current edition, harness is probably worth the small additional cost. Maintain the harness according

FIGURE 3-42 Hasty Hitch

The hasty hitch is an improvised harness that can be quickly attached to an ambulatory and conscious patient for lifting and lowering.

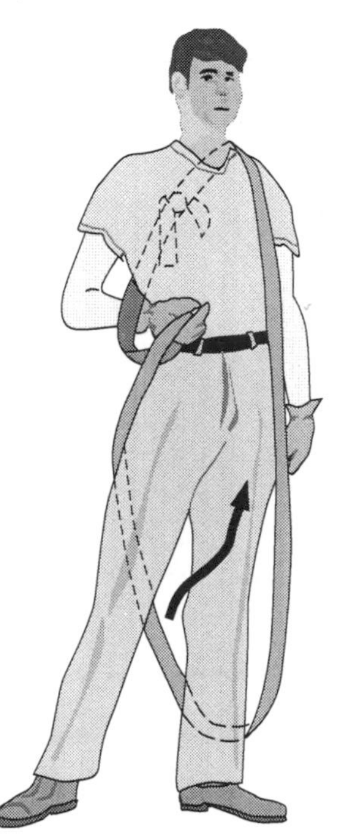

A The pre-tied loop is placed over the victim's shoulder and between shoulder blades—*not around back of neck!*

B Three parts of the webbing are brought forward: (1) over the shoulder, (2) under the arm, and (3) between the legs.

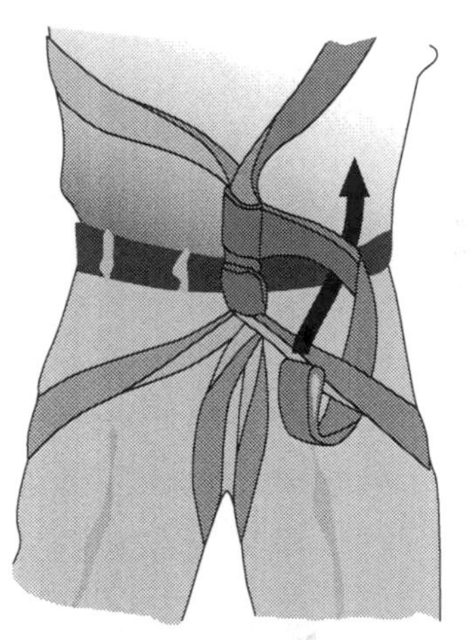

C All excess webbing is pulled out using the part between the legs. Make sure to pull the other two parts tight against the body. Wrap the long leg section around the other two sections. Finish with a downward wrap of one of the leg loops and a half hitch. This keeps all the connections low and away from the head.

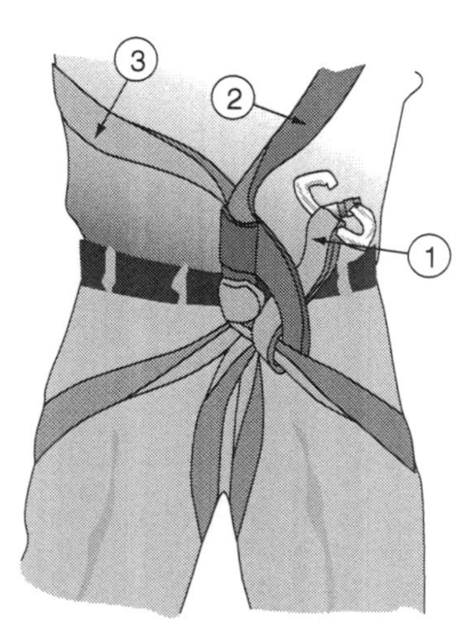

D Connect a carabiner to three attachment points.

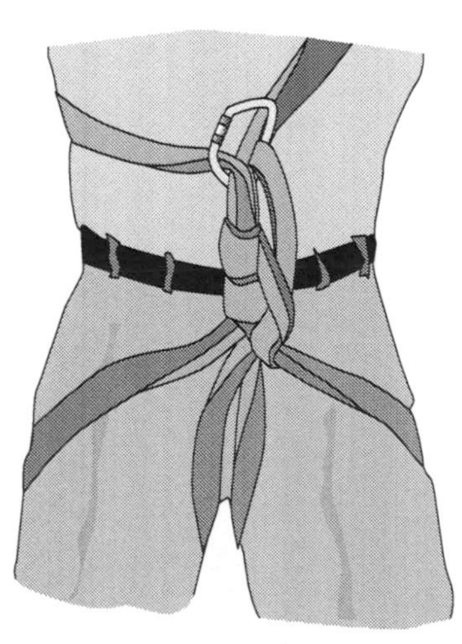

E Victim can now be carried hanging in the carabiner. **Important:** Keep victim's arms down.

FIGURE 3-43

A presewn Class III manufactured harness. (*Courtesy Yates Safety Equipment*)

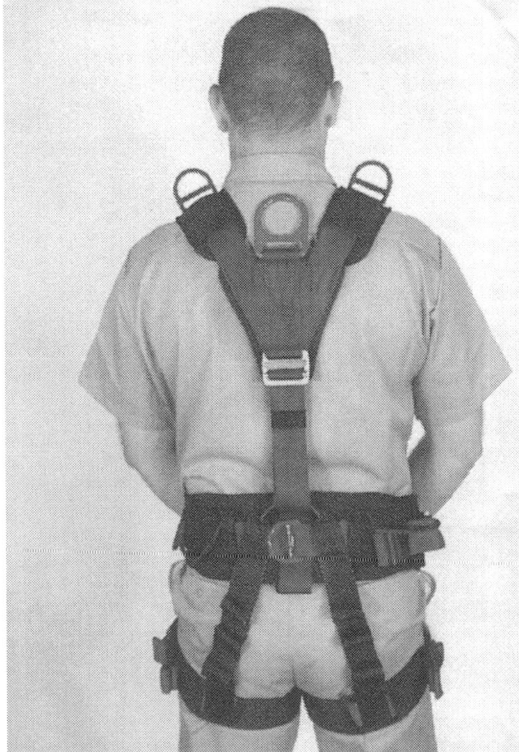

FIGUR 

A presewn Class II harness.
(*Courtesy Yates Safety Equipment*)

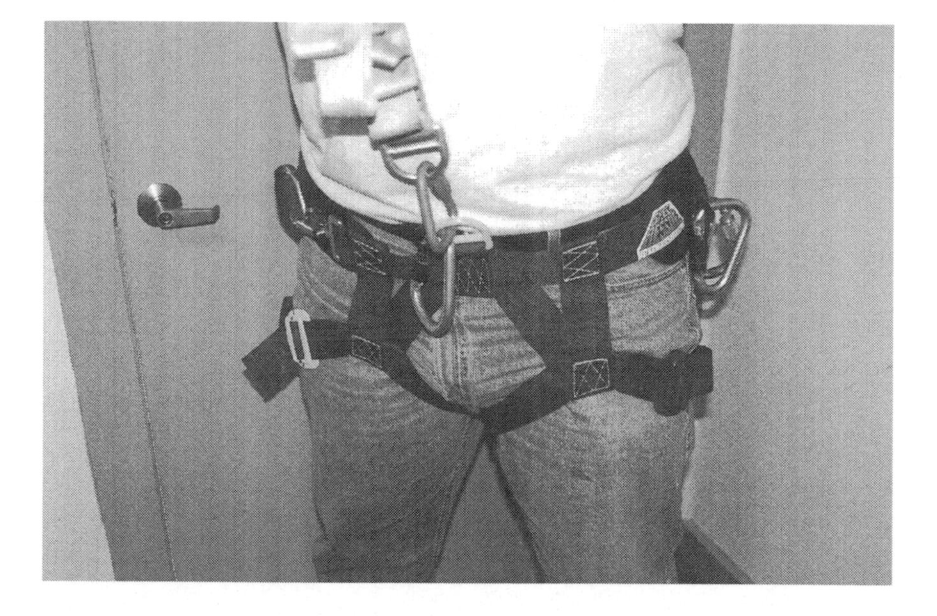

to the manufacturer's recommendations. Always keep it clean and away from chemicals, heat, and sunlight. With proper care, you should be able to get years of life out of your personal harness.

The belt, as defined in NFPA 1983 "is a system component; material configured as a device that fastens around the waist only and designated as a ladder belt, an escape belt, or a ladder/escape belt." New to NFPA 1983, the belt option allows firefighters and rescuers some latitude in escape and simple safety support uses (Figure 3–46A). Certified belts are commonly used in station uniforms, sewn into turn-out gear, and donned when climbing ladders (Figure 3–46B) for safety and support. Belts do not provide lower-body support via thigh and buttocks straps and can be particularly uncomfortable if loaded for more than a few seconds. Falls on belts are almost certainly fatal to the wearer's kidneys at best or, at worst, to the wearer.

► **prusik loop**

A loop tied using the double fisherman's bend to connect the ends of a single piece of cordage.

The **prusik loop,** more commonly called simply the prusik, is one of the most useful tools in a rope rescuer's tool box. The loop is tied using the double fisherman's bend to connect the ends of a single piece of cordage. The cordage is just a smaller diameter version of the low-stretch kernmantle rope we use for two-person loads. The diameter of the cordage is critical to the intended use of the prusik loop. Cordage diameter is listed by most manufacturers in millimeters (mm). It can be bought in 0.5 mm all the way to 9 mm diameter. Anything thicker is considered to be rope. The most common cord diameters used for rope rescue work are 7 mm, 8 mm, and 9 mm.

Common uses for the universal prusik loop are:

1. as a victim pick-off tool
2. as an anchor attachment
3. as a sacrificial element in a system to protect more expensive components
4. as a self-rescue device

FIGURE 3-45

(A) A presewn Class III harness. (B) A presewn Class III harness (back view). (*Courtesy Yates Safety Equipment*)

(A)

(B)

FIGURE 3-46

(A) "Last chance" emergency bailout uniform belt. (B) Firefighter ladder belt.

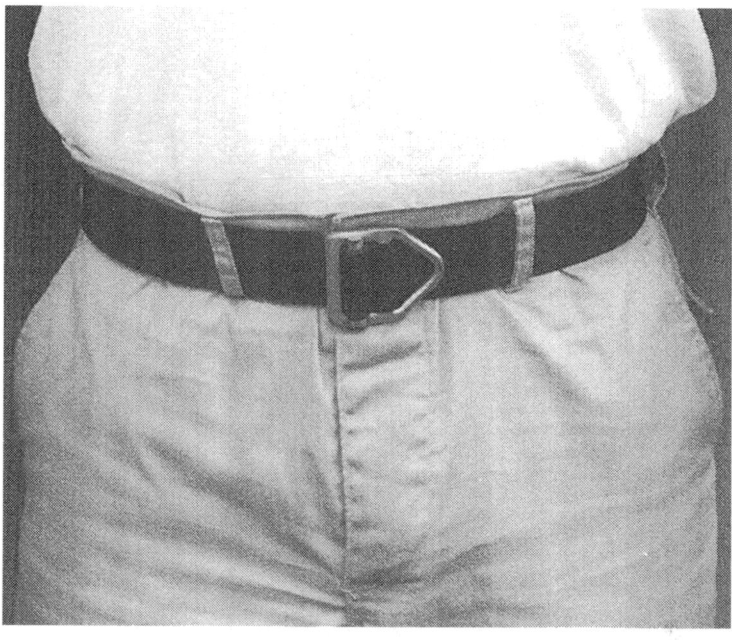

(A)

(B)

5. as a clutch device that slips and warns the rescuer if the system is receiving unexpectedly high forces
6. in the long loop (20 feet or greater) versions, as an excellent load-releasing device (see Chapter 7—hokie hitch)
7. as a component of the safety jig used on the dual-line highline

The simplest use for a prusik loop is as a connector between two people. For example, in the basic victim pick-off maneuver, a rescuer can attach a prusik from the Class II harness worn by the rescuer to a reasonable harness worn by the victim to safely extricate and rappel or be lowered to the ground. The prusik is an inexpensive version of the pick-off strap.

A prusik loop can be wrapped around a suitable anchor, a carabiner attached to both ends, and an anchor attachment made.

Calculations must be made to ensure that enough nylon is being used. For example, most 8-mm nylon cordage is rated at about 3,000 pounds. Tied in a loop and wrapped around an anchor, the single 8-mm cordage will be at least 4,500 pounds. This is very conservatively calculated by starting with the original manufacturer's listed breaking strength of 3,000 pounds. When doubling the cord, its strength does not necessarily double, so take half the original figure and add it to itself—3,000 + 1,500 = 4,500 pounds breaking strength at that anchor attachment.

Somebody, probably some bored sailor thousands of years ago, found out that if you wrap a smaller diameter piece of rope, a cord, around a larger diameter rope, a line, in just the right manner, the smaller cord will hold onto the rope very tightly and provide a very useful handle of sorts on the line. This configuration will hold to a point and then slip when the tension becomes too great. This hold/slip phenomenon has many uses in engineering modern rope rescue systems.

The prusik loop gets its name purportedly from Austrian music professor Karl Prusik, who used the loop and double fisherman's bend combination to fix musical instruments during World War I. He popularized the loop as a self-rescue technique for European alpinists in the early 1930s. This configuration is an adaptation from the Mangus or Arborheal hitch commonly used by tree surgeons. The use of the two- and three-wrap configurations on modern synthetic ropes has adopted and retained the term prusik loop.

Small prusik loops (Figure 3–47) are made from 5-foot lengths of cordage and, when the ends are tied together with three-wrap double fisherman's bends, they are 16 inches long. Medium prusik loops are made from 7-foot lengths of cordage and when tied, the loops are 28 inches long. Long prusik loops are made from 9-foot lengths of cordage and when tied are 36 inches long. Making your prusik sets consistent will avoid problems when you are rigging in the field. The short loop is small enough to just barely wrap around a 12.7-mm rescue rope three times and leave room for the double fisherman's bend to emerge and for a bight for a carabiner attachment. While it may take some practice to put it on the rope quickly

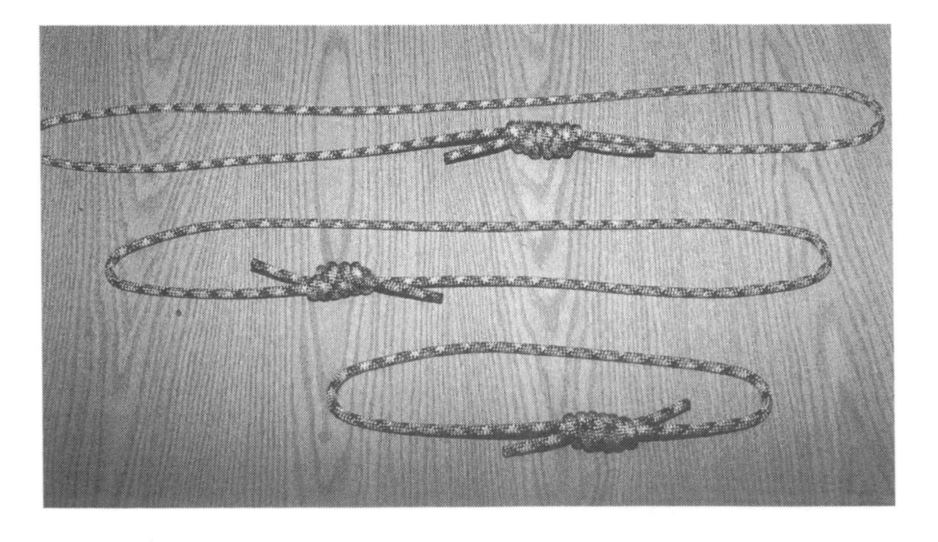

FIGURE 3–47

An 8-mm prusik set. One large, medium, and small prusik, starting size (before tying the double fisherman's bend) is 5, 7, and 9 feet. (*Courtesy Martin Grube*)

Building Prusik Loops

To understand the prusik loop and its applications, build some prusik loops with which to practice. A good starting supply of cordage loops can be made by obtaining a 115-foot length of 8-mm R/Q cordage. This will give you, before looping with a three-wrap double fisherman's bend, (3) shorts at 5 feet each single, (3) mediums at 7 feet each single, (3) longs at 9 feet each single, and (2) XXX longs at 25 feet each cordage loop (for load releasing devices).

In other words:

$3 \times 5 = 15$
$3 \times 7 = 21$
$3 \times 9 = 27$
$2 \times 25 = 50$
$15 + 21 + 27 + 50 = 113 + 2$ (for cutting and minor measurement errors) $= 115$ feet

Chapter 9 describes the use of XL (11 footers) and XXL (13 footers) for safety jigs on certain highlines. With 9-mm-diameter cord, add 8 or 10 inches to the single lengths before tying them. With 7-mm cordage, subtract 3 inches per cord.

and dress it out, you will find the small prusik to be your favorite for rope grabbing techniques. The medium prusik, when used with the small, provides a perfect set of tandem three-wrap, 8-mm prusiks. Always use the prusiks in tandem with the next larger size—the small with the medium, or the medium with the large. The former, as you will see, is the best set of tools for rope grabbing, absorbing force, and capturing progress.

An 8-mm prusik loop in single wrap or girth hitch around the host rope slides free along the length of the rope. A double-wrap girth hitch will grip quite nicely and slip, or slide down the host rope, between 2,500 and 3,000 lbf. A triple-wrap girth hitch grips even more and will slip around 3,000 to 3,500 lbf. The 3,000 lbf slippage is a very useful phenomenon for rope system engineers.

It is useful to understand the interface dynamics between the prusik loop fibers and the host rope (Figure 3–48). In effect, the automatic twisting action of the three-wrap girth hitch on the host rope acts to increase the friction coefficient of the loop/host interface. In the three-wrap girth hitch, the two legs of the loop are in tension from the center of the wraps in a rotational manner, twisting the rope, only to be balanced by the encompassing bight of the triple girth. This balanced action effectively cycles and focuses the linear tension applied to the prusik legs onto the host rope. This results in increased friction and, therefore, holding action, as force in the system becomes greater. At some point, the tension applied exceeds the friction coefficient of the loop/host interface and the loop slides down the length of the rope some distance. Used wisely, this grip and slip phenomenon makes a useful rope grabbing technique. It also can be used to release undesirable tension in the system and at least signal the rescuer or even prevent catastrophic failure.

FIGURE 3–48

Tandem three-wrap prusik set (one small and one medium) on 1/2-inch rope. (*Courtesy Martin Grube*)

The controversy over using prusiks derives from the myriad factors that affect their performance. For example, *all slippage ratings listed in this book are estimates based on experience and testing. Individual results may be different.* Unlike a mechanical ascender or rope grab that is stamped or machined to microscopic uniformity, a soft rope grab, has many shape variations. Also, unlike cordage, mechanical rope grabs are immune to absorbing water and humidity. At best, only approximations about grip versus slippage can be applied to both soft and hard rope grabs. However, since the metal part of a mechanical grab interface, consisting of a metal device and host rope, is finite, approximations of rope slippage can be guessed more closely. Metal rope grabs have a history of destroying rope at high forces due to the interface device, usually a cam configuration, and in particular, personal ascenders with conical toothed cams. Some ascenders with this type of cam bite the rope so hard that they can actually chew through the outer sheath of the rope under tension, usually between 1,000 and 1,500 lbf. Other ascenders have ribbed cams (Figure 3–49) that are much easier on the host rope but still damage the rope around 2,000 lbf. In an effort to address these issues, manufacturers of quality rescue equipment have been attempting to increase the surface area of the cam and spread the forces of the cam interface out over a larger number of sheath pics. Machined interfaces are usually curved, concave, with a matching geometry, convex, to the cam. Stamped interfaces are simply speed bumps stamped into the shell to minutely increase the surface area as well as the cost of manufacture. Other attempts at tempering the bite of mechanical ascenders, so they will slip rather than chew up the rope, involve coating the metal rope interface parts with a tough but slippery substance like polyethylene or Teflon®. In effect, the mechanical action of the device works up to a certain point, say 2,000 lbf, and then the coated cam and shell slip, releasing some tension and regrabbing the rope

FIGURE 3-49

Common cam designs for mechanical rope grabs. (*Courtesy Martin Grube*)

a short distance away. This is a step in the right direction, no doubt. As the coated rope grabs are relatively new, the jury is still out on the coating's ability to stand up to repeated use and slipping.

One heavy, stainless, industrial-type steel rope grab incorporates a very long, flat, ribbed friction plate that is pressed into the rope via the leverage action of the device. The plate is just long enough to contact each individual sheath pic in the rope via the natural rotation of the weave pattern in most R/Q ropes. Simply stated, if you were to draw a longitudinal black line down the center of a piece of 12.7-mm rope, and then trace a single bundle of sheath fibers around and down the rope, the black line would have to be 1⅜ to 1⅝ inches long in order to touch again the same spiraling fiber bunch. The friction plate on the rope grab is 1⅝ inches effectively pressurizing every fiber bundle. This concept practically eliminates rope damage from the mechanical interface and allows slippage. However, the slippage occurs in the 300-lbf to 1,000-lbf range on some ropes, which makes the rope-grab less than perfect for heavy hauls and tensioned rope systems in that range. Expect to see great strides in hard rope-grab technology in the next few years, where advantageous rope slip occurs at predictable tensions.

Factors that affect prusik loop and host rope interface are stiffness or looseness of fibers; type, age, and condition of fibers; manufacturing technique; rope coatings; shock-load versus slow-pull applications; correlation of cordage diameter to host rope diameter; vicinity of other equipment; angle and direction of pull with relation to host rope; weather conditions; hot, dry, wet, or humid rope; how the girth configurations are dressed; and various human factors.

After some experience, however, you will find a performance groove with certain combinations of prusiks that will reinforce your confidence in their use. In general, the perfect combination, or prusik set, of soft rope grabs on 12.7-mm low-stretch rope is two, three-wrap **(t3w)** 8-mm prusiks, preferably one small (16 inches) and one medium (28 inches). Using the prusiks in tandem reinforces component redundancy concepts. If one of them is improperly dressed or damaged, another one is in place and ready for action. Additionally,

► **t3w**

Tandem three-wrap prusik loop on a rope.

Size Correlations of Ropes

Size correlation may be expressed mathematically as a percentage of the cordage diameter relative to the host rope diameter. For example, 8-mm cordage is 63% of the diameter of 12.7-mm host rope. However, to determine the friction interface of the prusik loop to the host rope, the circumference of each rope must be calculated. For example,

8 mm × Π {3.14} = 8 × 3.14 = 25.12-mm circumference
12.7 mm × 3.14 = 39.8-mm circumference
25.12 mm ÷ 39.8 mm = 0.63, or 63%
8 mm is 63% of 12.7 mm

Additionally, an 8-mm three-wrap girth hitch on 12.7-mm-diameter rope spans about 57 mm, or 2.25 inches on the host rope.

57 mm × 39.88 = 2,280 square millimeters of surface interface

The size correlations are as follows:

Prusik loop diameter, mm	Host rope diameter, mm	Percent	Estimated slip, lbf
6	11.1	51	**3,200***
6	12.7	47	**4,500***
7	11.1	63	2,500
7	12.7	55	**4,700***
7	15.8	44	**5,000***
8	11.1	72	2,250
8	12.7	63	3,250
8	15.8	50	4,500
9	12.7	70	2,800
9	15.8	57	3,400
10	12.7	79	1,250
10	15.8	63	2,800

*Bold numbers indicate probable destruction of the cordage before slippage on host rope.

a bit more friction is obtained with the combined prusiks biting the rope, making the approximate slipping forces around 3,500 lbf on 12.7-mm rope.

Correlating the diameter of the cordage in a prusik loop to the diameter of the host rope is extremely important.

► **anchor slings**
Synthetic-fiber flat or tubular webbing configurations that wrap around anchors for the purpose of attaching system components.

Anchor slings (Figure 3–50 and Figure 3–51) are synthetic-fiber flat or tubular webbing configurations that wrap around anchors for the purpose of attaching system components. They come in different sizes and colors and with different performance characteristics. Anchor slings are used mostly to wrap around large, bombproof objects and to create a focus point to connect and anchor a rope system.

Conclusion

The number of equipment options available to you and your team are virtually endless. Obtain at least an intermediate, or operational, level of training before purchasing your cache. Remember, a lot of trainers will try to sell you a particular product, or in some

FIGURE 3–50

Synthetic anchor slings. (*Courtesy Martin Grube*)

FIGURE 3–51

Anchor sling around tree.

cases *any* product. Try to separate the two. Train and then train some more before making that big equipment purchase. Do not be afraid to seek expert advice.

From the beginning, document the history of your gear. Inventory and mark your gear so you can prove its use, lack of use, or even misuse. It will pay big dividends in terms of your credibility when the chief or an attorney inspects your equipment. The small amount of time invested in inspecting and documenting *to the manufacturer's recommendations* and the AHJ's policies, will reap big benefits.

Consider standardizing the majority of your gear. Develop a color-coding system for all the different lengths of prusik, webbing slings, and rope. Color-mark your hardwear, so it is easy to tell yours from other, maybe less well-documented gear. And remember, the manner in which you package, store, and transport your gear is a direct reflection on your own professional rope rescue pride.

■ SUMMARY

There are literally thousands of combinations of tools available for rope rescue work. Hundreds of manufacturers design, build, and sell equipment, all of which has its merits and demerits. The informed technical rope rescue team spends a tremendous amount of time researching and testing equipment until the right combination of useability, weight, strength, and expense is achieved. Understand early in the development, and/or maturity of your team, that equipment research is a living project that has no end, but that there are some equipment articles that have found their groove and are going to be hard to improve on.

■ KEY TERMS

Hardware
Metallurgy
Alloy
Aluminum
Anodization
Steel
Stainless steel
Descent control device (DCD)
Dulfersitz
Eight-plates
Brake bar rack
Hyper-bars
Closed-end rack/closed frame rack
Tube type DCD
Miscellaneous DCD
Carabiner
Dynamic roll out
Mechanical rope grabs and ascenders
Rigging plates
Swivels and snap links
Anti-torque devices (ATDs)
Block creel
High-stretch, or dynamic, kernmantle
Improvised harnesses
Emergency egress improvised harness
Modified Swiss seat
Hasty hitch harness
Presewn harnesses
Prusik loop
t3w (tandem three-wrap) prusiks
Anchor slings

■ REVIEW QUESTIONS

1. Rope rescue equipment is classified _____.
   a. as either metal or nylon
   b. according to use: urban, wilderness, or tactical
   c. according to task: rappeling, raising, lowering, or tensioned
   d. as either hardware or software

2. The term _____ is used to describe the combination of two or more different elements.
   a. symbiosis
   b. alloy
   c. recombinant
   d. casting
3. The most commonly used aluminum alloys in the manufacture of rope rescue hardware are _____, _____, and _____.
   a. 2024, 6061, 7075
   b. 1983, 1006, 1670
   c. Type 300, Type 400, Type 700
   d. hardcoat, softcoat, unfinished
4. Most stainless steel used in rope rescue equipment is of the AISI Type 300, _____, _____, and _____ stainless steel.
   a. nonmagnetic, tough, ductile
   b. semimagnetic, tough, brittle
   c. nonmagnetic, soft, ductile
   d. rustproof, nonmagnetic, iron free (nonferric)
5. DCDs can be classified in one of four categories:
   a. eight-plates, brake bar racks, tube type, and miscellaneous
   b. manual, automatic, human interactive, and miscellaneous
   c. closed-end, open-end, with ears, or without ears (deaf)
   d. wilderness, rural, urban, tactical
6. The preferred choice for the bars on a brake bar rack are _____.
   a. hollow aluminum
   b. high carbon steel, hollow
   c. titanium shell, aluminum core, solid
   d. tubular stainless steel stock
7. Six brake bars are added to the brake bar rack in a specific order to ensure proper rope loading, they are (from top to bottom) _____.
   a. training groove, straight slot, and four slanted slot bars
   b. straight slot, slanted slot, and four straight slot bars
   c. training slot, straight slot, slanted slot, and four slanted slot bars
   d. stainless steel and four aluminum straight slot bars
8. The _____ is the best device available for engineering rescue load-lowering systems.
   a. rescue eight with ears
   b. motorized hydraulic descent device (MHD_2)
   c. tensionless wraps around a standpipe or a bombproof tree
   d. closed-end rack

9. The best method of unlocking a stuck carabiner gate lock is to _____.
   a. load it to approximately 75% of its minimum breaking strength and attempt to unscrew it with a flexible rubber pad
   b. wrap two turns of cordage around the gate lock, have someone hold the carabiner, and pull the cordage so the lock action is counterclockwise
   c. place it on a hard surface and attempt to rotate the lock counterclockwise between your boot and the surface of the ground
   d. use the carabiner wrench supplied by the manufacturer, and do not use one manufacturer's carabiner wrench on another's carabiner
10. Nylon Type 6.6 707 has a melting temperature of about _____.
    a. 350°F
    b. 400°F
    c. 450°F
    d. 550°F
11. Nylon loses a considerable amount of strength when saturated with water, as the hydrogen in the water causes some of the hydrogen bonds in nylon to have a weaker molecular attraction, causing it to lose up to _____ of its tensile strength when saturated.
    a. 50%
    b. 25%
    c. 15%
    d. 5%
12. Rope constructed without knots or splices in the yarns, ply yarns, strands or braids, or rope is called _____.
    a. shuttle loom
    b. block creel
    c. continuous filament
    d. kernmantle
13. In tubular webbing, _____ is the preferred construction method as it is somewhat stronger and cannot be unwound by cutting the edge seam.
    a. shuttle loom
    b. braid on braid
    c. chain lock
    d. block creel
14. A prusik set of small, medium, and large loops can be made with three sections of 8-mm cordage. Before tying, the length of those sections of cordage should be _____.
    a. 28 inches, 36 inches, and 42 inches
    b. 9 feet, 11 feet, and 25 feet
    c. 42 inches, 54 inches, and 66 inches
    d. 5 feet, 7 feet, and 9 feet

CHAPTER

4 Rigging for Rappeling

OBJECTIVES

Upon completion of this chapter, you should be able to:

- identify suitable anchors for rope rescue operations.
- define and identify a pseudo-anchor.
- define the difference between a knot, a hitch, and a bend.
- explain the dynamics of rope configurations, what works, and what causes failure.
- identify and correctly construct the twelve basic rope configurations.
- engineer a three-point static rope rescue system.
- engineer a three-point dynamic rope rescue system.
- engineer a dynamic belay system.
- identify leading edge concerns and proper edge management techniques.
- discuss the major safety points of using a helicopter as a rope rescue platform.
- discuss the advantages and disadvantages of using helicopters as an adjunct for rope rescue.

A DAY ON THE ROPES

The highline between the two buildings was complete. At an elevation of only about 165 feet and a span of almost 150 feet, this was a relatively mundane highline. The purpose was to teach technical rescue teams some of the techniques of moving the occupants of a hazardous building to the relative safety of a close and nonhazardous building via near-horizontal rope systems. I had an opportunity to be the rider and enjoy that floating sensation that you can only get with a near-horizontal rope system. Approximately mid-span, I heard that super-cool monotone of one of my team members, that could only suggest something was really going wrong, say over the radio in my chest harness, "Mike, DON'T MOVE!" My body froze instantly while my mind switched into overdrive, screaming the possibilities that might cause the DON'T MOVE command. With a resulting catenary angle in the highline of about 100 degrees, that made me a little over 120 feet in the air, not quite high enough to reach terminal velocity, but gave me enough time to do some serious thinking on the way down.

Dean and Charlie were wrestling with the massive twin 5/8-inch track lines on which my life depended (Figure 4–1A, Figure 4–1B, and Figure 4–1C). While the 16-inch steel I-beam anchor to which the ropes were attached was completely secure, part of the anchor system was failing enough to cause some serious concern to my team and some major concern to me, and there was not a thing I could do about it. I would have to rely on my team to get me out of my situation. The anchor on this highline was a massive steel I beam. The anchor system involved some delicate maneuvering of the rope around obstacles on the roof and down into the building via the removal of an exhaust fan attached to some duct work on the roof. As with most training scenarios, if we wanted to use the building again, we would have to make sure we left no marks. Therefore, the anchor system consisted of the I-beam, the twin 5/8-inch ropes, some rope padding, and four 4 × 4 × 18 number 2 pine wood crib blocks stacked to raise the ropes over the delicate duct work. The "don't move" command was called because the weight and shear force on the crib blocks caused by the ropes had caused the cribs to shift, or roll out, and the track lines were now lying directly on the sharp edge of the duct. Somehow, the duct was holding the downward and outward pull of the big 5/8ths balanced precariously in perfect equilibrium. The ducting had not collapsed, which would have been better, as the highly tensioned track lines appeared to be lying on the blade of a makeshift knife!

It was a good thing I did not know all of the details, as I would have been even more worried. Living out on a string has a way of drawing thoughts out of you like a blade. Every stitch in your harness, every dropped carabiner, and every uninspected rope becomes suspect. Technique and retechnique runs through your mind like a videotape on fast forward with the sound going too fast to understand the words. A rescuer has some time in the don't move mode to do some divine reconsideration. It is like getting a theology degree in thirty seconds.

Dean informed me over the radio that they would have the little problem worked out in a minute. Harold was some distance away but was able to give me the approximate details of the event. I felt some comfort knowing the horizontal control lines were standing ready, as always, in the *unlikely* event of trackline failure. I'd seen dozens of tests, even ridden some failed trackline tests, and had complete confidence in the backup technique. Somehow it was different now.

Charlie, wisely, was able to work a 10-foot 4×4 under the ropes and, using some spare cribbing as a fulcrum, developed a Class I lever to raise the ropes off the duct. Dean carefully rebuilt the crib system and together they transferred the load off the sharp ducting back onto the crib station.

FIGURE 4–1

(A)

(B)

(C)

(A) Ventilation duct removed to access bombproof anchor under roof covering. Wood cribbing is being used to elevate rope over metal ductwork. (B) Steel I-beam anchor inside building. Note rope padding. (C) Anchor ropes extended to the leading edge focus point. (*Courtesy Martin Grube*)

NOTE: A rigger I grew to respect very much told me he always envisioned system component failures in his mind when building his rope rescue systems. While that is a bit gloomy for me, his mental tool helped him have a pre-plan in mind should any failures ever occur.

Three interesting points from this near-incident help to highlight this chapter on rigging for rappeling. Number one, have a backup system that *automatically* reacts in the event of system or human failure. Number two, be extremely careful of steps you might take during training that are different from steps you might take during a real incident. Had this occurred in a real emergency instead of training, there would have been no need to protect the duct by adding the wobbly crib station. We would have simply bashed it down with a hammer and padded whatever was left. And number three, always stick with good fundamental rigging techniques. Do not cut corners, EVER. If or when something does go wrong, basic, proven, redundant rigs will pull you through to a safe finish.

ANCHORS

► **rappeling**
The act of sliding down a rope in a controlled fashion.

► **anchors**
The items to which rescuers attach rope system components.

► **bombproof, bombing, or bombing off**
Terms that have been applied to guaranteed foolproof anchors.

► **rope system**
The sum of all the equipment components that are engineered into a rescue tool for the purpose of saving lives and transporting people and equipment to areas that are otherwise not accessible.

Rappeling is simply the act of sliding down a rope in a controlled fashion (see Chapter 5). To rappel requires a transportation system—a rope or ropes that are sufficiently anchored to assure a rescuer's safety while working on the rope. **Anchors** are the items to which rescuers attach rope system components, which are solidly in place to eliminate any possibility of unwanted system movement. **Bombproof, bombing,** or **bombing off** are terms that have been applied to guaranteed foolproof anchors. In theory, even if a bomb goes off, the integrity of the anchor is not in question. Using a quality, bombproof anchor allows you and your team to work with much greater focus, confidence, and concentration by removing any worries about the foundation of the system. Your team should always consider the anchor, its attachments, and all of its necessary components as an anchor system. Each of the components is interdepen-dent and important to the success of anchor security. In short, do not make the mistake of underengineering part of the anchor system just because there is one really bombproof anchor. The crib system used in the example at the beginning of the chapter should have been as stable and bombproof as the anchor and its attachments or eliminated altogether. When making rigging assignments, consider appointing a mini-team of two or three people to establish the anchor system. When it is time to attach the rest of the **rope system,** the anchor team can point with confidence to the best place to attach to the anchor.

Locating good anchors takes some experience, but in general the following items work well as bombproof anchors:

1. Major load-bearing structural members, steel I-beams and columns, and rated maintenance anchors (pad-eyes) (Figure 4–2A, Figure 4–2B, and Figure 4–2C) are almost always perfect anchors. In urban and industrial locations, they are plentiful and make for an almost sterile rigging environment. Make sure that these structural components are, in fact, load bearing and solid. Decorative facades that appear massive often disguise 2×2 inch galvanized channel steel supports and are not good anchors. Get in the habit of routinely looking for anchors and analyze them for being bombproof.
2. Heating, Ventilation and Air Conditioning (HVAC) footings offer excellent low, heavy anchors. Be particularly careful around running machinery, electrical parts, and fan belts. Always secure, by locking out and tagging, electrical parts of any anchorages to HVAC units.
3. Entire buildings can be used simply by padding the corners and wrapping a rope in true tensionless fashion around the structure. Be aware that many buildings have a styrofoam-like surface that can crush easily, damaging the building and perhaps your rope. Elevator towers (Figure 4–3) can be great anchors and excellent high-anchor points to get an elevated leading edge.

FIGURE 4–2

(A) Bombproof reinforced concrete anchor with rope system attachments. (B) Steel boxbeam anchor. (C) Maintenance pad-eye anchor on rooftop. (*A, Courtesy Martin Grube; B, Courtesy Chase N. Sargent; C, Courtesy Martin Grube*)

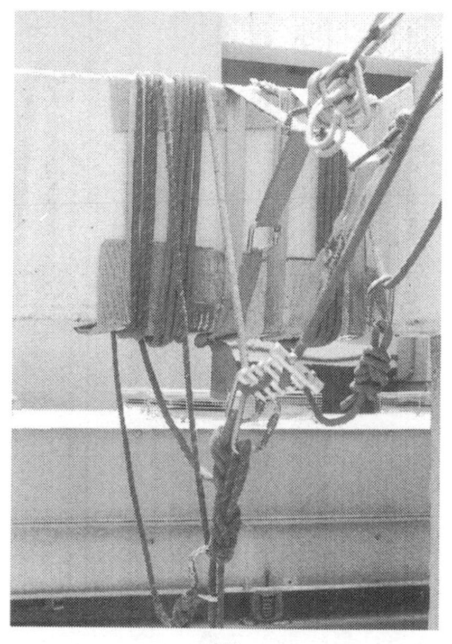

(A)

(B)

(C)

4. Grouping windows and door frames or roof scuppers can create good anchors (Figure 4–4A and Figure 4–4B). Simply use a large, long rigging strap or a section of rope to wrap the

Machinery and Equipment Hazards

By its very nature, rope rescue involves lots of ropes, strings, webbing, and loose equipment. Running machinery, especially around urban rescue sites, is incredibly dangerous. Drive shafts, fan belts, hydraulic rams, and similar machinery turn on and off without notice. An elevator motor that has been inactive for hours can suddenly switch on, driving huge cables and drive-wheels. Electrical cables and devices can be activated automatically, manually, and remotely, changing a stable condition to one of extreme hazard.

Always look for machinery hazards. Assume any machine or machinery part that is idle will suddenly start moving. Assume all electrical parts and anything metal may be energized. Secure all electrical and mechanical hazards. The building engineer should be consulted to assist in analyzing any area in which you intend to work. Lock, chock, double blank, and tag any equipment that may affect your operations. Minimize your hazards by wearing the appropriate personal protective gear. Rescuers, particularly rope rescuers, tend to have lots of gear hanging from their bodies. Eliminate loose gear. Make sure harness webbing and attached equipment is tight and secure to the body. Zippered boots are much safer than boots with laces. Laces flop about and can come untied and cause tripping hazards. A loose boot lace or a piece of webbing that is hanging from a harness can get caught in a piece of running machinery and injure or kill a rescuer in a second. Never store equipment around the neck. Sometimes people stash emergency prusik loops around their necks, which is highly dangerous. Make sure your team's helmets are fastened snugly and the chin straps used. A helmet that falls off a rescuer's head onto people or a patient's head below can kill just like a dislodged rock from above. Most people prefer lightweight helmets with automatically detachable chin straps, which means that if a rescuer somehow gets caught by the helmet, the chin strap will break free. Remember, specialized rescue requires specialized personal protective equipment.

FIGURE 4–3

An elevator tower wrapped with a rope can be used as an anchor and as en elevated leading edge. (*Courtesy Patrick Anderson*)

FIGURE 4–4

(A) Grouping roof scuppers for anchors. (B) Roof scupper anchor attachment. (C) Scupper rigging. (*A and B, Courtesy Martin Grube*)

(A)

(B)

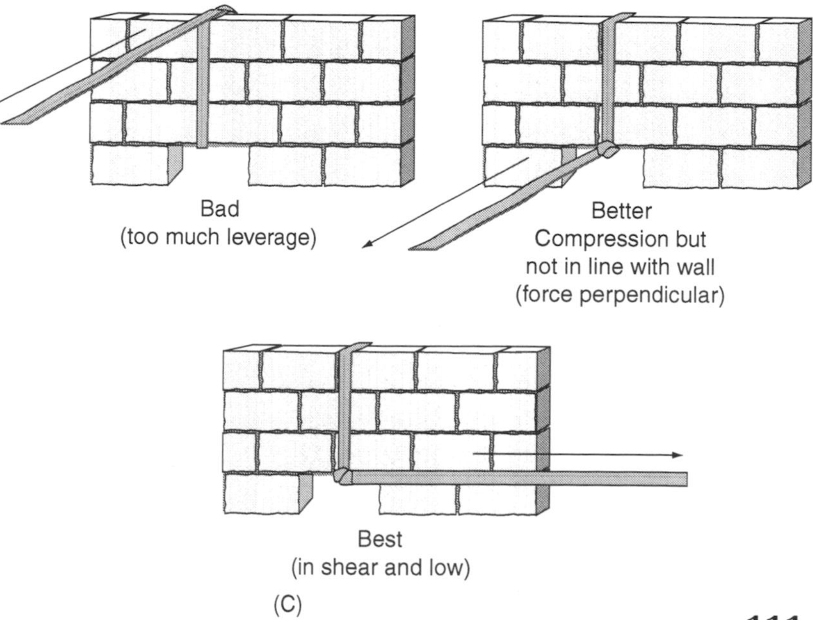

(C)

FIGURE 4–5

Steel staircase used for an anchor. (*Courtesy H. Dean Paderick*)

frames or scuppers together. Take special care to watch for sharp glass and metal flashing that might cut the rope. Pad all edges liberally.

5. Building standpipes and stairs in stairwells make good anchors (Figure 4–5). Check them for signs of hidden rust or poor maintenance and always try to rig them in shear with relation to their attachment to the building (Figure 4–6A and Figure 4–6B). Group stair rail components in a load sharing arrangement when in doubt.
6. Mechanical friction bolts attached to a building or to rock faces (Figure 4–7) according to the manufacturer's recommendations can make excellent anchors.
7. Trees can be good anchors if they are alive, sufficiently heavy, and well rooted. Most leaf-bearing trees have a root system roughly one-third the size of the canopy (Figure 4–8).
8. Large rocks and outcroppings (Figure 4–9A and Figure 4–9B) can be wrapped and used as good anchors in many circumstances. Always make sure the rock is sound before using it as an anchor. Beware of boulders that may be resting on loose surfaces like mud, pebbles, and moss that can cause them to shift. Always treat rocks and trees with respect. Try not to deface or damage them in any way; you or someone else may need them again someday.

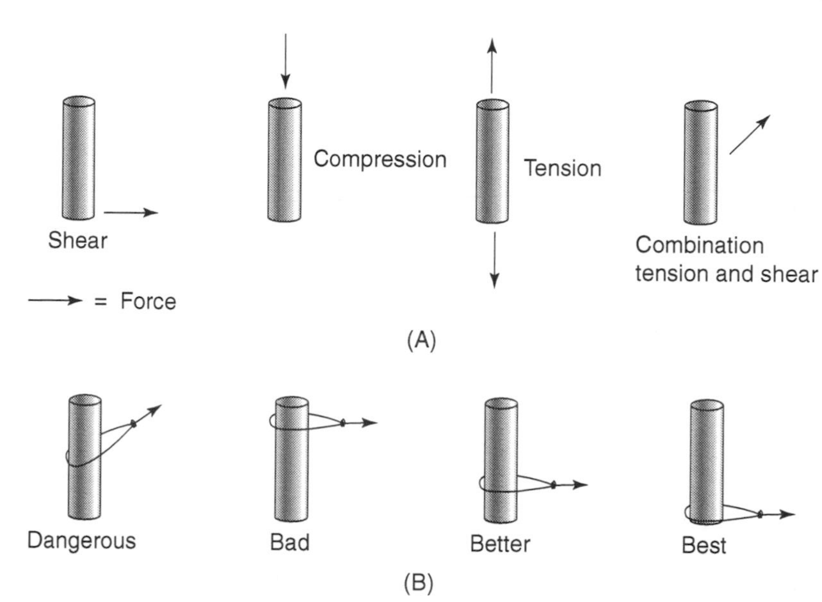

FIGURE 4–6
(A) Forces on anchors.
(B) Leverage considerations on anchors.

FIGURE 4–7
Friction or expansion bolt anchor and hanger. (*Courtesy Chase N. Sargent*)

▶ **anchor systems**
Groups of anchors that work in unison to provide either more conveniently located anchors or backup anchors if one of the anchors fails.

▶ **tie backs**
These tension two in-line anchors together to incorporate the strength from both anchors simultaneously.

▶ **load sharing anchor system components**
Sections of rope or webbing configured to transfer the rope system load to two or more anchors.

Anchor systems are groups of anchors that work in unison to provide either more conveniently located anchors or backup anchors if one of the other anchors fails. Tie backs and load sharing bridles are the most common and logical anchor systems. **Tie backs** (Figure 4–10) tension two in-line anchors together to incorporate the strength from both anchors simultaneously. If you can imagine a three-point picket system where a rope is attached from the bottom of a rear anchor to the top of a front anchor and then tension is applied to hold them together, then it is easy to build a tie-back anchor system. A 1/2-inch section of R/Q rope can be configured between the two anchors using a trucker's hitch, or an adjustment jig. Using a pulley and t3w prusiks to hold the rope tension back on itself also works well.

Load sharing anchor system components are sections of rope or webbing configured to transfer the rope system load to two or more anchors (Figure 4–11A). This shares the load between multiple anchors instead of just one. Some systems can even be constructed that

FIGURE 4-8

Deciduous (leaf-bearing) trees have a root base approximately equal to one-third of their canopy size.

FIGURE 4-9

(A) Rock anchor. (B) Rock anchor and lowering system. (*Courtesy Chase N. Sargent*)

(A)

(B)

FIGURE 4–10

Pulley system tie back anchor.

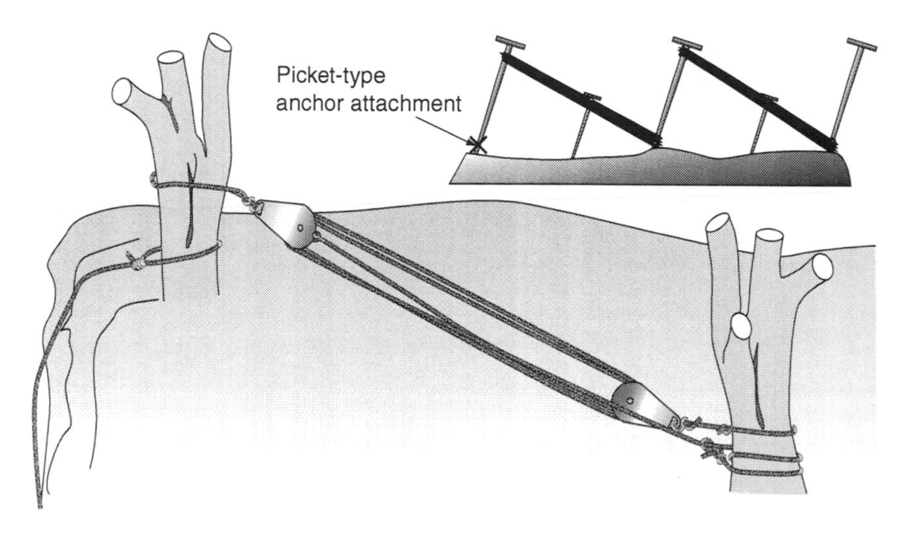

will partially equalize the load on the anchors. One of the most common load sharing anchors is created by tying the double bight eight load sharing bridle (Figure 4–11B and Figure 4–12). (More on the double bight eight later in this chapter.)

▶ **pseudo-anchors**

Items that can be used as anchors but may not always be considered bombproof.

Pseudo-anchors can be defined as those items that can be used as anchors but may not always be considered bombproof. Pseudo-anchors have applications, but great care must be taken when using them in rescue operations. In general, they are great tools, but they must not be used as the sole anchor point for any rope system. Some pseudo-anchors include the following:

1. Bipods, tripods, quadrupods, gin poles, and scaffolding are mainstays in confined space (Figure 4–13A) and many rope rescue applications because they provide an elevated leading edge. However, if a bomb were to go off, it is doubtful any of these items would remain intact. When using any of these devices, be sure that a belay line is being used that is completely free from the device (Figure 4–13B). In addition, if there are any forces other than straight down, such as in a manhole, make sure the device is tied to something substantial so it cannot be pulled off the edge (Figure 4–13C). For example, the rope angling off the side of a building and a tripod is being used to elevate the leading edge.
2. Heavy rescue rigs and fire trucks are excellent anchors (Figure 4–14A and Figure 4–14B) but for obvious reasons cannot be considered bombproof. There is always the possibility that someone could get in and drive away. However, assuming that the rig is in good condition and has manufacturer-certified attachment points, *and* that the rig has been thoroughly secured—keys (if it has them) removed, emergency stop activated, locked in park (or gear), emergency brake applied, wheels chocked, security person with machine gun stationed on the rig, and so on—it can be an excellent anchor. Additionally, if the building is completely devoid of anchors, you can always anchor to a ground rig

(A)

(B)

FIGURE 4–11

(A) Load sharing anchor using rigging straps. (B) Two rope load sharing (but not equalizing) double bight eight anchor and rigging plate focus point. (*Courtesy Martin Grube*)

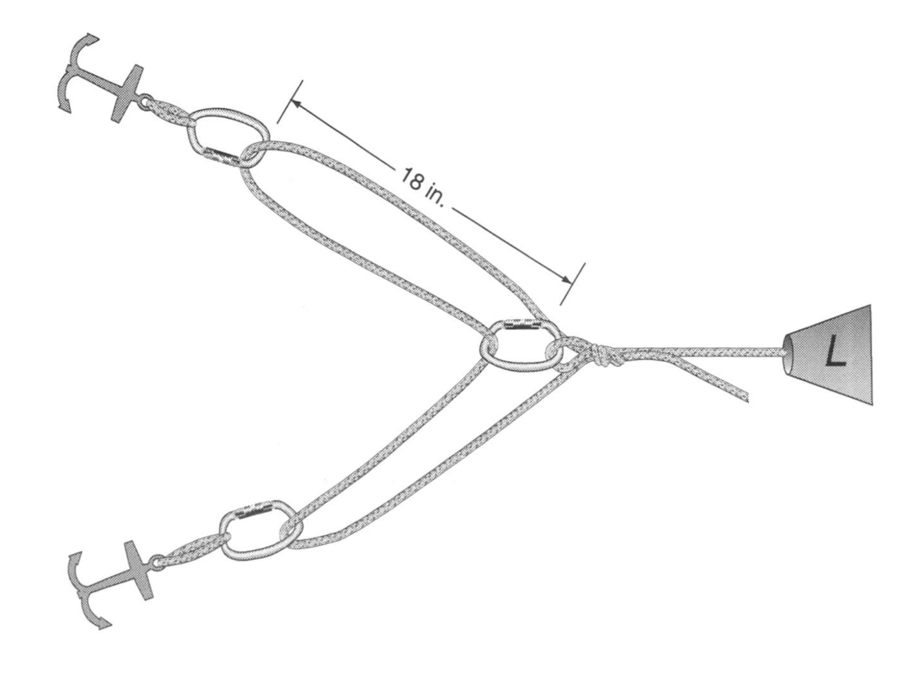

FIGURE 4–12

Load sharing and equalizing two-point anchor bridle tied with a double bight figure eight.

FIGURE 4–13

(A) Tripod pseudo-anchor used to elevate rope and load off the edge. (B) Tripod pseudo-anchor used to elevate rope and load off the edge. Belay rope kept separate from tripod. In case the tripod is pulled over, the belay rope is ready to hold the load from falling. (C) Tripod used to elevate leading edge. Note tripod tied to an anchor to secure it to the building. (*A, Courtesy Martin Grube; C, Courtesy Patrick Anderson*)

(A)

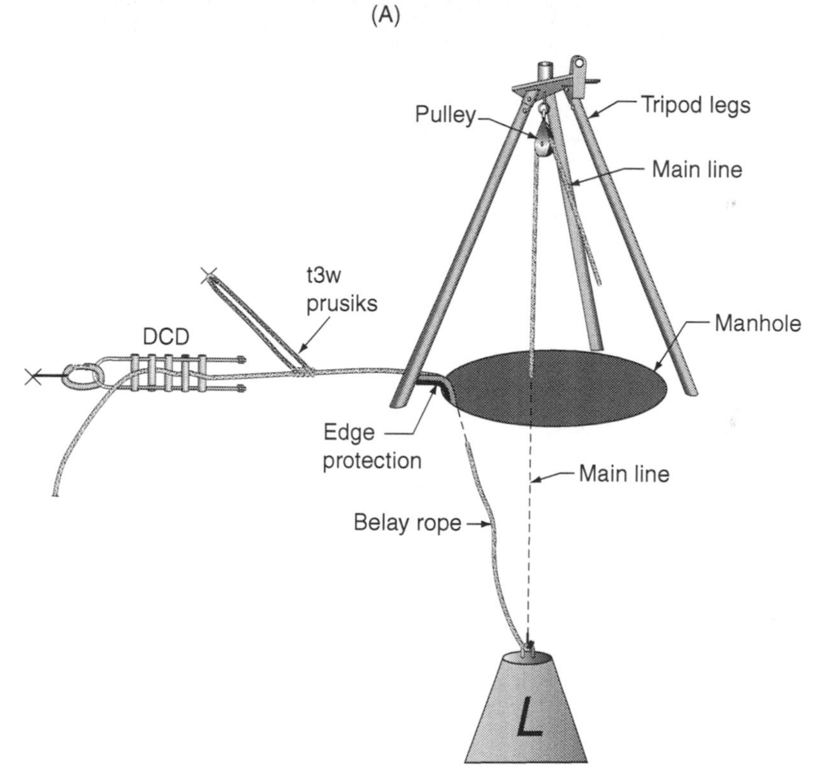

(B)

(C)

FIGURE 4–14

(A) Heavy equipment pseudo-anchor. (B) Heavy equipment pseudo-anchor (surf rescue unit, LARC V). (C) A building void of adequate anchors can still be used by extending ground anchors to the roof. (*A, Courtesy Martin Grube*)

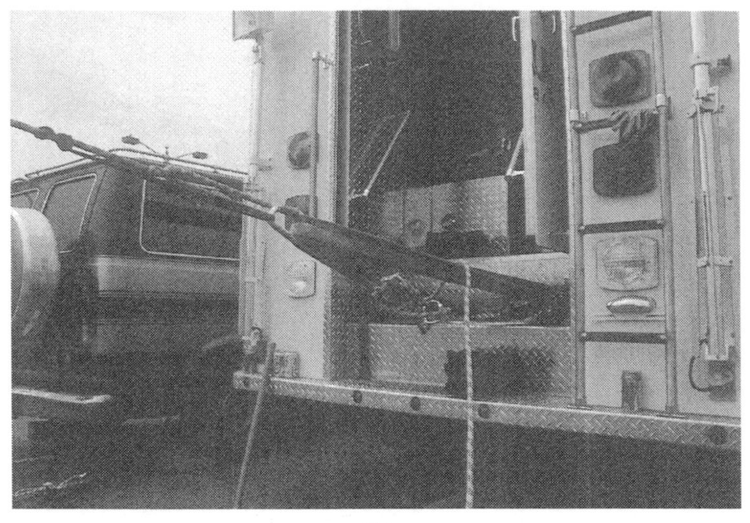

(A)

(B)

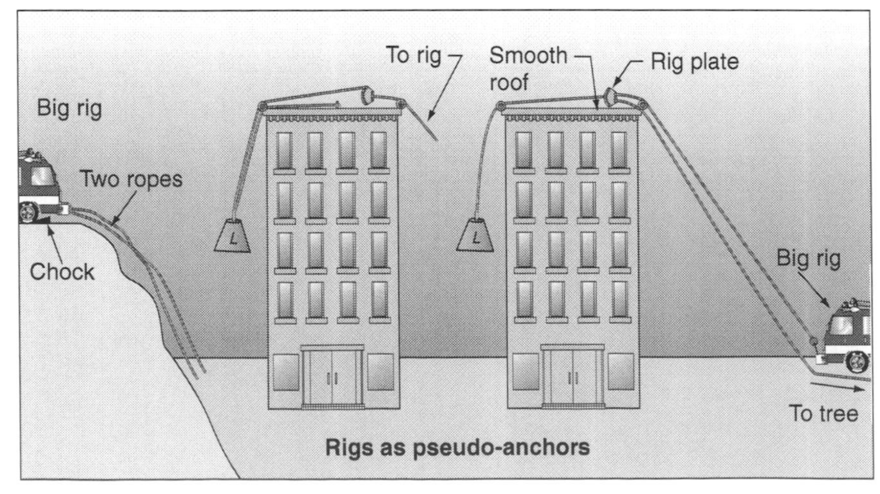

(C)

NOTE: The Question: A wonderfully simple analytical tool that you and your team can use to quickly design and check a rope system is "The Question." Ask yourselves the following: If this system component disappears (explodes, disintegrates, dissolves, and so on), where will the load be transferred to? If the answer is (1) I don't know, (2) to the ground, or (3) complete failure, then the system has failed "The Question." In some applications, it becomes easy to rig the whole system to a single anchor or rigging plate, or to use only one rope. It can even become habit to rig to single, super-reliable components.

► **super bombproof anchor (SBA)**

A bombproof anchor in which the rig master feels so confident that no other anchors are used in the system.

► **training to fundamental correctness**

Sticking to the rules by not cutting corners when building rope rescue systems. Building a system using a series of rules and philosophies one day and then bending those rules the next day makes for irregularities that are confusing to a team and can cause a general lack of focus on complicated and stressful rescue calls.

► **primary anchor**

The main anchor that a load depends on.

► **secondary anchor**

The backup, alternate, or belay anchor.

(Figure 4–14C) and bring the rope up the side of the building, effectively extending the anchor to the roof from the ground.

3. People have been used as anchors on occasion and unfortunately will be again. During high-hazard emergency operations that are absolute life and death situations, the weight and strength of a number of people can be used as an anchor. For the purposes of this book, however, no recommendations for using people as anchors will be made. The obvious risk, of course, is pulling the anchor people over the edge to certain death. Also, when people are used as anchors, they usually become friction control devices, mechanical advantages, and the like. They might get tired or develop an attitude, and let go of the rope and the load. When people are the anchors, there may not be anyone left to safe the edge and make rope calls. Finally, when people drop people, emotions are affected as well as bodies.
4. Small plants, grass, roots, and the like can be grouped together to make a surprisingly strong anchor. Webbing and cordage can be wrapped around a series of small plants using round turns (Figure 4–15A). Using load sharing anchor systems (see Chapter 7), the collective strength of the anchorettes (Figure 4–15B) can be gathered when absolutely nothing else is available.

ANCHOR ATTACHMENTS

As important as good anchors are, the attachment method is equally important. Always analyze the direction of the force that the rope system creates on the anchors. You can create or reduce an incredible amount of force on the anchors simply by understanding and applying the system forces in a safe manner (see Figure 4–6A and Figure 4–6B).

All rope systems require a minimum of two anchors, preferably three. Some rig masters fall into the trap of relaxing with just one **super bombproof anchor (SBA).** Under most circumstances, the SBA will do just fine. However, two problems can occur with this thinking. First, the rig master might have misjudged or simply assumed the anchor was an SBA when it was not. Rig masters are human and make mistakes. It is not wise to build a rope rescue system that is entirely reliant on a single anchor. In short, do not put all your apples in one basket. Second, **training to fundamental correctness** *always* helps ensure rope system security. Using two anchors sometimes and only one anchor other times leads to irregularity and the possibility of indecision on the team. There may be obvious choices for second anchors that the rig master, who has focused in on a single SBA, has not noticed or even refuses to acknowledge. Make a minimum of two anchors the rule, with no exceptions.

The main anchor that your load depends on is called the **primary anchor.** The backup, alternate, or belay anchor is called the **secondary anchor.** A third anchor is referred to as the **tertiary anchor.**

FIGURE 4–15

(A) Many small anchors can be grouped to form one usable load sharing pseudo-anchor. (B) Small trees grouped with rigging straps for load sharing pseudo-anchors.

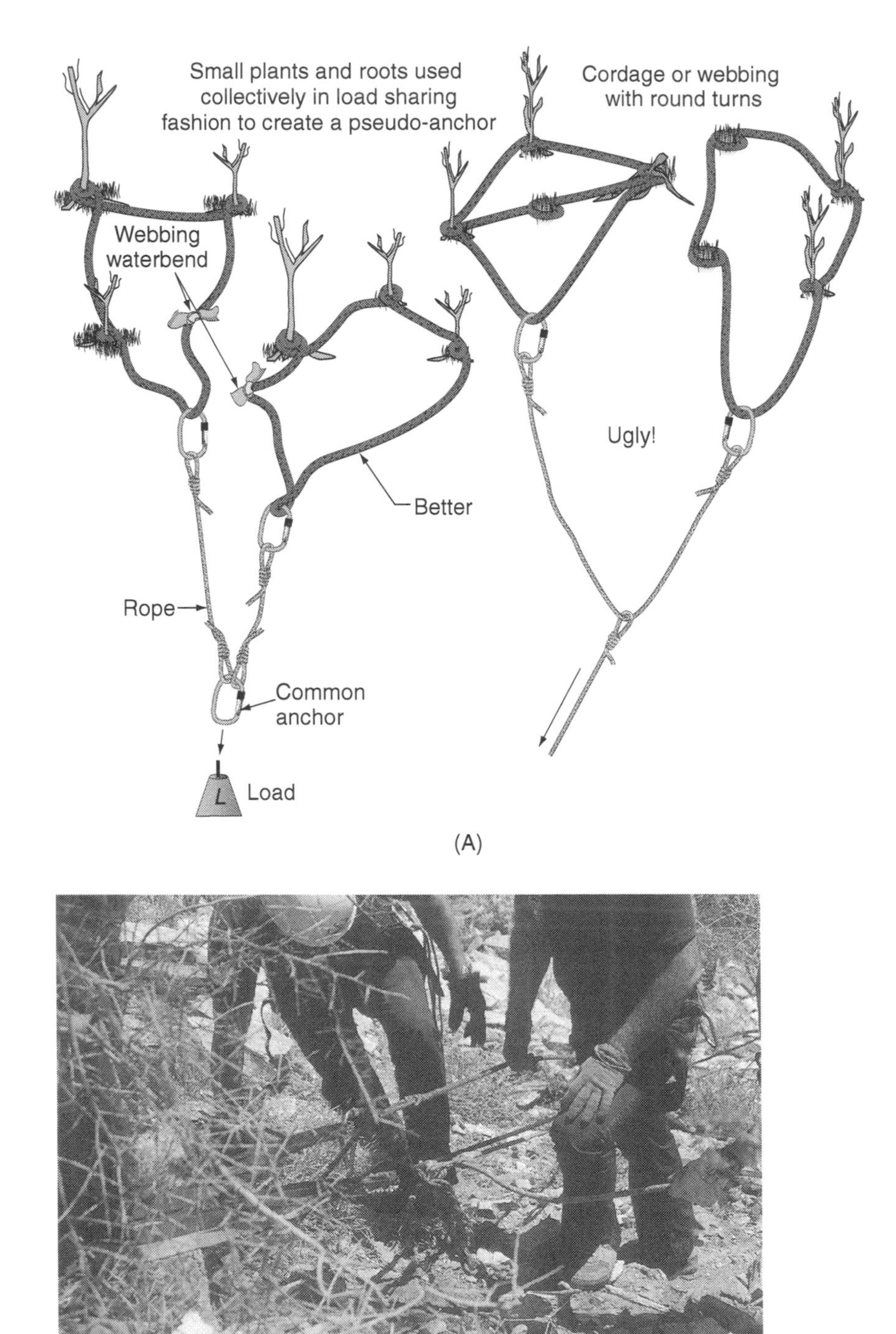

▶ **tertiary anchor**
The third anchor.

KNOTS, HITCHES, AND BENDS

Rope, particularly synthetic fiber rope, is one of the few tools on earth that can be molded, bent, and tied into different configurations that alter the way the tool performs. If you need extra grip on an axe, it is impossible to tie the handle into a loop to make it easier to grip. A lawn mower is a lawn mower, and only with extreme modification can it be turned into a go-cart. A rope, however, is a flexible linear medium that is malleable and moldable into an infi-

NOTE: For the purposes of this book, considerable effort has been made to avoid colloquial terminology and to coincide with NFPA 1983, NFPA 1670, NFPA 1006, and an international effort to standardize common terms. The double fisherman's bend, for example, is also commonly called the grapevine knot, the barrel knot, or the fisherman's knot. It all depends on where you learned to tie it and who taught it to you.

NFPA 1983

Standard on Fire Service Life Safety Rope and System Components

NFPA 1670

(Proposed) Technical Committee for Technical Rescue Operations and Training

NFPA 1006

(Proposed) Standards for Rescue Technicians Professional Qualifications

The History of Knots and Their Mathematical Implications

Knots have been used for thousands of years to make rope into more useful configurations. Many of the knots we consider useful today are adaptations of ancient natural fiber knots used in sailing rigs. It is thought that those sailors, who had many idle hours to fill, spent their time experimenting with ropes and knots. Long before there were accurate clocks to help navigate the seas, sailors used knotted ropes to calculate the approximate speed of a ship. To do that, a rope was knotted in precise increments. One sailor would play this rope into the sea and count off the knots as they passed between his fingers, another sailor would time the passage of knots using a sandglass. If eight knots passed between the sailor's fingers when the top of the sandglass was empty, the speed was said to be eight knots. This knowledge provided valuable time and distance information that could be used to estimate arrival at some faraway location. It is also the origination of the nautical (and aeronautical) designation for speed called knots, or about 1.16 miles per hour. Many sailors were completely illiterate and, I think, developed a sort of psychomotor spatial reasoning of knots and their applications. They could read their knots and imagine practical applications the way you might read a book. I believe that every aspiring rope rescue rigger might want to look beyond knots, hitches, and bends as just configurations and try to gain an understanding of the dynamics of rope interacting upon itself and the rest of the equipment.

Physicists and mathematicians sometimes study knots as mathematical objects, like numbers. They try to ascertain if they are equal. For mathematicians, all knots must have their two ends connected, visually, artificially, manually, or mathematically, to create a foundation premise for an equation. Mathematicians envision knots as closed loops or routes that can be traced with a finger, as if the two free ends had been welded together. Classification of knots can then be mathematically qualified as topologically equivalent or topologically nonequivalent. The zero loop, for example (Figure 4–16A), can be drawn in planar form, twisted around an infinite number of ways, and still be a zero loop. Trefoil knots, so named because they favor the shape of a clover leaf (Figure 4–16B), may look similar and can be twisted in any number of ways, but they are topologically nonequivalent. To add to the confusion, knots can be added together, subtracted, multiplied, and divided ad nauseam. There are even recent discoveries of deep interconnections between knot theory and a branch of physics that studies the fundamental particles and forces that are the building blocks of the universe. DNA, the molecular structure for human reproduction, has knotted configurations that, when deciphered, may help to explain some of the riddles of life.

nite number of configurations, some of them particularly useful, and some of them completely useless.

Understanding knots, hitches, and bends means looking at the interaction of individual fibers in given configurations and the factors that make some of them self-destruct at very low forces and others be almost indestructible. All knots weaken the rope somewhat. It is common to refer to a knot by the strength remaining in a well tied and dressed knot when tensioned to break. For example, a bowline knot is usually considered to be a 75% knot, meaning that if the rope is rated at 9,000 pounds (new), 9,000 × 75% = 6,750 pounds estimated remaining strength in the rope. When tensioned, the rope and

FIGURE 4–16

(A) The zero knot. Understanding knot dynamics begins with the most basic configurations. Knots can be identified mathematically and may be compared by being considered equal or not equal. (B) The trefoil knot. Similar but not equal.

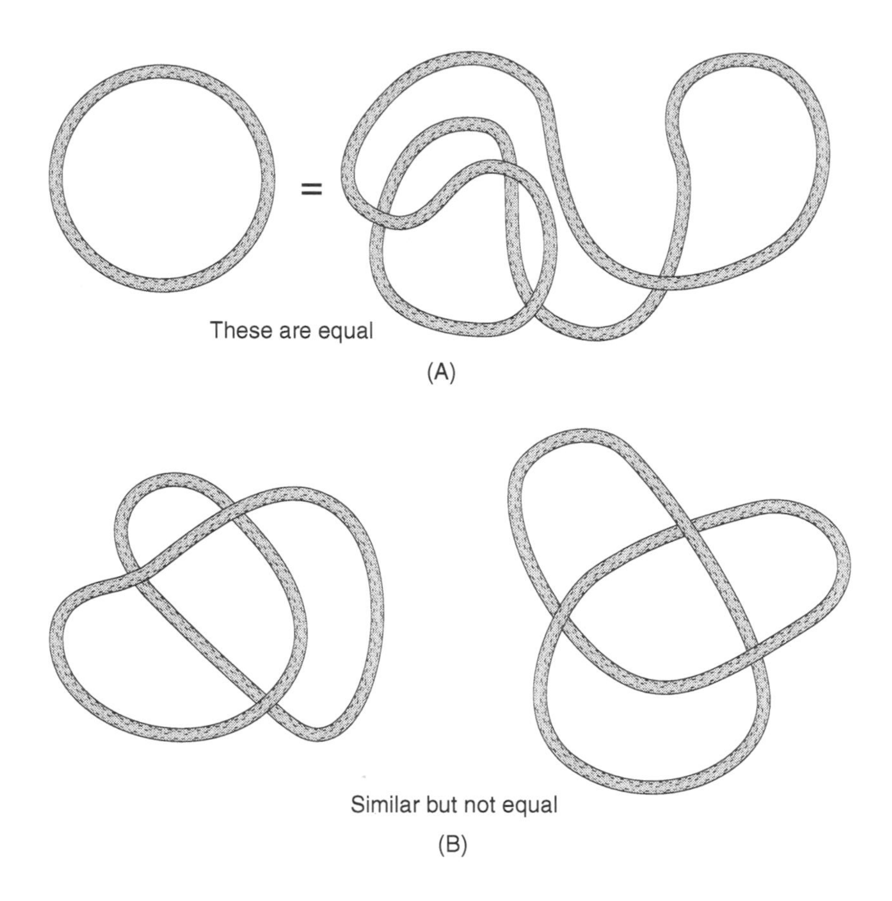

knot configuration should break in the knot at about 6,750 pounds, provided the rope is in good condition and dry and the knot is well dressed. Other knots have differing strength ratios and, based on many variables and test methods, the numbers can vary significantly.

Knots break when they are subjected to enough force to cause the internal rubbing of fibers, or friction, that cause them to melt and break. Nylon and polyester fiber knots break by pinching and literally melting themselves apart when under applied force (Figure 4–17). Natural fiber ropes, such as manila, sisal, jute, and cotton, commonly break from radius bending long before the fibers can heat to the point of melting. It is common to hear people mistakenly refer to radius bending as the major cause of knot failure with nylon and polyester ropes. Throughout the 1970s and 1980s, manufacturers and many instructors told of the horrors of bending ropes too tightly around objects, causing them, at times, to demonstrate less than 50% of their advertised tensile strength. Subsequent testing has suggested repeatedly that it is the fiber itself moving against adjacent fibers that creates enough friction to melt the fibers. For example, tie a well-dressed figure eight on a bight in 12.7-mm rope and connect the bight to a standard 12-mm rod stock steel carabiner. Attach the running end of the rope to a *large* round testing bollard and apply force. The rope will eventually break in the knot and not around the carabiner, even though the carabiner has a *smaller* diameter than the rope. Under extreme pressure, the rope flattens out against the carabiner like nylon webbing, maintaining much of its original strength. Again, the *knot* self-destructs.

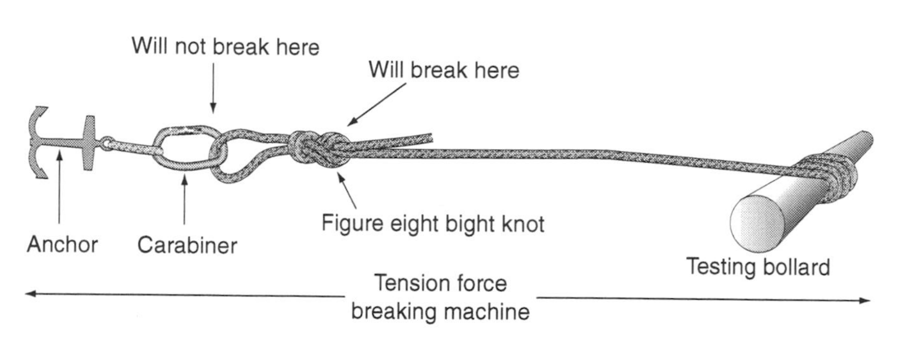

FIGURE 4–17
Most knots break because of internal friction and not over associated equipment components.

► **inherently tight or inherently loose configurations**
Two categories of knotted rope and webbing configurations. Inherently tight knots, such as the eight on a bight, double bight eight, and the butterfly, stay in position, do not come untied by themselves or during use, and do not require safety knots as backups. Inherently loose knots, such as the clove, girth, and münter hitches, have a history of coming loose or capsizing (changing shape) during some uses and require backup safety knots to make them hold their shape.

► **dressing**
Manipulating the parts of a knot so they are all parallel, neat, and not crossing.

► **pretension**
To pull hard on the knot components in the direction of the intended load to work out any looseness that could cause the knot to be dressed differently when carrying a rescue load.

► **standing part**
The anchored side of the rope.

► **running part**
The working end of the rope.

► **line**
A rope that is in use.

► **knot**
A rope or webbing configuration comprised of bights, loops, and round turns that maintains its shape independent of outside factors.

Therefore, rescuers should use configurations that tend to self-destruct later in operation rather than earlier and that break at much higher forces rather than lower forces. The most common way to achieve a strong knot is to spread the frictional forces out over a greater surface area. The figure eight on a bight does this better than most knots. Another way is to create a configuration with opposing components that, when tensioned, cannot pull through each other, like the double fisherman's bend. Particularly weak configurations have a very small surface area contact and tend to pinch themselves apart, like the square bend or the girth hitch.

Rope and webbing configurations lend themselves to being **inherently tight** or **inherently loose.** The stretching capacity of nylon tends to tighten some configurations and to loosen others as they are being worked. The figure eight on a bight, the butterfly, and the double fisherman's bend are inherently tight configurations. The bowline, the clove and half hitches, and the square bend all tend to be inherently loose, meaning they can come untied under certain operational conditions. It is not necessary to use a safety backup knot on inherently tight knots. However, it should be mandatory to use either an overhand knot or a two-wrap single fisherman's knot to safety all inherently loose knots. If in doubt, safety them.

The knots you tie are your own personal rigging signature. Take pride in every one of them. **Dressing** a knot means manipulating its parts so they are all parallel, neat, and not crossing. This means spending a minute or more on some knots. Well-dressed knots are stronger and are easier for other teammates to identify in a hurry. Always **pretension** knots in their manner of function. This will set the knot previous to tensioned lock-up and help ensure that it loads in the intended manner. Pretension knots (Figure 4–26) by pulling on alternate and opposing legs, or hang them in a safe place and load them with your body weight.

When rigging a rope system or even tying a simple knot, it helps to know which part of the rope is going to be anchored and which part is going to be working. The anchored side of the rope is called the **standing part.** The working end is referred to as the **running part.** A rope that is in use is commonly referred to as **line.**

All rope configurations are comprised of some combination of bights, loops, and round turns (Figure 4–18A, Figure 4–18B, and Figure 4–18C), collectively called parts, legs, and tails. They are tied in various ways to interact and either slip or hold their shape in a useful manner. A **knot** is a rope or webbing configuration comprised of bights, loops, and round turns that maintains its own

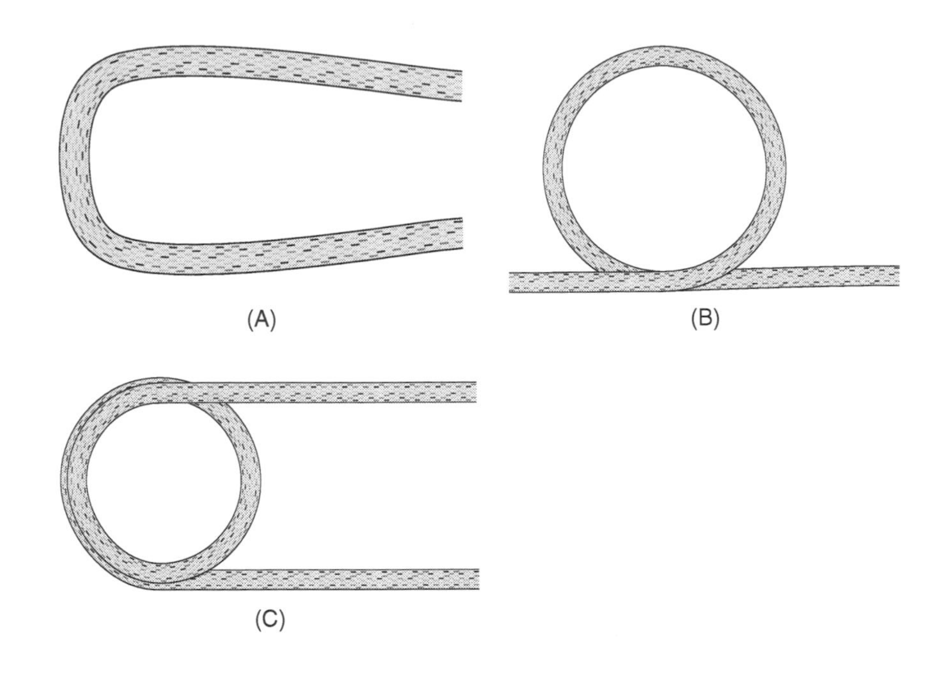

FIGURE 4–18
(A) Bight. (B) Loop. (C) Round turn.

► **bend**
A configuration where two ends of a rope are joined.

► **hitch**
A configuration that is tied around an object.

► **overhand knot**
The most basic knot; a good safety backup to inherently loose knots.

► **half hitch**
A simple rope configuration that is made by passing the running end around an object, around the standing end of the rope, and then back through the resulting loop. It is inherently loose and is often misused as an adequate safety knot for inherently loose knots.

► **single figure eight**
A knot that is a turn and a half around itself.

► **figure eight on a bight**
A preferred, inherently tight and relatively strong knot that makes a bight in the end of a rope for connecting people or equipment components.

shape independent of outside factors. A **bend** is a configuration where two ends of a rope are joined. A **hitch** is a configuration that is tied around an object.

The family of eights is the most important of the basic knots. A team should develop a thorough knowledge of the family of eights before moving on to more complicated configurations. They probably comprise 80% of all the knots rescuers will need under most circumstances.

The most basic knot is the **overhand knot** (Figure 4–19). It is a good safety backup to inherently loose knots and learning it first helps to understand the family of eights. Do not confuse the overhand knot with a **half hitch** (Figure 4–20). The overhand knot actually rotates around itself 360 degrees, giving moderate surface contact and much better friction interface than a half hitch. A common mistake is to substitute a half hitch as a safety backup for inherently loose knots. Learn to identify quickly the difference between the two.

While the half hitch is a single turn around itself, the **single figure eight** is a turn and a half around itself (Figure 4–21). The single figure eight makes a wonderful stopper knot to prevent ropes from pulling out of pulley systems and other equipment. It is also a good identifier for people to feel as the end of the line approaches while in use.

The **figure eight on a bight** is the best choice for a majority of static, nonmoving, rigs. It is well suited to be used as a terminal, end of the rope, anchor attachment. The figure eight on a bight is tied by first forming a bight in the rope. The bight wraps around its two legs a complete turn and a half (Figure 4–22), similar to the single figure eight but now with two parallel parts. This is probably the most-used knot in rope rescue, because it is quick and simple to tie, is inherently tight, and only weakens the rope about 15%. It can, therefore, be listed as an 85% knot.

When using 1/2-inch (12.7 mm) diameter rope, always leave about a 6- to 8-inch tail in the knot. Synthetic fibers have an inherent lubricity to make them flow through the braiders more easily.

Overhand knot.

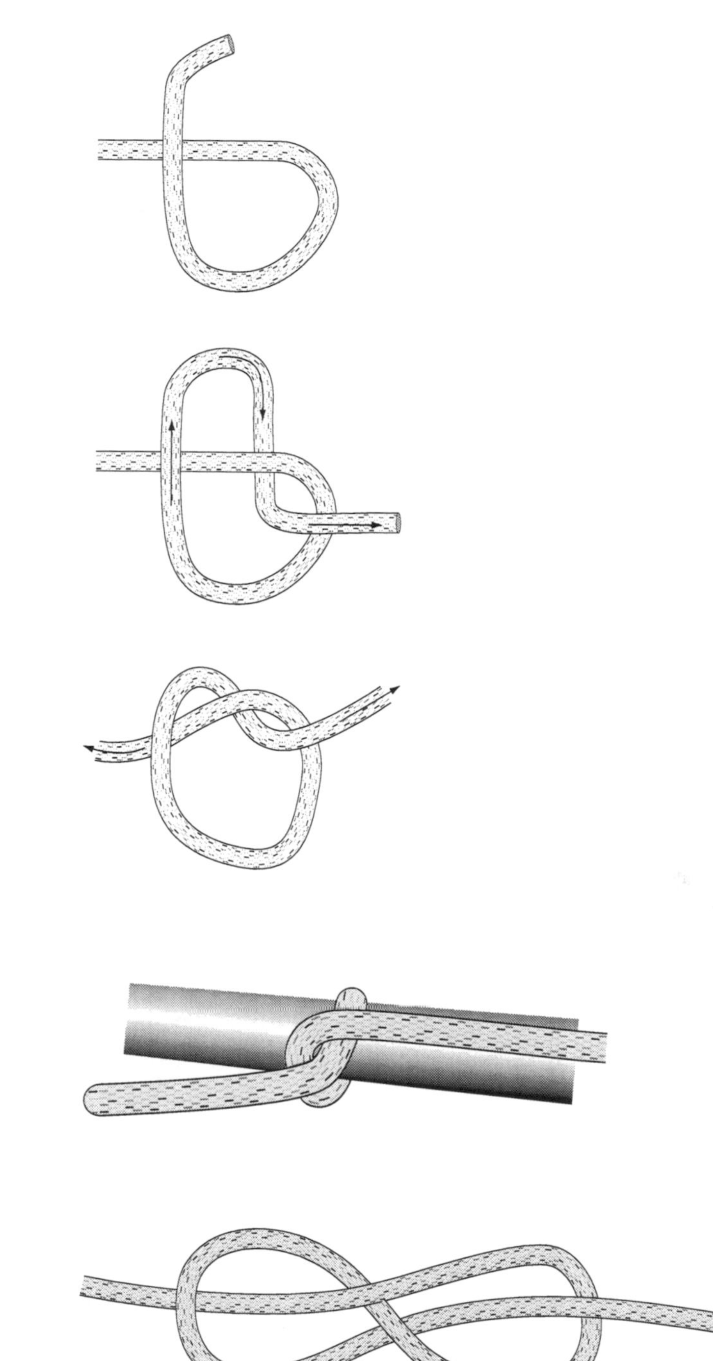

Half hitch.

Single figure eight (stopper knot).

This makes the rope "travel" somewhat under tension, causing the tail end of the knot to work towards the knot. Some knots are so loose that, under tension, the tail can travel completely through the knot and effectively and dangerously untie the knot. On the figure eight on a bight knot this is not a big problem since it tends to lock into place under moderate tension. The 8-inch tail ensures lockup before it can untie. Tails that are much longer than 8 inches can get caught in other pieces of equipment on a rope rig. The proper length tail mean a safer system and no wasted rope.

FIGURE 4–22

Figure eight on a bight.

Make the bight only as big as necessary for the application. To just clip a couple of carabiners into the bight, make the bight only about as big as the "okay" sign you can make with your fingers. Overly long bights (Figure 4–23A and Figure 4–23B) tend to waste rope and, when the load is almost to the top of the edge, it is too far away to safely grab and raise over the edge. A similar situation is encountered when using a tripod over a manhole (Figure 4–23C). If the bight is very long, the knot will run into the top pulley before the load is far enough out of the manhole. A very small bight allows the load to be raised almost as high as the pulley at the top of the tripod. This makes a tripod tilt maneuver very simple, and it is much easier on rescuers' lower back muscles.

A figure eight on a bight, or figure eight follow through, can also be tied around a stationary object (Figure 4–24). This is achieved by tying a single figure eight with a long tail; wrapping the tail around the object; and tracing the original knot from the standing end of the knot, which is the side closest to the standing end of the rope, through to the running part. As with all knots, bends, and hitches, it is dressed out and pretensioned before use. This is an excellent knot to use on rounded columns, standpipes, and trees. It also saves on rigging webbing and carabiners. Like all eight on bights, it has the disadvantage of being inflexible when under tension. If this was your main anchor attachment and you had to switch to a lowering operation, you would be hard pressed to complete the lower with this piece of rope.

► **figure eight bend**

Quick and simple method of connecting two ropes of equal diameter.

A **figure eight bend** (Figure 4–25) is a quick and simple method of connecting two ropes of equal diameter. The figure eight bend is made by tying a single figure eight near the end of the rope and tracing the original single eight backward with the second piece of rope. This is similar to tying the eight on a bight around an object, except that you start tying at the running end of the knot instead of the standing end. The tails should come out opposite sides, like on a bow tie, instead of parallel, as in the eight on a bight. Make sure

FIGURE 

(A) Make figure eight bights as small as possible, so the load stays within reach. (B) Long bights can make the load too far away to handle easily. (C) On tripods, long bights can leave the load still partially in the void. The raising operation is stopped by the knot in the pulley, and the load is still not out of the hole.

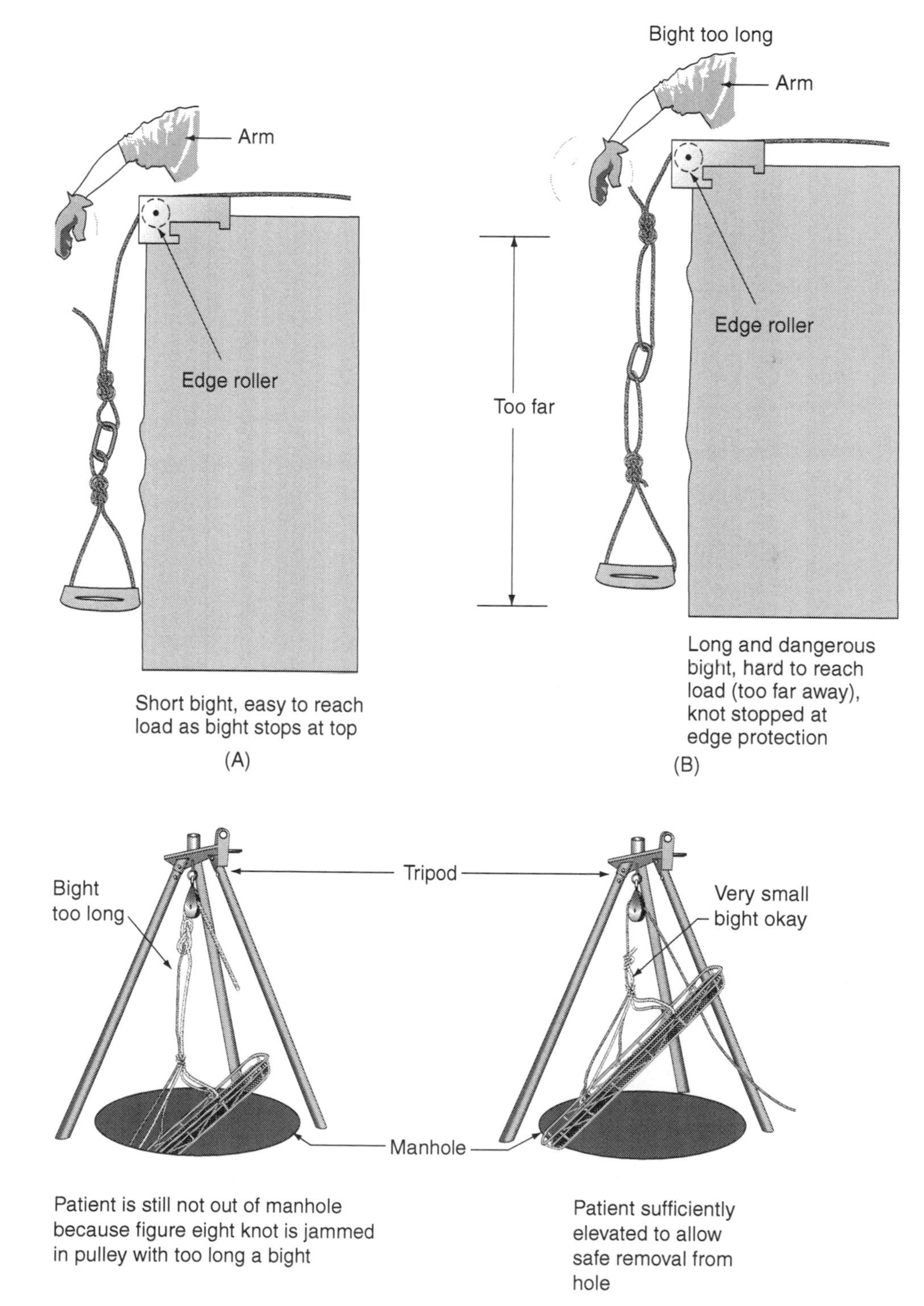

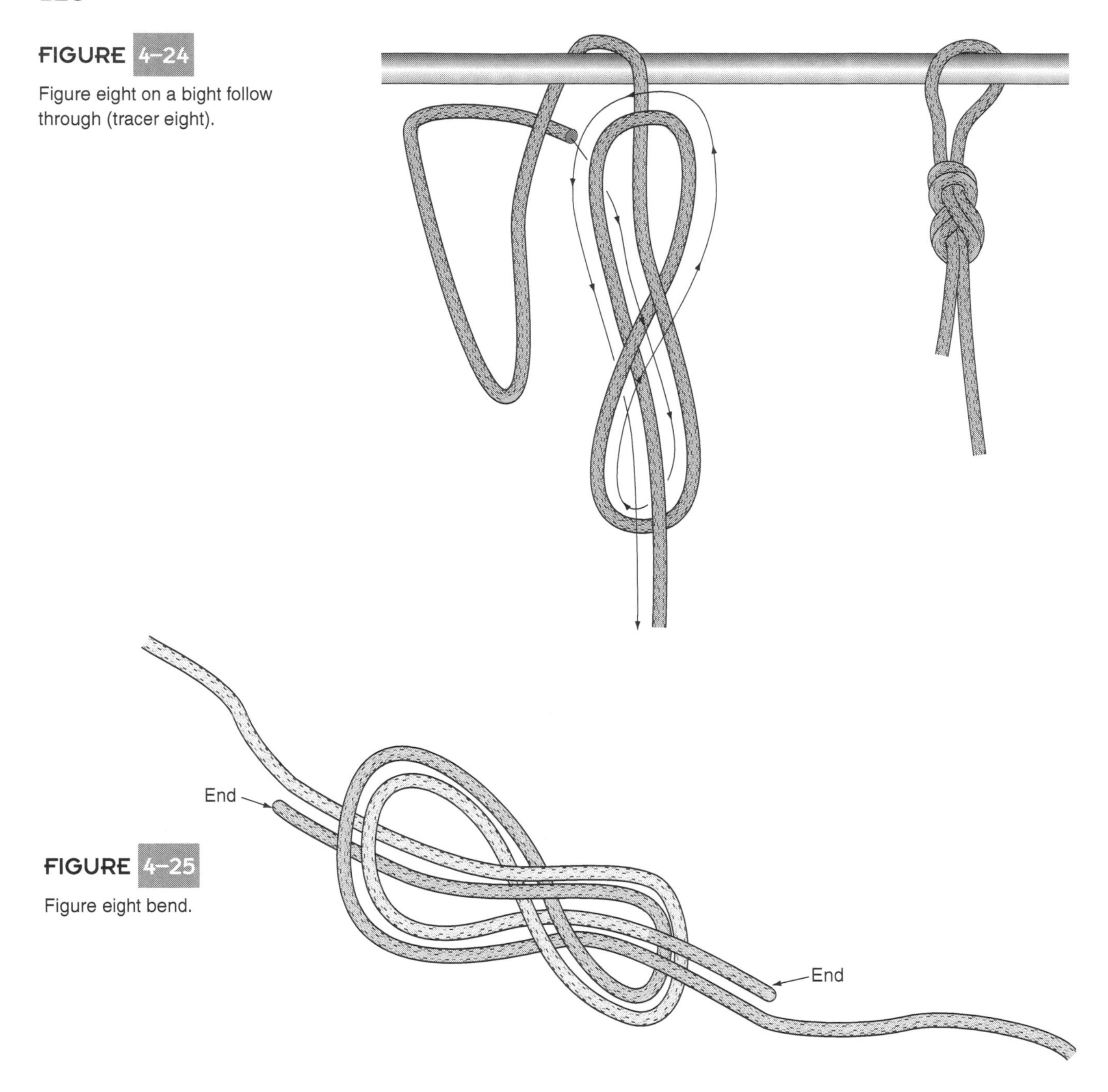

FIGURE 4–24

Figure eight on a bight follow through (tracer eight).

FIGURE 4–25

Figure eight bend.

to dress and pretension the knot, leaving 8- to 10-inch tails. It is easy to pretension (Figure 4–26) any of the double-part figure eights by pulling *hard* on opposing legs and tails, then switching to the remaining leg and tail. This helps set the knot before lockup under tension.

► **double bight figure eight**

An inherently tight and enormously strong rope configuration that provides two bights in the end of the rope for connecting people or equipment components.

The **double bight figure eight** (Figure 4–27A) is great for multipoint attachments and load sharing anchors. Due to the added surface area of the knot, it is slightly stronger than the figure eight on a bight. When tied, the two bights can be pulled out equally, or one bight can be pulled out much further than the other (Figure 4–27B). The small, equal bights are useful as twin attachment points for carabiners. It is debatable whether there is any real value in using

FIGURE 4–26

Pretension figure eights by pulling hard on opposing legs.

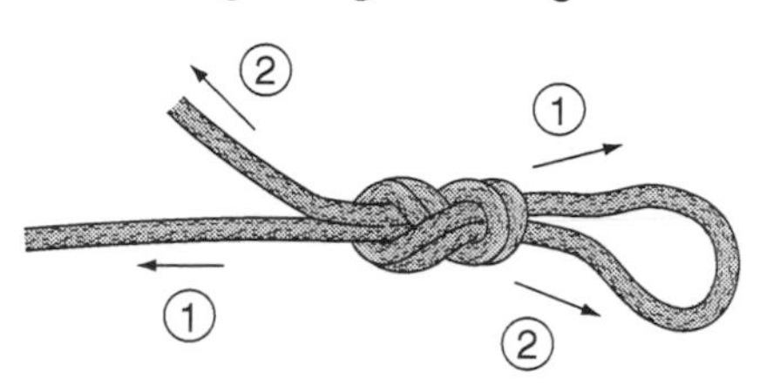

Pull opposing legs or parts 1 then
Pull opposing legs or parts 2

FIGURE 4–27

(A) Double bight figure eight with equal length bights. (B) The top of the double bight eight *has two* parts; *the bottom has three.* One of the three parts is common to both bights and can be rotated to adjust bight lengths. (C) The load sharing double bight eight is made by exaggerating one of the bights and making the other very small. They are connected in the middle with a carabiner.

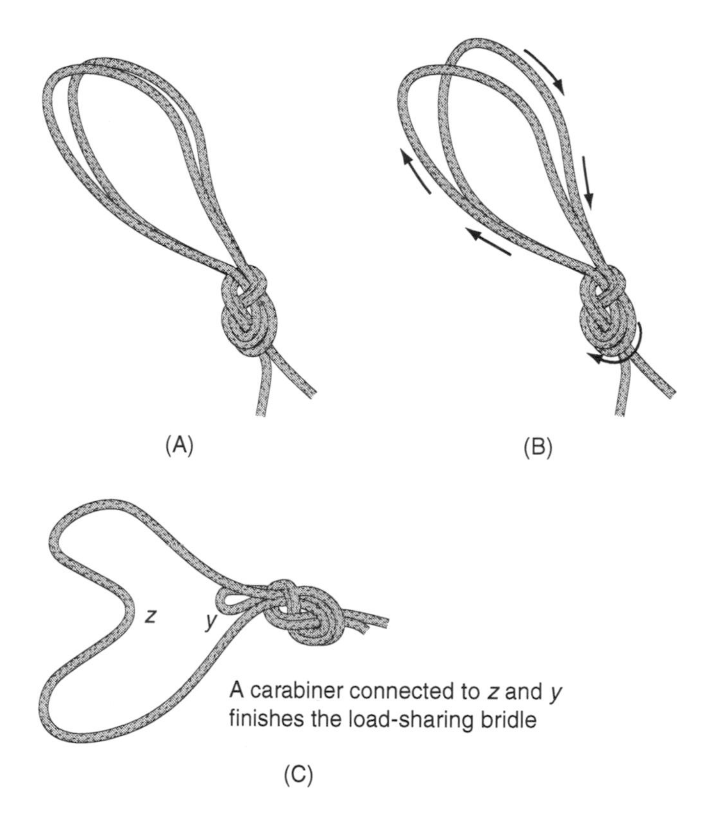

both bights in the same carabiner. The additional rope bight increases the surface area that is in contact with the carabiner, transferring some of the force towards its weaker gated parts and away from the stronger spine side of the carabiner. The most practical use of the double bight eight is as a load sharing bridle (Figure 4–27C). Tying the double bight eight begins like tying the single bight eight, but instead of passing the bight through the knot to finish, both legs of the bight are passed through, leaving the bight out. The legs become the double bights of the double bight eight. What would have been the single bight, had it been passed through to conclusion, is instead passed over the entire knot, encompassing the double bights and the rest of the knot configuration. From the tip of the *two bights* down, there are four legs, then two diagonal knot strands, and finally *three lower, parallel knot-strands.* The single bight that you passed

NOTE: To understand the different uses of the butterfly and the eight on a bight, it is important to see the way tension affects both knots. Tie, dress, pre-tension, and examine both knots. Pull on the standing (long part) and running end (tail) of the line that ties the eight on a bight in opposite directions. The tension tries to pull the knot apart or capsize it to destruction. This action weakens the figure eight on a bight down to approximately 60%, versus about 85% when pulled in-line. A knot in a 9,000-pound rope that is pulled in this manner will break at about 5,400 pounds of applied force, all other factors being equal. The butterfly, on the other hand, is a bit weaker but is designed to have tension pulled through the knot without capsizing it under load. Pull on the standing and running legs of the butterfly. The knot's dynamics cause it to tighten, not pull apart. Understanding the way the two knots work allows the rescuer to decide instantly on the proper rigging application and the appropriate knot.

► **butterfly knot**

Used instead of eight on a bight where a strong mid-line, rather than a terminal end attachment, is desired.

► **double fisherman's bend**

Two opposing fisherman's knots tightened against one another to make the common prusik loop.

over the knot to finish it, makes the third strand near the bottom. It is this strand that is shared by both working bights. It is used to size the bights to either make them fairly equal or purposely make them unequal for making the load sharing bridle (Figure 4–27C).

The load sharing bridle is easily crafted out of the double bight eight. First, make unusually long bights in the double bight eight, and then maneuver the third strand around so one of the bights is very small and the other very long. Both anchors are attached to the long bight. A carabiner is brought onto the big bight between the anchors and drawn down to and connected to the smaller bight.

The **butterfly knot** (Figure 4–28) is used in lieu of the eight on a bight where a strong (75%) mid-line, rather than a terminal end attachment, is desired. It is one of the easiest and most fun knots to make, because it looks complicated when tied, but can actually be tied very quickly and does not need much dressing. Since it is a mid-line knot, it has no tails to flop around or run through the knot and come untied. Practice in rigging systems will help you learn just where to start the knot and, because it is so quick and easy to tie, do not be afraid to tie it and test its location relative to the other components of the system. If it is not positioned exactly right, it is easy to relocate it in seconds. The butterfly is tied by picking the line some distance from the ends. Lay two complete and relatively large loops in your hand (Figure 4–28A). This means you will have three parts, or strands, of the knot in your hand. Take the front part, closest to your fingertips, and lift it back and over the remaining two parts (B). Release the first part and move the new front part up and back over the two parts in your hand. Finally, take the new front part and bring it up and back over the two parts in your hand, so it is in the rear. Now pull it under, through, and in front of the remaining parts (C). This becomes the bight and attachment. Pull the bight out to the desired length (D) and then set the knot by pulling the running and standing parts of the line apart (E). The knot usually dresses itself to some extent and takes on the vague appearance of a demented butterfly.

The **double fisherman's bend** (Figure 4–29), as the name implies, connects the ends of two ropes that have the same diameter. This knot consists of tying two **fisherman's knots** over one another so they oppose, meaning the tails come out in opposite directions. Dynamically, the knots work in combination to tighten on one another and to try to pull past one another, which becomes increasingly impossible as the knots tighten. This creates an extremely tight bundle of fibers, resulting in an enormously strong and inherently tight bend. A double fisherman's bend is very difficult to untie once it has been loaded to a rescue load, greater than 600 lbf. It is estimated that the properly tied double fisherman's bend is about a 90 to 95% configuration. Compared to the figure eight bend, it is about 5% to 10% stronger and maintains a somewhat smaller profile. It is the bend of choice for the prusik loop. However, the eight bend is easier to tie, dress, and inspect, and it might be a better choice for beginners and awareness level rescuers.

FIGURE 4–28 The Butterfly Knot and Attachment Point

The butterfly is tied by picking up the line some distance from the ends.

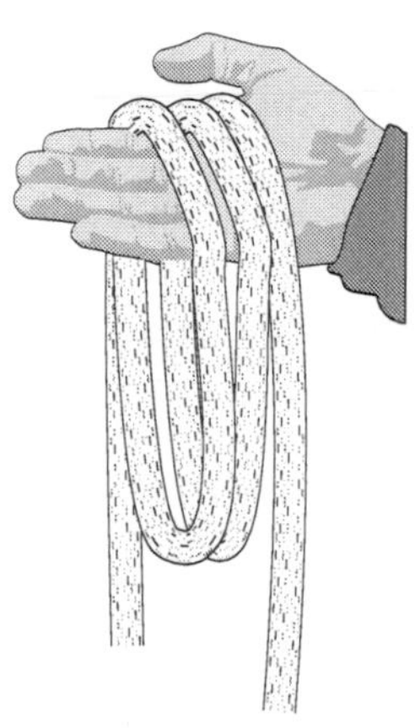

A Lay two complete and relatively large loops in your hand.

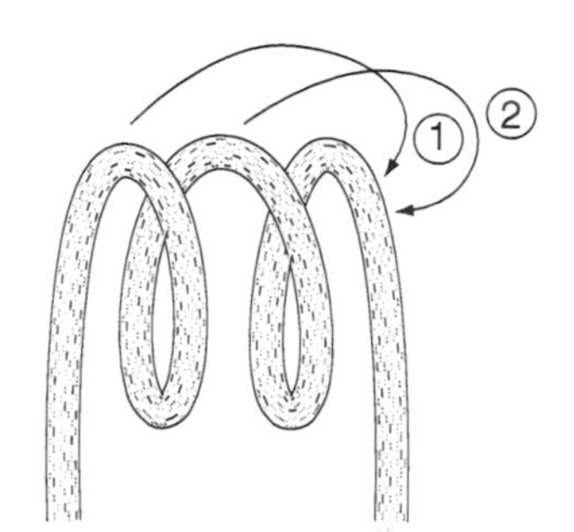

B Take the front part, that closest to your fingertips, and lift it back and over the remaining two parts. Release the first part and move the new front part up and back over the two parts in your hand.

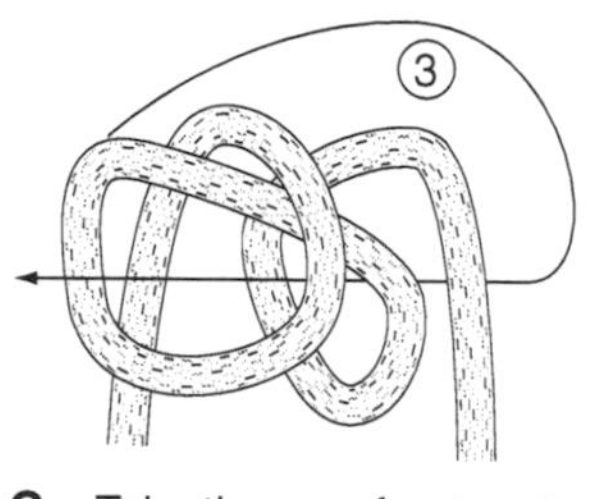

C Take the new front part and bring it up and back over the two parts in your hand, so it is in the rear. Now pull it under, through, and in front of the remaining parts. This becomes the bight and attachment.

D Pull the bight out to the desired length.

E Pull running and standing parts to set the knot.

FIGURE 4–29

Double fisherman's bend.

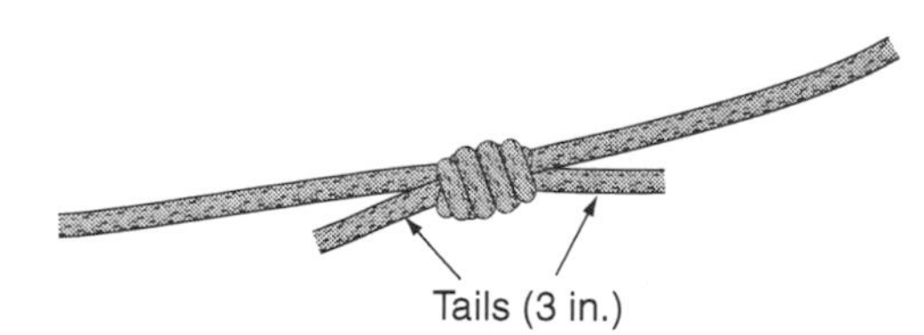

► fisherman's knots

A series of inherently tight and very strong rope configurations used to secure equipment to the end of the rope and as safety knots for inherently loose knots.

The three-wrap double fisherman's bend is tied (Figure 4–30) by overlapping the two ends of line. It is helpful to lay your finger along the length of the receiving end, the end being wrapped. This will create a little tunnel to send the tail of the knot back through. Begin by back wrapping, or wrapping the rope back over itself, with *three* consecutive wraps (A). Remove your finger and push the tail

FIGURE 4–30 Tying the Double Fisherman's Knot Using the Index Finger as a Tracing Tunnel

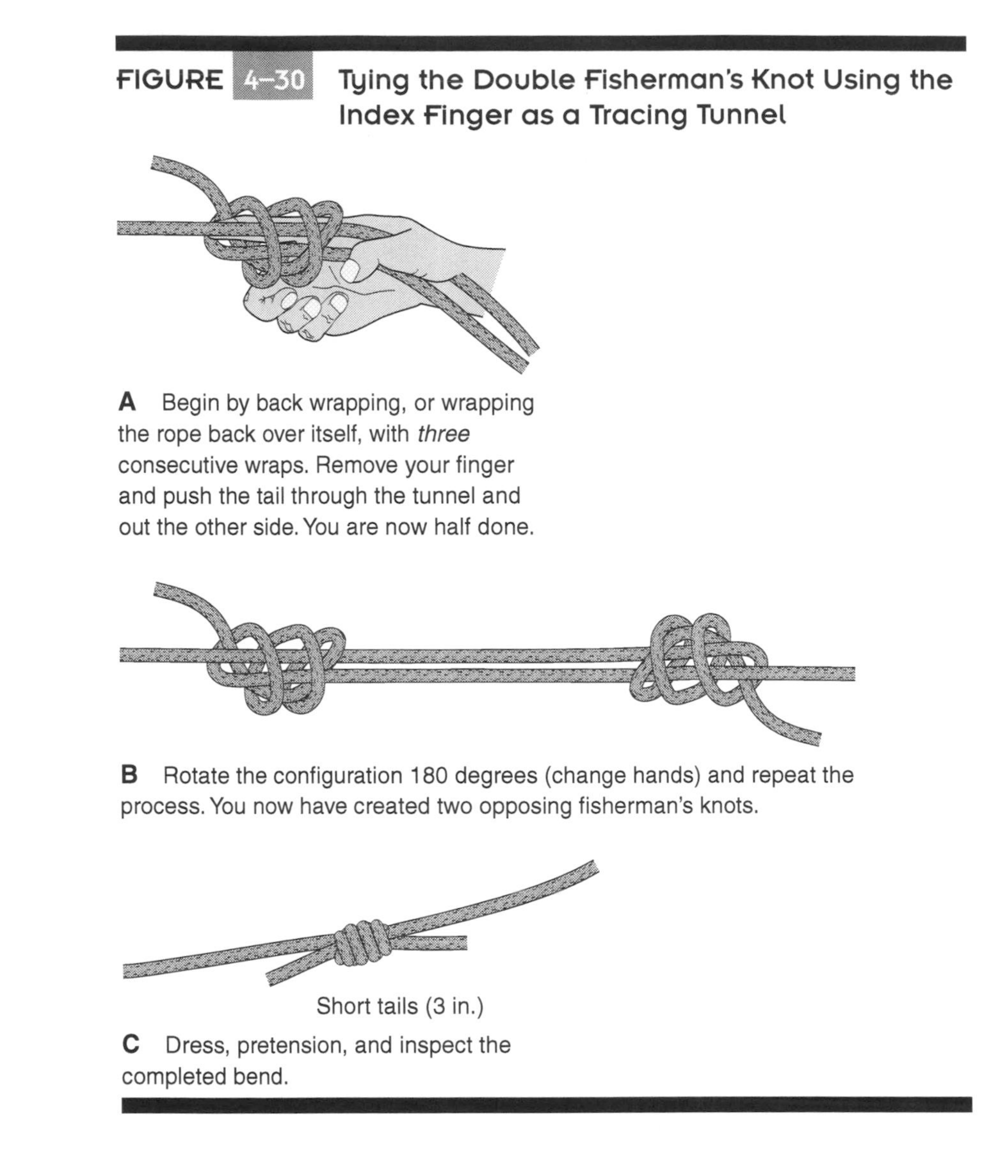

A Begin by back wrapping, or wrapping the rope back over itself, with *three* consecutive wraps. Remove your finger and push the tail through the tunnel and out the other side. You are now half done.

B Rotate the configuration 180 degrees (change hands) and repeat the process. You now have created two opposing fisherman's knots.

C Dress, pretension, and inspect the completed bend.

through the tunnel and out the other side. You are now half done. Rotate the configuration 180 degrees and repeat the process. You now have created two opposing fisherman's knots (B). Dress, pretension, and inspect the completed bend (C). Make sure the tails are opposing and, for 12.7-mm rope, about 8 to 10 inches long. The tails should be about 3 inches long in 8-mm cord (see Chapter 3).

► **cheater prusik**
Single prusik used in a variety of ways to enhance the flexibility of a system.

A prusik loop makes a wonderful sacrificial element in a rope system that will see a lot of use. As you will see a bit later in this chapter, a **cheater prusik** is a single prusik used in a variety of ways to enhance the flexibility of a system. For example, an anchored three-wrap prusik hitch on a rappel rope (Figure 4–31) will hold all of the weight of the rappelers throughout a day of rope sliding and will never load any other parts of the system. In effect, the prusik does the work instead of the more expensive equipment. If a shock load to the system occurs, the prusik bite will slip, absorbing energy and loading the rest of the system, which would normally be loaded anyway. At less than $0.50 per foot, a long prusik loop costs about $4 and can be used for years in this capacity. A set of prusiks, including a 16-inch

FIGURE 

Applying the two-wrap prusik loop to a rope.

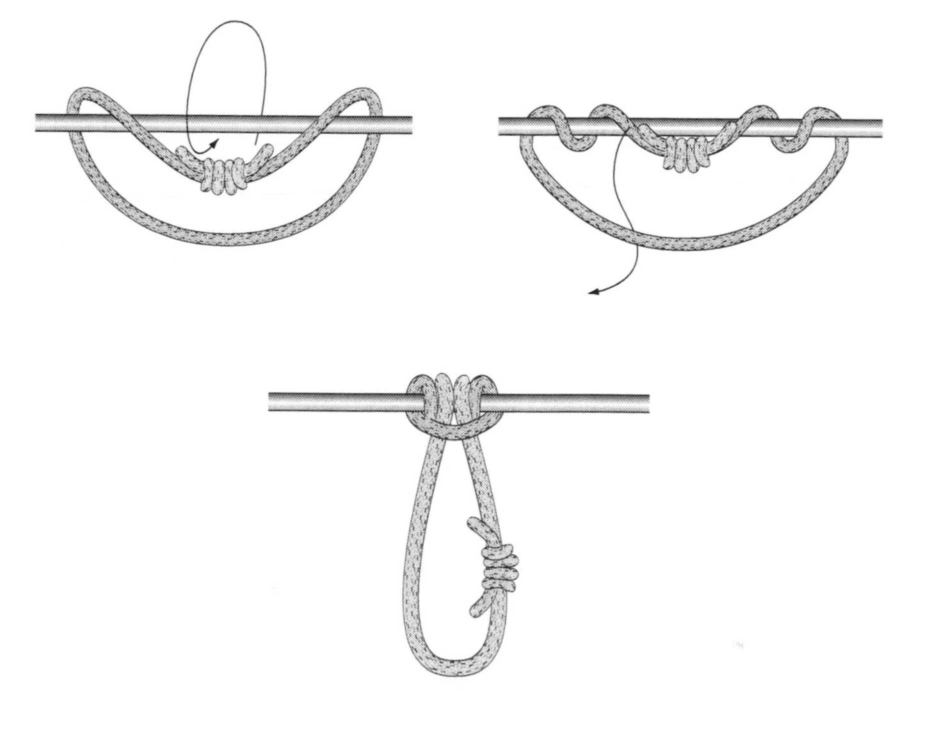

short, a 28-inch medium, and a 36-inch long loop (Figure 4–32A and Figure 4–32B) can be made for less than $10. When a prusik wears excessively, or if it receives a shock load, simply discard it and use another. The majority of the components, except of course the rope that is being rappeled on, stay like new.

Webbing, or flat rope, is a completely different breed of software as you know from Chapter 3. The dynamics of knots in webbing are completely different from the same knots in round rope and, therefore, require some special attention. The best way to connect the ends of webbing sections to make harnesses or rigging slings is by sewing, but sewing is a special discipline that requires a measure of talent and experience. Life support equipment must have the elements of doubt removed. Use a respected manufacturer's pre-sewn software unless you know the techniques and equipment that are best for sewing your own gear.

► **water bend**

Joins the ends of the same piece of webbing to make a sling or connects two different sections of webbing.

In lieu of sewn software, there are a couple of configurations that can be tied that work very well. The **water bend** (Figure 4–33) is excellent for joining the ends of the same piece of webbing to make a sling or for connecting two different sections of webbing. To tie the water bend, make a loose flattened overhand knot near the end of the webbing (A). With the other end of the webbing, trace the original loose overhand knot, starting at the tip of the webbing in the first knot (B and C). In effect, you have created a mirror image of two knots that together make the water bend. Make sure there are no folds or twists in the bend, dress the knot so it looks nice, and pretension it well (D). After it has been thoroughly tightened, you should still have about a 3-inch tail in 2-inch webbing and a 2-inch tail in 1-inch webbing. If you do, be comfortable with the knot. Backup safeties will not help secure the water bend further, but make checking the tail lengths part of your routine inspection program.

(A)

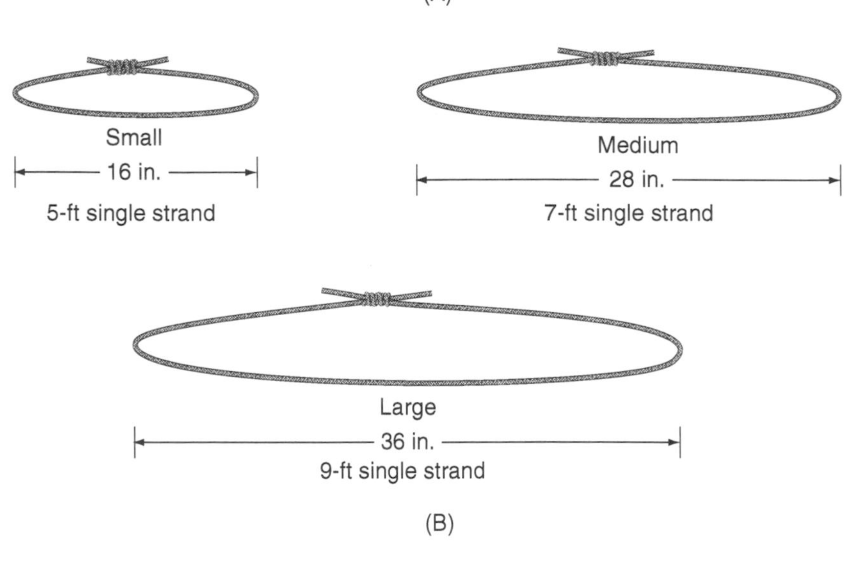

(B)

FIGURE 4–32

(A) A set of tandem three-wrap (t3w) prusiks. (B) A three prusik set: small, medium, and large. Eight-mm cordage should be cut in five, seven, and nine-foot lengths, tie with three-wrap double fisherman's bends to make small (16-inch loop), medium (28-inch loop), and large (36-inch loop). (*A, Courtesy Martin Grube*)

► **internal water bend**
Variation of the water bend.

The **internal water bend** (Figure 4–34) is an interesting variation of the water bend. Its advantage is that the tails do not flop around and get caught in other equipment. Its slight disadvantages are that it is somewhat harder to tie, and the inspection of the tail lengths must be done completely by feel. The internal water bend must be tied with tubular webbing. When using 2-inch webbing, begin by tying a loose and flat overhand knot about 18 inches from the end of the webbing. Open the tubular end, forming a tunnel down toward the overhand knot. Slide the other, non-knotted, end section into the tunnel until it almost touches the overhand knot. Now, simply work the overhand knot towards the end of the webbing. This will, in effect, tie a water bend since the internal webbing mirrors the original knot. Work the bend into position so that the tails are about 4 inches long before pretensioning and about 3 inches long after pretensioning. Make sure the sections are flat and the bend is well dressed. You will have to feel through the outer webbing to tell

FIGURE 4–33 Tying the Water Bend in Webbing

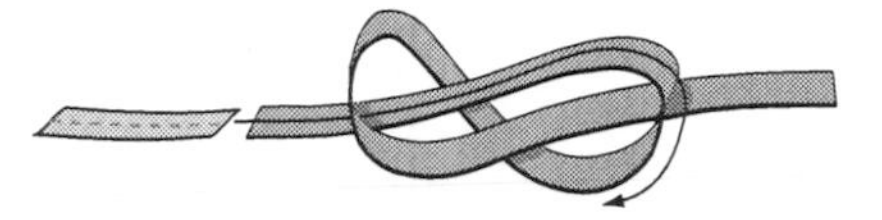

A Make a loose flattened overhand knot near the end of the webbing.

B **C**

With the other end of the webbing, trace the original loose overhand knot, starting at the tip of the webbing in the first knot.

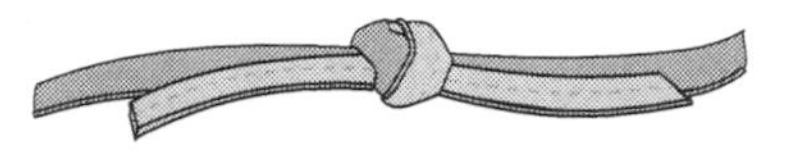

D Make sure there are no folds or twists in the bend, dress the knot so it looks nice, and pretension it well.

FIGURE 4–34

The internal water bend.

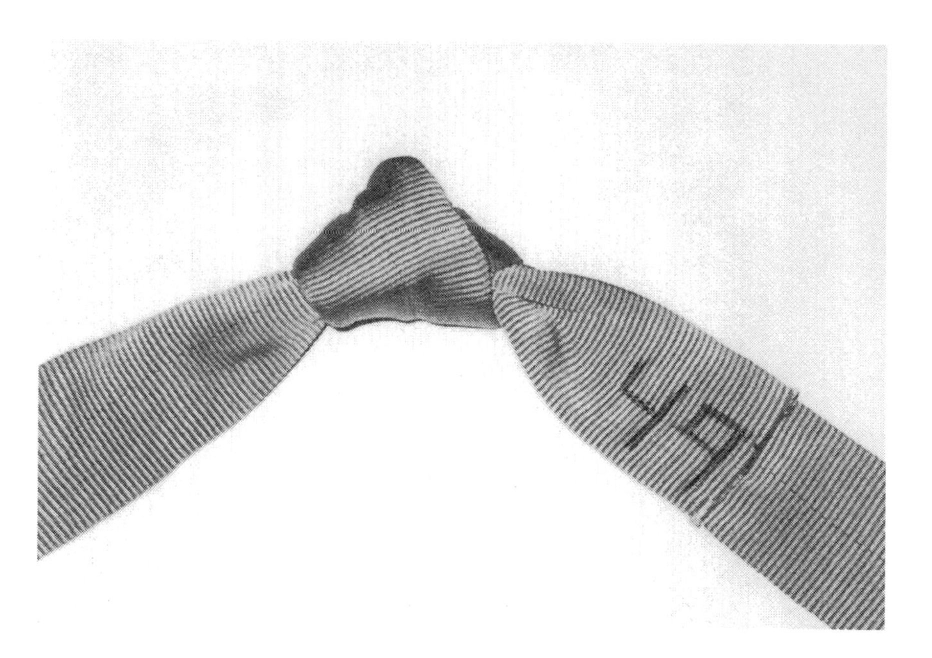

► **overhand on a bight**

Makes a connection point in the terminal ends of webbing.

► **square bend**

Recognized as a square knot; can also be used to join webbing.

how long the inner tail is. With some practice and patience, you will be able to tie them more quickly.

The **overhand on a bight** in webbing (Figure 4–35) makes a great connection point in the terminal ends of webbing. Its dynamics are similar to the water bend in webbing. Umbilicus hoses that incorporate airlines, cordage, and an intercom line inside two-inch webbing are terminated with the overhand on a bight.

The **square bend** (Figure 4–36A), which is recognized as a square knot, can also be used to join webbing. It is quick to tie and

FIGURE 4–35

Overhand knot on a bight.

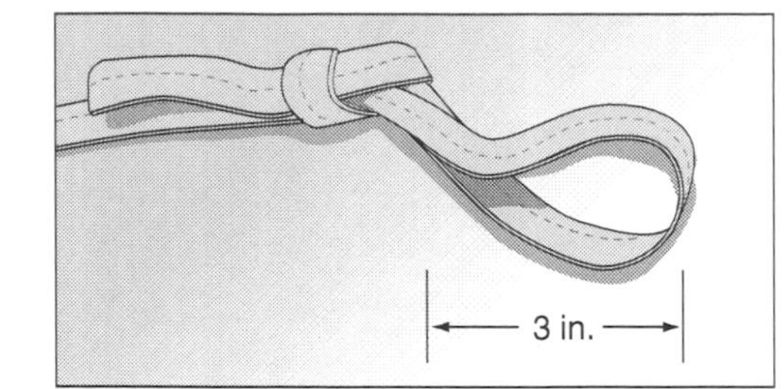

FIGURE 4–36

(A) Square bend in webbing.
(B) Square bend requires overhand safeties because it can capsize *and* it is inherently loose. It is fast, however, and useful in emergency egress situations. (*Courtesy Martin Grube*)

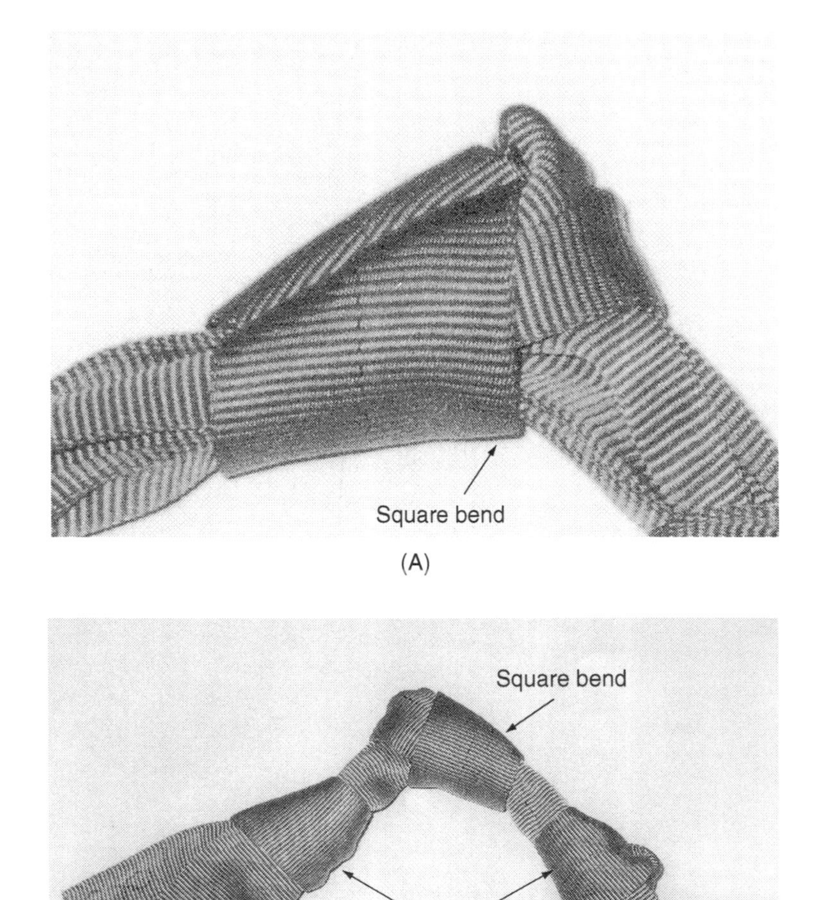

allows for tightening around the waist, unlike the water bend, and is therefore recommended for the emergency egress improvised webbing harnesses. The disadvantages are that it is easy to mistie it and form a rolling granny knot, it is hard to tell whether the knot is a proper square in webbing, and the square bend is notoriously weak, about 50%. The square bend is tied by taking both ends and overlapping them, first right over left and then left over right. The tails should come out on the same side of the bend, like a bow tie, and not

FIGURE 

Single-line static rappel system.

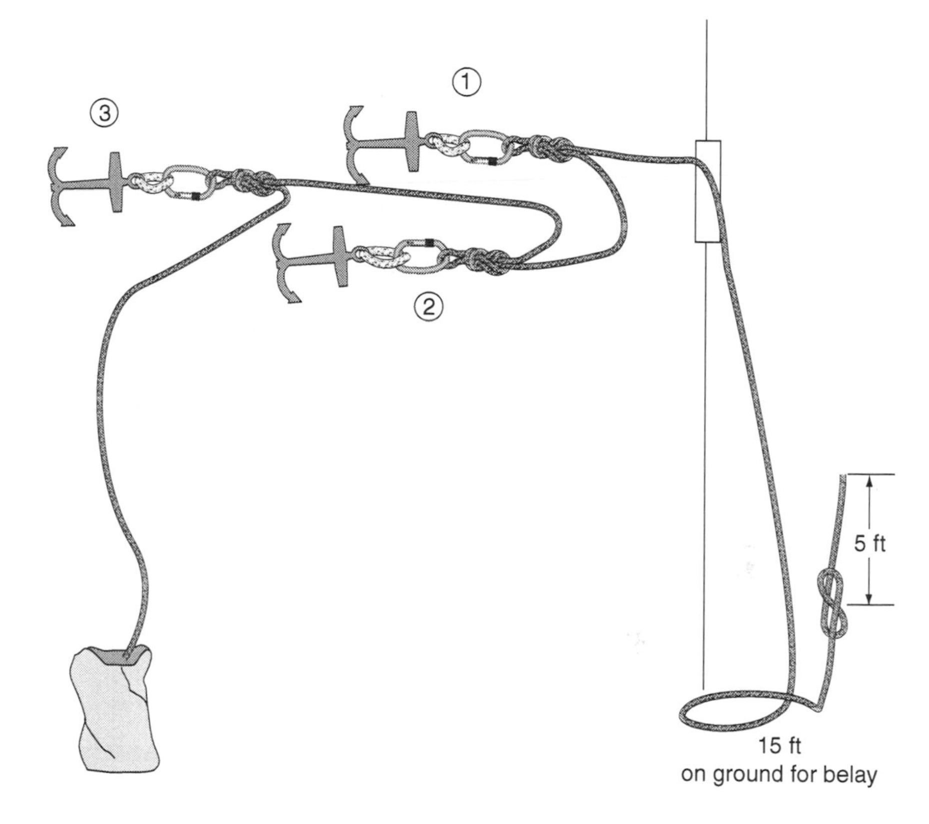

on opposing sides, diagonally. The resulting configuration forms two bights that interlace. In webbing, you should be able to identify a pouch formed by the bend. Because this knot is inherently loose, it requires at least a half hitch safety and preferably a complete overhand knot (Figure 4–36B). The purpose of the safeties in the square bend is to hold the two bights in position. Even a properly tied and tensioned square bend can be easily capsized by pulling on the running end, one of the tails, thereby straightening out one of the bights. The other bight then slides right off the end. All this concern about the square knot is justified. Limit its use to emergency egress and use a safety on each end to realize its full potential.

► **single-line static rappel line**

A single rappel rope that has been tied off to anchors (static) and that cannot be moved while a rappeler's weight is on the rope.

► **leading edge**

The part of the edge of a high place that can be considered a danger to fall from and the immediate area surrounding the edge, usually considered to be about 10 feet back from the edge; the area where fall protection equipment is mandatory; the loading zone for loading a rope rescue system.

Rigging the **single-line static rappel line** (Figure 4–37) is simple and quick. Assume that you are working with a relatively sterile rappel site and have decided on the single rope system.

1. Make your edge safety (Figure 4–38) and anchor decisions. If you are working on an unprotected edge, that is an edge with no handrails or parapet wall, rig edge tethers for everyone near the edge. Proclaim a "no-go zone" and mark it with fire line tape or even surveyor's ribbon so all personnel know where it is safe and appropriate to go and where it is potentially dangerous and off-limits. Limit the number of people in the edge danger zone. Take whatever steps are necessary to elevate the **leading edge.** Elevating the leading edge allows those who get on the rope to have a feeling of security before they are actually out of the safety zone. It also helps, when raising the load, to have such a landing zone to

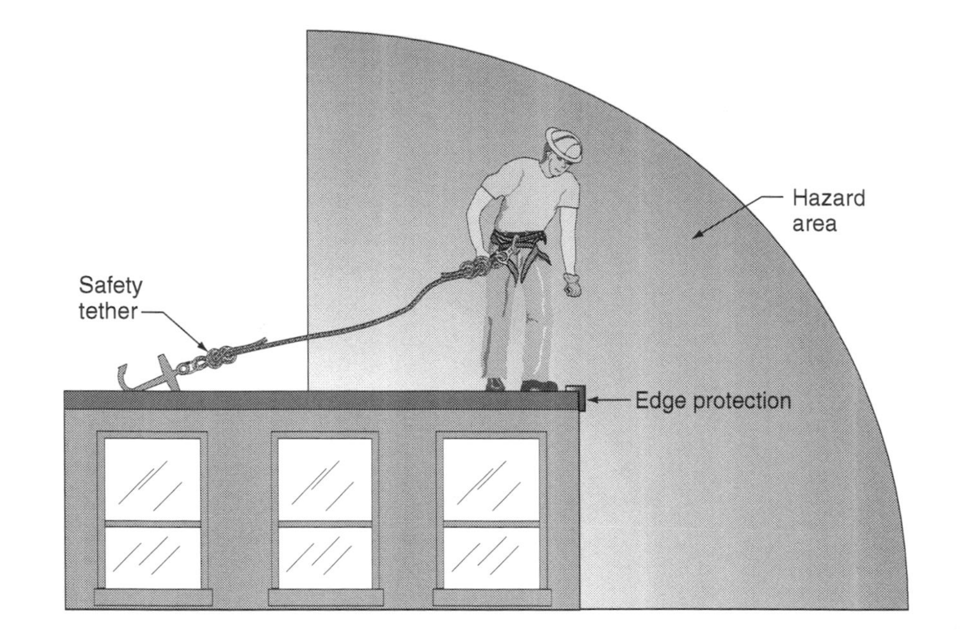

FIGURE 4–38

The hazard area near an edge, usually calculated as about 10 feet. Anyone in this area must be secured to an anchor.

safely capture the load, the patient and/or rescuer. Calculate the angle in the rope created by your leading edge elevated attachment point, from anchor to edge. Try to keep it as wide as possible (see Chapter 7), and make sure the elevated attachment is a solid anchor in itself. Make sure the rope is protected wherever the loaded rope touches the edge. Remember, loaded ropes can cut very easily. It is estimated that 90% of all rope failures are a result of inadequate edge protection.

2. Tie a single figure eight stopper knot about 5 feet from the terminal end of the rope. The stopper knot identifies the end of the rope by feel and prevents the bottom belayer from backing off the end of the rope. If entanglement in foliage, rocks, rubble, or building components is a problem, the stopper knot can be tied later, but do not forget to do it.
3. Lower the rope and the stopper knot to the ground and deploy an additional 15 feet of rope onto the ground. The extra ground rope provides a safe bottom belay. Avoid the urge to drop rope bags over the edge unless there is some tactical reason for fast deployment of the rope. Because dropped rope bags almost always get damaged from the force of the fall, it is best to keep the extra rope on the top with you. Some rope bags also have other equipment compartments in them. Hardware will cut the rope and the bag when it hits the ground and will usually damage the hardware as well. Finally, a dropped rope bag with any remaining equipment in it can kill a person who is standing on the ground.
4. Decide on a primary anchor site. Tie a figure eight on a bight and connect it to the anchor using a steel carabiner. Which figure eight on a bight you decide to use will depend on the

NOTE: It is helpful to learn how to estimate lengths of rope in the field. One method is to measure your arm span. You can quickly pull off and count arm spans of rope and multiply them by your arm span measurement. Most people have an arm span equal to their height. If you are 6 feet tall, your arm span will be about 6 feet in length. If you count off ten arm spans of rope, you will have field measured about 60 feet. Practice this with a tape measure. You may be surprised at how many times it becomes useful.

FIGURE 

Single-line static rappel system using a cheater prusik to carry all the load, serving as a shock absorber and sacrificial element leaving the rope knots unloaded.

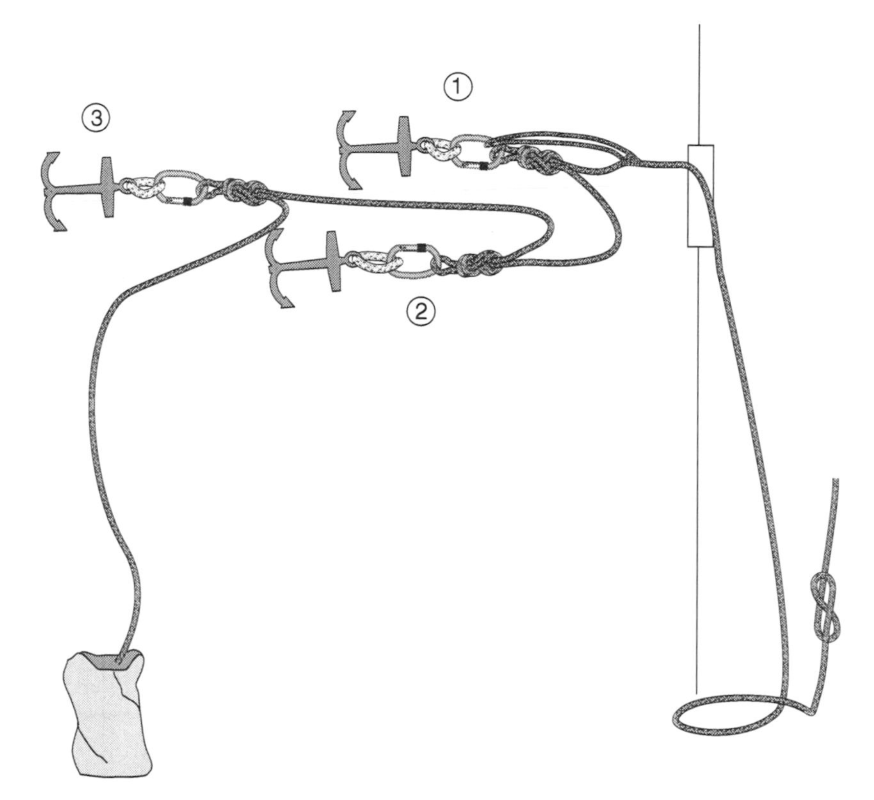

anchor attachment, and experience will help you to make this choice. For emergency applications where there is no system redundancy, your rig is ready for rappel.

5. Tie the secondary eight on a bight and attach it to the secondary anchor. Remember, in a non-load-sharing system, the primary anchor is the one that will be holding most of the load. The secondary anchor is the backup. While the primary anchor is commonly the one closest to the edge, it does not have to be. It might be the anchor farthest from the edge. Make sure there is minimal slack in the rope between anchors. A long piece of rope between anchors causes dangerous acceleration time if one of the front anchors blows. Even worse, it could cause the load to careen into a wall object or the ground.
6. Tie a third eight on a bight, if you are using three anchors, and attach it to the tertiary anchor.
7. If you have time, you can put a cheater, or single, prusik in here (Figure 4–39). It is not life or death, but it is handy and keeps a lot of wear and tear off the other parts of the system. If you are using a 12.7-mm rope, attach a prusik loop to the rope using the prusik, three-wrap, hitch. Anchor the prusik onto the primary anchor. Pull some slack through the prusik hitch so, when loaded, the system forces are on the prusik and not on the primary's eight on a bight knot. Again, this allows the figure eight knot to remain unloaded. At the end of the operation, it is easier to untie, and the rope has not been damaged by loading a knot all day. In the event of some kind of shock loading, the prusik will slip some expending

energy via the heated destruction of the prusik. If it slips enough, the load eventually will rest on the eight on a bight, where it would have been without the shock load.

8. As with all rope systems, *test them before you load them.* Assemble three or four of your crew members in a safe location. If you are forced to work near the edge, make sure everyone is on a previously tested safety tether system. Have the crew members pull on the rope system in the direction of anticipated force. Try to simulate at least the amount of force you are anticipating on the system. Observe all the equipment components under the artificial load. Has the cheater prusik set properly, and is it holding the entire load? Release the cheater and watch the rest of the system perform under the load. Are the knots well dressed and tensioning properly? Is the equipment lining up the way you anticipated? Are all the carabiners locked and not loading sideways? If you made any attachments of carabiners to webbing or slings, are all of the parts of the webbing and slings in the carabiners?
9. Finally, have someone else walk through the system and look for problems. This person will be responsible for certifying the system for loading with live loads. Encourage discussion about any rigging questions. Have the certifier hand touch all the components. Hand touching helps to eliminate highrise hypnotism that can draw the focus away from important details. Have the certifier acknowledge verbally that the system has been checked and certified and is ready for action.

► **single-line dynamic rappel system**

A single rope system that has been rigged through a descent control device so the rope can be lowered or moved while a rappeler's weight is on the rope.

Rigging the **single-line dynamic rappel system** (Figure 4–40) gives you a much safer and more versatile system than does the single-line static system. If you train a large number of people, this system will give you a much greater feeling of security than any static rope system. In short, the system can be released while in use, and any load can be lowered to the ground. Anyone on the rope who is having problems can simply be lowered to the ground quickly, as opposed to attempting a pick-off. Remember, a static system is used when rigging time is critical to operational success, emergency egress, or quick patient access. Static systems are fixed, or static, to the anchor sites and are quick, light, cheap, and potentially dangerous. In the single-line *dynamic* rappel system, however, the rope can move, and therefore anything on it can be lowered to the ground. All dynamic systems require some sort of DCD to be anchored. Instead of rappeling along a fixed rope, a person stays fixed on the rope and is lowered *with* the rope.

1. As with all rope systems, check all edge safety considerations, including edge protection, an elevated leading edge, and personnel safety tethers.
2. This discussion uses an eight-plate as a DCD. Locate a suitable anchor site for the eight-plate and secondary and tertiary anchor sites. Remember, the rope will *always* attempt to be perpendicular to the edge. Try to select anchors that

FIGURE 4–40

Three-point dynamic rappel system. Rappeler can be lowered through the anchored DCD if necessary.

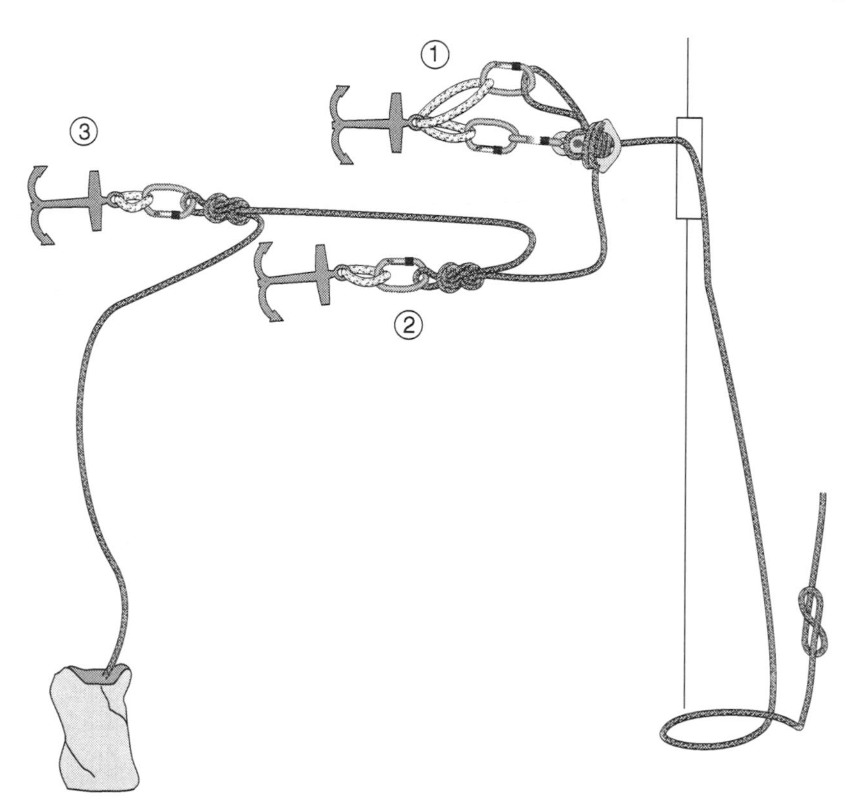

align with your target. If they do not, use change of direction (COD) system components (Chapter 7).

3. Tie a single figure eight stopper knot near the end of the rope and feed the rope over the edge. Make sure the rope stays in the edge protection and play the rope toward the primary anchor attachment. Reeve the eight-plate and attach it to the anchor with a carabiner. Have someone create artificial tension in the rope, pulling toward the edge and simulating a loaded rope. This will help you properly lock off and tie off the eight-plate. Take a double wrap across the top of the eight-plate, making sure the rope snaps down both times between the rope exiting the top of the eight-plate and the eight-plate itself (Figure 4–41A and Figure 4–41B). Continue with the tie off of the eight-plate, using a doubled overhand knot. Secure the bight of the tie off with a carabiner to the anchor (Figure 4–41C).
4. Attach the rope to the secondary and tertiary anchors with figure eight on a bight knots. Make sure to leave a minimal length of rope between the anchors to avoid acceleration time in the event of primary anchor failure.
5. Attach a prusik cheater (Figure 4–42A) in front of the DCD to carry the burden of the loaded rope.
6. As with all rope systems, *test them before you load them.* Again, assemble three or four crew members in a safe location. If you are forced to work near the edge, make sure everyone is on a previously tested safety tether system. Have the crew members pull on the rope system in the direction of

FIGURE 4–41 The Anchored Eight-Plate DCD Lock-Off

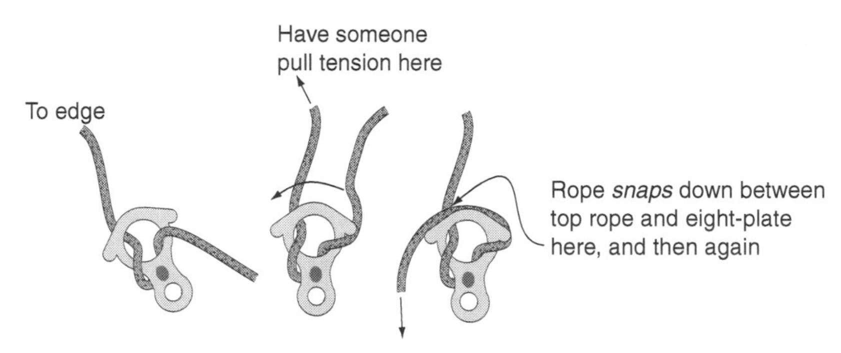

A The eight-plate DCD lock-off is started by lifting the running part of the rope up and over the top of the plate.

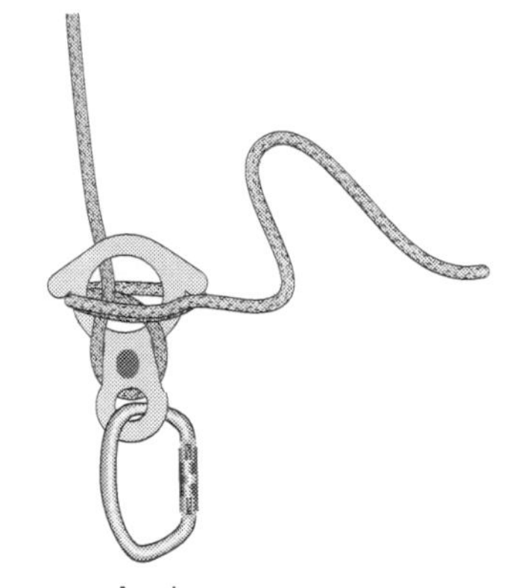

B The running part of the rope passes over the top of the eight and is snapped down between the standing part of the rope and the top of the eight-plate again.

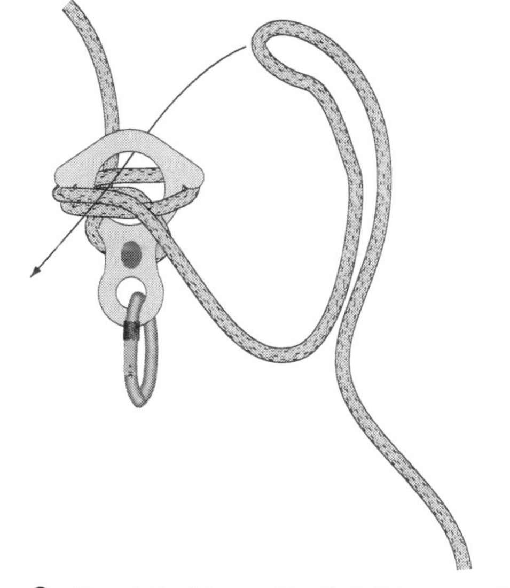

C The eight-plate must be tied off to ensure that the rope does not come loose while other activities are being performed. Take a large (arms length) bight off rope in the running end. Pass this bight through the large top "O" in the eight-plate, and pull out any additional slack. This long doubled piece of rope will be used to tie off the rope above the eight-plate.

This bight will be
secured back to the
anchor with another
carabiner

D Simply use the doubled piece of rope and tie a doubled overhand knot onto the standing part of the rope. Dress the knot down tight to the top of the eight-plate. If the belayer can pull on the rope and the knot comes out, you have probably tied a half hitch and will have to try again to complete the total doubled rope overhand knot.

anticipated force. Try to simulate at least the amount of force you are anticipating on the system. Observe all of the equipment components under the artificial load. Has the cheater prusik set properly, and is it holding the entire load? Release the cheater and watch the rest of the system perform under the load. Are the knots well dressed and tensioning properly? Is the equipment lining up the way you anticipated? Are all the carabiners locked and not loading sideways? If you made any attachments of carabiners to webbing or slings, are all of the parts of the webbing and slings in the carabiners?

7. Finally, have someone else walk through the system and look for problems. This person will be responsible for certifying the system for loading with live loads. Encourage discussion of any rigging questions. Have the certifier hand touch all of the components and, again, verbally certify the system.

The dynamic system behaves very much like the static rope system during normal rappeling operations, since the prusik cheater does most of the work. The real beauty of the system comes when you want to lower the rope and the load to the ground.

1. First, release the cheater from its grip on the rope. This can be done by building a hokie hitch or other load releasing device (see Chapter 7) into the cheater prusik, and then releasing the device until the load is transfered onto the DCD. Another technique is to gather people around the rope and simply have them pull up on the rope, taking enough slack in the cheater to allow someone to release the grip. *Make sure all the people who are pulling are safely tethered to an appropriate, and preferably separate, anchor system.* Depending on edge protection and leading edge friction, four or five people, that is a 4:1 or 5:1 mechanical advantage pulley system, can pull a one-person load up enough to release the cheater. A two-person load and lots of edge friction may require seven or eight people, creating crowding on the edge. Anticipate this when possible, and rig in the hokie hitch beforehand. The final alternative for loosening a dead-locked prusik is to cut it in a controlled manner, preferably using scissors instead of a knife. *Always* try to keep the relaxed cheater on the rope to be lowered. All lowering systems require a safety brake (see Chapter 7) in case the load starts moving too fast or someone needs to take a break. If the original prusik cannot be salvaged, three-wrap another prusik to the main line and anchor it before cutting the original.
2. The majority of the load is now in the anchored eight-plate, but some of the load always remains in the edge protection and/or elevated leading edge component. You can now untie and unlock the rope from the eight-plate. The person assigned to perform the lower must be ready to accept the load and lower it safely to the ground. The eight-plate will reduce the load, via friction, by approximately ten times. Therefore, a one-person, or 300 lbf, load will feel like about 30 lbf. Try to make sure the rope moves at a continuous rate. Watch the pics and patterns in the rope as it travels through the eight-plate and aim for a consistent descent. This makes for a smoother, less shock-inducing ride.
3. A person should always be assigned to monitor the loading and lowering operation. Verbal or radio communications make long and/or complicated lowers much safer.

NOTE: For the purposes of discussion, all systems in this book, unless otherwise designated, are two-line systems. Emergency egress helicopter operations and sterile teaching environments with closely managed belays are the exceptions.

Standardize your rope calls. Tension, slack, stop, slower, and faster will handle 90% to 100% of your rope movement needs. (See note in Chapter 5—Rope Calls in this text.)

SINGLE- VERSUS DOUBLE-LINE SYSTEMS

The single-line dynamic rappel system, while versatile and safer than the single-line static system still only uses one rope. Modern nylon and polyester ropes are remarkably strong and reliable, but we still have to ask The Question. "What happens to the load if this part, the rope in this case, fails?" The answer is that someone is going to fall! Really sterile teaching environments with lots of edge protection, elevated anchor attachments, and multiple anchor sites are important as learning sites because all of the extraneous system components are absent. Therefore, the single-line system is easy to learn.

▶ **double-line dynamic rappel system**

Two rope rappeling system in which the rappeler's descent control device slides on a fixed rope and the second rope, or belay rope, is attached to an anchored DCD on the top and fed through the DCD as a backup in the unlikely event of mainline failure.

The **double-line dynamic rappel system** (Figure 4–42B) answers "The Question" nicely, however it uses more equipment, more people, and more time. These are all important considerations to take into account when engineering rescue systems. In addition, this system is the basis for all of the lowering systems you will use later in the book. As you work your way through the steps, envision your team performing them simultaneously rather than in the step-by-step fashion in which you are reading them. A good rig master is thinking ahead and making assignments to small teams or groups of people so the system is assembled from multiple points. You are building a life support rope system, and time makes a difference.

1. Secure the edge with safety tethers for workers near the edge and clearly mark any hazard zones. Brief your personnel on your plan and point out all safety concerns. Check and recheck everyone's personal safety equipment.
2. Obviously, you have to determine where the target is, and therefore where your lines will have to be positioned. Locate bombproof anchor sources. Look for at least four anchors, two for the main line and two for the belay line. If the anchors and the target do not align, look for change of direction anchors offset from the edge and can be used to bring the rescue lines in line with the target.
3. Anchor a DCD, a closed-end brake bar rack for elevations of 50 feet or more or an eight-plate for shorter drops, to the primary anchor. Play out the rappel rope to the ground, including 15 feet for the bottom belay that you may choose to use during the rappel operation. Work the rope toward the primary anchor and DCD. Reeve and lock the DCD.
4. Protect the ropes from edge problems. Pad, eliminate, delineate, or preferably elevate whatever is necessary to keep the rope off sharp edges.
5. Position the belay line so the terminal end is near the place where the rappeler will go over the edge. A figure eight on a bight will connect the belay rope to the rappeler. Walk the belay rope to its primary anchor, which is separate from all

FIGURE 4-42

(A) The dynamic single-line rappel system with t3w prusiks. (B) The two-line dynamic rappel system with prusik brakes in place.

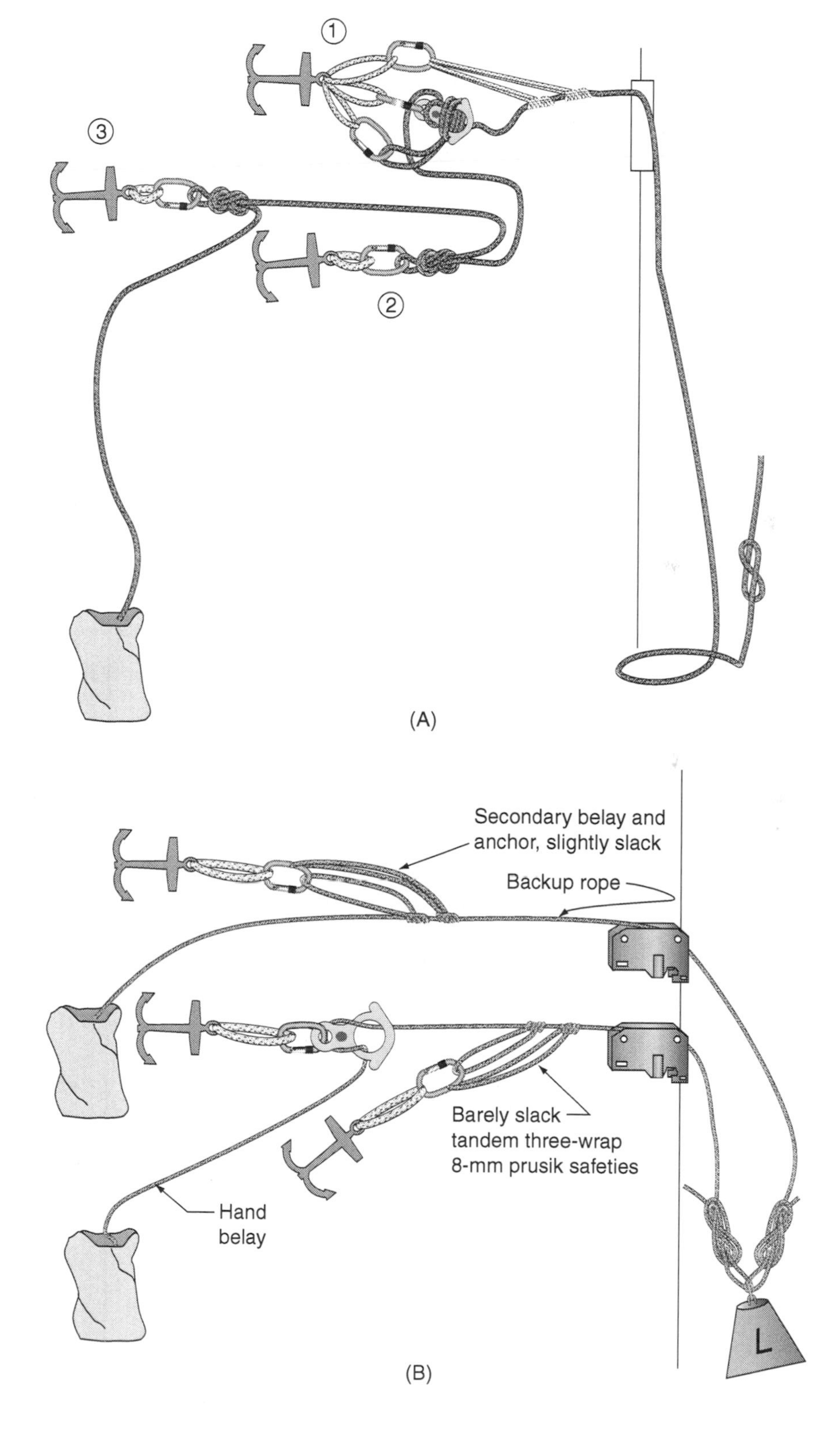

other anchors. Attach tandem three-wrap (t3w) 8-mm prusiks to the rope. These will act as emergency stops should the rappeler start falling too fast.

6. Work the belay rope back to its secondary anchor. Reeve it lightly onto a DCD. When using an eight-plate, it should be watched to keep the rope loose and flowing and not

inhibiting the rappelers actions on the main rappel rope. The purpose of this DCD is to offer a backup mechanism for a lower in the unlikely event of a mainline failure.

7. Work the primary rappel rope towards its secondary anchor and secure it with a terminal end knot. The eight on a bight will work nicely.
8. The rappel and top belay lines can be connected to a tertiary anchor now if it is thought to be necessary. Keep the top belay rope dynamic, or moveable, in case a lowering operation becomes necessary.

Operating the Lowering System

You have rigged a dynamic rappel system for the primary purpose of sliding a rope to your target for medical assessment and stabilization, to perform a pick-off, or for reconnaissance reasons. The dynamic part is there for versatility. First, if something happens to the rappeler, you can safely lower him to the ground from above. Second, as a tactical maneuver, it provides quick access via the rappel. While the rappeler is tending the patient, the remaining lower operations can be performed safely up above.

Operating the system is relatively easy, since gravity does all of the work. Once a lowering operation is decided on, the following actions will be required:

1. The belay rope has probably not been used and remains slightly slack. Someone must watch the belay prusik brakes, making sure they do not grab unless it is necessary to stop a falling load. The DCD behind the prusiks has minimal friction and is ready for lowering should something happen to the main line. As the lowering operation proceeds, keep the belay line slightly slack so *if* the main line contacts something sharp and separates, the belay line is not subjected to the same cutting action under tension. Too much slack, however, equates to acceleration time in the event of main line failure. Experience is the best teacher for belay line slack, but the following suggestions might help. (a) You should be able to shake a lazy sine wave into the belay line with your hand, but any kind of dip in the belay line is too much slack. (b) The belay line should stay beside the main line and not wind around branches or other obstructions.
2. Ready the main line DCD for the lowering operation. Untie and unlock the DCD and untie the secondary anchor attachment so the DCD is in a position to accept the full load.
3. The prusiks in front of the DCD must be loosened before the load can be lowered. If you have wisely built in a hokie hitch or other **load releasing device (LRD),** this is as simple as releasing the hitch and applying the load to the DCD and operator. Make sure to swap the untied hitch with a new one or, if you are proficient, retie the LRD and reset the prusiks in

► **load releasing device (LRD)**

A system component that will allow the release of another loaded component, used to transfer forces from one component to another or when passing knots.

the loosened, ready-to-brake position. If you did not build in an LRD, you must *raise* the load a bit in order to release the prusiks. This can be accomplished by having several rescuers safely pull up on the main line. You can also attach a pulley system to the main line with a couple of prusiks and raise the load mechanically until the prusiks can be released. Make sure to also compensate slack in the belay rope at all times.

4. Apply the load to the DCD. It is always best to apply maximum friction to the device keeping the lower static until you get a feel for the load. With a brake bar rack, as always, all of the bars will have been activated before being tied off. With the eight-plate, a double wrap will allow control over two-person loads. With either device, make sure the DCD operator understands the most open and most closed position. Opening the angle of the rope in relation to the DCD reduces friction and allows the load to move faster. Closing the angle of the rope in relation to the DCD increases friction and slows or stops the load.
5. When all of the components and operators are ready, give the command "SLACK." Make sure the person giving the calls can see the load or is in contact with the person on the rope via radio or hand signals. Anticipate a delay of a second or two between giving the command and seeing a response in the system. People have to hear and comprehend the order, and nylon is relatively elastic. Try to maintain a steady lowering speed. Do not allow the lower to stop and go, thereby shock loading the system, anchors, and the rescuers and patients on the rope.

Helicopters

NOTE: The following discussion of the potential interface between helicopters and rope rescue equipment and providers is intended for orientation purposes only and not as a conclusive instructional guide.

Rotor wing aircraft, more commonly known as helicopters (Figure 4–43), can make very useful rope rescue platforms. They combine two modes of transportation, gravity as the down engine, and turbofans as the up engine. This incredible balancing act of a screaming assemblage of parts looking for a place to become coordinated, and the gravity, the human factor, friction, and rope rescue equipment makes for an almost unbelievably useful rescue tool. The versatility of helirescue is unparalleled, which, paradoxically, is what makes people concerned about using it. The variables in using rope components for rescue combined with the variables that make helicopters fly, and not fly, are just too much to stomach for some rescue managers.

There are many facts about helicopters that a good rescue team should analyze before considering their application. First, the negative side:

1. Short of the Space Shuttle, helicopters are about the most expensive pieces of equipment to own and operate anywhere in the world. Helicopters big enough to carry several team members and lots of gear, and that have two engines are

FIGURE 4–43
Helicopter rescue team. (*Courtesy Chase N. Sargent*)

prohibitively expensive for most fire and rescue departments. Powerful helicopters are most often manufactured for military applications (Figure 4–44) and for military budgets. Smaller, more affordable helicopters are very limited in their payload and usually have only a single engine. The redundancy factors are therefore limited to only the most basic transportation duties. Risk managers/rope team managers often insist on minimum safety ratios of 10:1 or maybe even 5:1 with their equipment applications. Smaller, more affordable helicopters with a rated payload of about 1,000 pounds (one-ninth that of rescue rope) leave very little in the way of safety margin. Assuming a great flying day, lightweight passengers, one pilot at 200 pounds, one rescuer at 200 pounds, one patient at 200 pounds, and 100 pounds of equipment the payload still adds up to 700 pounds. The 1,000-pound payload divided by 700 pounds of people and equipment equals 1.42 or a safety margin of a little less than 1.5 to 1. As the helicopter approaches its maximum rated payload and the applied torque on the engine, transmission, and rotor assembly approaches 100% of rated capacity, mechanical failure becomes a greater possibility.

2. Helicopters add exponentially to the rules, regulations, policies, procedures, standards, laws, regulations, jurisdiction, and nosey people that affect a rescue program. There are too many to name here, but in addition to the rope rescue rules, the Federal Aviation Administration (FAA), the National Transportation and Safety Board (NTSB), and local and state police all have some impact on your program. If you are able to wade through the myriad regulatory and enforcement agencies, there may be some hope of using the

FIGURE 4–44

Military helicopters can be the perfect rescue platform. (*Courtesy Chase N. Sargent*)

local helicopter for a rescue platform. Finally, the manufacturer should approve whatever you are planning to do with their aircraft, an impressive task in and of itself.

3. Experienced rotor wing pilots trained in rescue missions are hard to find. It is sometimes hard to tell the difference between really good pilots and marginal ones. Routine transport pilots have busy schedules and have to make time to train and qualify in the special operations arena. Additionally, it is hard for pilots, just like rescuers, to turn down missions, especially emergency missions, for which they are not qualified.
4. The aerial performance of a helicopter-based rescue operation is only half the program. Ground support training is often compromised or neglected altogether. Helicopters have two exposed rotors that turn at tremendous speeds and are often within touching distance of people walking around on the ground. In particular, the tail rotor can spin so fast it becomes almost invisible. With all the commotion coming from the main rotor overhead, the tail rotor is often unnoticed and people can walk into it. This primary safety consideration is sometimes overlooked by team members who are focused on other rescue objectives. Even worse, untrained team members and well-meaning volunteers walking around the landing zone are a huge problem in heli-rescue programs that have weak ground-support discipline and training.
5. Other safety considerations are powerlines and other elevated obstructions, loose objects on the ground, rotor wash, hovering operations, detrimental weather conditions, fuel and smoking, and loose objects attached to the craft, such as rope, slings, stretchers, and so on. These should all be considered when analyzing a potential heli-rescue program.

NOTE: There are two important safety considerations when using the short haul system. A mechanism must be in place to allow the instant release of the rope or cable attached to the aircraft. First, a helicopter leashed to the ground by a rope or cable will crash catastrophically if not released. If the helicopter were to go down, the release of the load by a pilot or rappel master in the craft may save the people on the rope or cable from double jeopardy—the fall plus entanglement with the aircraft. Second, the main rotor on a helicopter effectively makes the craft a huge electrical alternator. Static electricity travels down the cable or rope very efficiently, especially when wet, looking for a ground. If you are standing between the rope or cable and the ground, and it touches you, you will receive a nasty shock.

Positive factors to consider about heli-rescue include the following:

1. Helicopters are excellent observation platforms and transportation devices for search and rescue missions (Figure 4–45 and Figure 4–46). There is no rescue tool that can cover more area more quickly and more thoroughly than a helicopter. With proper communication and a safe landing zone, the helicopter is the perfect solution for transporting lost and injured people to safety. Also, under the right conditions, helicopters are the best way to quickly transport rescue team members and equipment into and away from rescue scenes.
2. When marginal and single-engine helicopters are being used as rope rescue platforms, short haul systems are usually the

FIGURE 4–45

Heli-rappeling is a fast way to insert rescuers into almost any location. (*Courtesy Chase N. Sargent*)

FIGURE 4–46

Helicopters make great observation and search platforms. (*Courtesy Chase N. Sargent*)

only viable rescue option. The *short haul system* (Figure 4–47, Figure 4–48, and Figure 4–49) involves a rescue rope or cable rigged to hang beneath the aircraft. The length of the rope or cable is variable depending on the helicopter, the pilot, the mission, and, of course, policy. Sixty-five feet is practical for most smaller aircraft. The purpose of the short haul is to keep the craft in the air and some distance away from the rescuer and/or patient and ground obstacles. The rope or cable allows a rescuer or a basket-type device to be lowered into places where the craft could not safely go, such as between trees or houses and into water for swift-water or

FIGURE 4–47

Short hauls can be the tools of choice for departments with single engine helicopters, especially with ambulatory patients. (*Courtesy Chase N. Sargent*)

FIGURE 4–48

Litters can be "flown" nicely with forward moving helicopters. (*Courtesy Chase N. Sargent*)

FIGURE 4–49

Short hauls are a fast way to transport people and rescue gear. (*Courtesy Chase N. Sargent*)

NOTE: An important safety consideration with heli-rappeling operations is that belay is practically nonexistent. Dual-line rappels are particularly cumbersome and create an opportunity for line entanglement that could dangerously slow a descent. The aircraft becomes unbalanced if one rappeler gets off the rope quickly and the other rappeler is stuck untangling a DCD. Top belay with a second rope presents even more problems than a dual-line rappel. Finally, the mandatory bottom belay for routine single line rappels is questionable, because more people are committed to the helicopter danger zone, and pulling down on the rope to belay a rappeler can unsettle the craft from its delicate hover pattern.

ocean-water rescue. Preferably, and providing the patient is ambulatory, no rescuer is attached to the bottom of the rope or cable, and the patient can climb unaided into a net or basket-type arrangement. This keeps payload to a minimum by eliminating the on-rope/cable rescuer, and it also removes the possibility of injuring a rescuer. The option still remains to add a rescuer to the short haul line to tend patient injuries and for patient packaging as necessary.

3. Properly rigged and rated helicopters make excellent rappel platforms for unparalleled access. Heli-rappeling (Figure 4–50 and Figure 4–51) can place rescuers into almost any location where the helicopter cannot land. It can place firefighters on the top of a burning highrise, and it can place law enforcement agents on buildings that have criminals in them, to name but a few. Heli-rappeling is a study in agility, balance, and on-rope finesse. The equipment used for heli-rappeling is virtually identical to that used for a single line rappel from a water tower or any other free-hanging location. It includes a single 1/2-inch rope between 100 and 150 feet long, a rescue eight-plate DCD, several carabiners, and a good harness. Loading the rope is different, however. Pilots do not appreciate surprises and, for a helicopter pilot, a shock load to the craft is particularly unsettling. Unless you are using a powerful twin engine craft, the pilot and rappel master will require the simultaneous balanced loading of twin ropes. This means that one rappeler must be on each skid, carefully

FIGURE 4–50

Heli-rappeling requires finesse, coordination, and superb communication among the pilot, the rappel master, and the rappelers. (*Courtesy Chase N. Sargent*)

FIGURE 4–51

Heli-rappelers should stay together to help keep the helicopter balanced. (*Courtesy Chase N. Sargent*)

bringing their weight onto their ropes and away from the craft at the same time. To avoid shocking the aircraft, an inverted rappel must occur until the rope comes in contact with the skid, at which time the rappelers can move into the classic position for the rappel to the ground.

SUMMARY

This chapter discussed how to securely anchor a rope transportation system. In general, it is as simple as tying a rescue quality rope to a bombproof anchor, ensuring a backup, testing the system and personal protective equipment, and over you go. Try not to complicate the system with extraneous equipment, and always pick up your extra equipment and cache it in a safe place before rappeling. Helicopters are magnificent rescue tools providing one has the right craft and the right endorsements, and can afford the operational costs of the helicopter and the training that goes into a good heli-rescue program.

KEY TERMS

Rappeling
Anchors
Bombproof, bombing, or bombing off
Rope system
Anchor systems
Tie backs
Load sharing anchor system components
Pseudo-anchors
Super bombproof anchor (SBA)
Training to fundamental correctness
Primary anchor
Secondary anchor
Tertiary anchor
Inherently tight or inherently loose configurations
Dressing
Pretension
Standing part
Running part
Line
Knot
Bend
Hitch
Overhand knot
Half hitch
Single figure eight
Figure eight on a bight
Figure eight bend
Double bight figure eight
Butterfly knot
Double fisherman's bend
Fisherman's knots
Cheater prusik
Water bend
Internal water bend
Overhand on a bight
Square bend
Single-line static rappel line
Leading edge
Single-line dynamic rappel system
Double-line dynamic rappel system
Load releasing device (LRD)

REVIEW QUESTIONS

1. In rigging, it is important to always have a backup system that _____ reacts in the event of main system failure.
   a. quickly
   b. automatically
   c. over
   d. gently

2. An anchor system that involves simply tensioning two inline anchors together to incorporate the strength from both anchors simultaneously is called _____ .
   a. a linear anchor
   b. a shared anchor
   c. a tensionless anchor
   d. a tie back
3. Items that can be used as anchors but may not always be considered bombproof are classified as _____ anchors.
   a. potential
   b. crummy
   c. pseudo-
   d. insignificant
4. The main anchor that a load will be dependent upon is called the _____ .
   a. front anchor
   b. super bombproof
   c. pseudo
   d. primary
5. You tie a well-dressed figure eight on a bight in 12.7-mm rope, connect the bight to a standard 12-mm rod stock steel carabiner, and attach the running end of the rope to a large, round testing bollard. Upon applying force, the rope will eventually break _____ .
   a. in the knot and not around the carabiner
   b. at the carabiner and not in the knot
   c. midway between the carabiner and the knot
   d. at the point the rope enters the rope
6. Natural fiber ropes, such as manila, sisal, jute, and cotton, commonly break due to _____ .
   a. the rotting of internal fibers
   b. radius bending
   c. the internalization of the fiber bundles
   d. melting when overloaded
7. When in use, a rope is commonly referred to as a _____ .
   a. hawser
   b. lanyard
   c. halyard
   d. line
8. A _____ is a rope or webbing configuration comprised of bight loops and round turns that maintains its own shape independent of outside factors. A _____ is a configuration where two ends of a rope are joined. A _____ is a configuration that is tied around an object.
   a. hitch, bend, knot
   b. knot, bend, hitch
   c. bend, hitch, knot
   d. bend, knot, loop

9. The _____ is the best choice for a majority of static, nonmoving, rigs.
   a. figure eight on a bight
   b. butterfly knot
   c. bowline
   d. double fisherman's bend
10. Rigging the single-line dynamic rappel system provides _____ system than the single-line static system.
   a. a much less secure
   b. a more complicated and difficult to rig
   c. a much safer and more versatile
   d. a safer but more equipment intensive
11. In addition to routine rope rescue policies, standards, and regulations, the _____, the _____, and local and state police all have some impact on a helicopter rescue program.
   a. FCC, FTSB
   b. FAA, NTSB
   c. DOT, FEMA
   d. OSHA, FAA
12. The _____ involves a rescue rope or cable rigged to hang beneath the aircraft.
   a. static line system
   b. hoist line configuration
   c. belly sling component
   d. short haul system
13. The equipment for heli-rappeling is _____ that used for a single-line rappel from a water tower or any other free-hanging location.
   a. virtually identical to
   b. remotely similar to
   c. completely different from
   d. much more expensive than
14. An important safety consideration when considering heli-rappeling operations is that _____ practically nonexistent.
   a. chances of finding an experienced pilot are
   b. safety margins are
   c. belay is
   d. federal funding is
15. During heli-rappeling, _____ particularly cumbersome and create(s) a(n) opportunity for line entanglement that could dangerously slow a descent.
   a. loose objects on the ground are
   b. dual-line rappels are
   c. untrained ground support people are
   d. rotor wash is

CHAPTER

5 Rappeling

OBJECTIVES

Upon completion of this chapter, you should be able to:

- define operational edge safety.
- safely transfer from an edge tether to the rappeling rope.
- define the communications cycle and the importance of feedback.
- define and list the major rope calls.
- identify safe bottom belay techniques.
- explain how to rappel and lock off using an eight-plate.
- explain how to rappel and lock off using a rappel rack.
- explain how to emergency rappel and lock off using a münter hitch.
- explain how to effectively rescue a person who is wearing a harness, using pick-off techniques.
- explain how to effectively rescue a person who is not wearing a harness, using pick-off techniques.

A DAY ON THE ROPES

The on-site industrial fire brigade and rescue team had more training than most municipal teams. In fact, they had responded to seven bonafide technical rescues in the huge refinery on the river since May. Four were confined-space rescues, but only *two* were real rescues. The other two were body recoveries. Two were pick-off rescues from a frac tower on the same call. One of the patients was unconscious and required spinal immobilization and a tricky two-line rope lower. The other person rappeled down by himself with some encouragement. Finally, there was a machinery extrication job on somebody's leg that was caught up in an electric motor axle that was said to be spinning at about 10,000 rpm. He should have had zippers in those steel-toed boots instead of laces!

So there was no unnecessary excitement when the call went out for "worker fallen," zone four, tower A-4. It was a 150-foot fractional distillation tower that had been out of service for about a month for extensive repairs. Upon arrival, the rescue team chief met with the site supervisor to discuss the situation. One of the newer workers had removed one too many bolts from the OSHA railing over his head, and it had pulled free from the tower. It struck him in the head and sent him over his own railing, dangling by his Class III safety harness.

To the rescue team chief, it seemed like a pretty quick fix. They would secure site safety, attach a couple of lines to the myriad anchors above the patient, and make quick access to secure the patient, both medically and physically. Then they would lower another rescuer and a Stokes litter down to the ground, which was only about 120 feet below.

Two particular hazards presented themselves to the team, though. First, it was the third shift, and it was dark even under the shadowy yard lights. Second, the noise from the cogenerator and its associated steam pipes was distracting, at best. The rescue team chief reviewed the plans and hazards with the team of nine rescuers, and they went to work. The main line and belay line for the first worker was established, and the victim was reached by an EMT on a standard rappel and lower combination in less than ten minutes. The stretcher and attendant line took more time. The rig master had to plan for a change of direction pulley to allow for some space between the original rescuer and the stretcher and attendant lines. Nevertheless, the stretcher and attendant were ready for the lower about twenty-seven minutes after positioning on the side of the tower.

As the stretcher and attendant lower began, there was some confusion about who was doing the rope calls. The rescue team chief turned the edge calls over to another member who had a voice that could be heard over the roar of the cogenerator. The lower began smoothly with a loud "SLACK" from the edge man. Then, almost in the same breath, as the stretcher attendant reached a critical point, he shouted "WHOA!" Over the din of the cogenerator, the lower team heard "GO" instead of "WHOA." Certain that the "WHOA" from the edge man was a "hurry up and GO!" command, the lower team let out approximately four feet of rope just when the attendant had started to get some purchase on the wall. The stretcher and the attendant fell about ten feet before the lower team could arrest the sliding ropes, coming to a rest inches above the first rescuer and the victim. Fortunately, no one was seriously hurt. There were some bruises and some lost pride, but good fundamental rigging and a sticky set of 8-mm prusiks saved the day for at least three men. During the critique, the lowering team swore they heard a demanding "GO," while rescuers near the edge, some eighteen feet away, heard "WHOA." A simple communications blunder almost cost some otherwise excellent rescuers their lives.

Communication

One of the most common errors team members make when communicating is failing to respond to questions and commands. Feedback is the most frequently neglected component of the communications cycle and the reason for many team mistakes. All communication involves five crucial elements (Figure 5–1):

A *sender*—someone or something sending a message.

A *message*—information that originates with a sender and is directed to a receiver.

A *medium*—some method of communicating the message: voice, radio signals, hand signals, hard-line intercom, semaphore, rope tugs, Morse code, and so on.

A *receiver*—someone or something that receives a message. It is important to note that the intended receiver may not be the actual receiver. Messages often get intercepted and revised or dismissed. Therefore, the medium used by the receiver may or may not be the same medium as that used by the original sender.

Feedback is used to acknowledge the clear receipt of the sender's message or to question the original message due to interference in the medium. This might include a dropped radio signal or excessive ambient noise, or maybe the original message simply does not sound right and clarification is needed. It is estimated that 80% of all communications failures are a result of inadequate feedback.

With feedback, the sender knows the message was adequately received and goes on to other work. Without feedback, valuable time is wasted while the sender waits to see if the message was received and whether the receiver is acting on it. An efficient team rarely has to call back to the receiver with another message, such as, "Did you copy my last message?" In effect, this places a huge responsibility on both the sender and the receiver to properly perform their respective communications duties. Senders must select a workable, tested medium that is reasonably capable of getting the message through to the receiver. The message must be clear, concise, and as short as possible to convey the message. This is a real art that takes time and effort to master. Once the message has been received, the receiver has the responsibility to respond back to the sender to acknowledge receipt. Failure to do so is common and dangerous. Simply saying COPY or OKAY and repeating back the message is the preferred method of responding to the sender. Repeating the message allows the original sender to correct any problems in the message that may have been obstructed by interference.

FIGURE 5–1

The communications cycle.

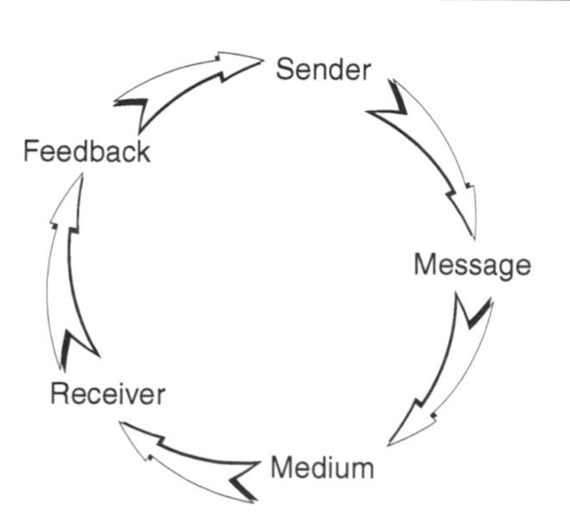

ROPE COMMANDS

► **rope calls**
Standardized words used as communications tools to eliminate mistakes and confusion while a team is operating a rope system.

Rope calls are standardized words used as communications tools to eliminate mistakes and confusion while a team is operating a rope system. Try to standardize and enforce a common set of words to be used under all rescue situations. Encourage the use of the word STOP for anyone on the team to use at any time. The word STOP is recognizable everywhere, and it is hard to confuse with other commands, like WHOA, HOA, HO, and WOHA. Encouraging liberal use of the STOP command gives all your team members a sense of security by empowering them to have a functional role in the series of events. It is like positioning warning sirens around the rescue scene. Should someone see something amiss, saying STOP can be a life saver. Explain that saying STOP does not automatically give them control over the entire operation. It gives everyone an opportunity to reassess a potentially bad situation and make corrections. The rig master, rescue team manager, or operations officer will then have the authority to alter something, to continue, or to abort the operation.

TENSION and SLACK are the words to use for movement in a rope system. They are hard to confuse with other sounds or other actions and they can refer to any direction.

SLOWER, FASTER, and LOCK are other common words that should be used to enhance safety and efficiency. SLOWER and FASTER are commands used to tweak a system's performance after TENSION and SLACK have initiated an action. LOCK simply means that the system is going to stop for some time, and the operators can lock their devices and rest or await another directive.

Preparing to rappel has a unique and almost universal set of commands. The person preparing to rappel must notify the rest of the team, particularly his belayer, that he is ready to load the rope. This command is a loud "ON ROPE." ON ROPE is a good command to use whenever a load could be introduced to the rope. The rappeler waits for the command response, "ON BELAY," from his top or bottom belayer before actually leaving the safety of his edge tether or parapet wall. Once the ON BELAY command has been received, the rappeler again confirms his intentions by saying, "ON RAPPEL." The belayer, now fully positioned and ready to account for the rappeler's weight in the belay system, says, "BELAY ON."

RAPPELING

Rappeling is the act of sliding a rope. The rope is the road, the chosen DCD is the vehicle, gravity is the engine, and friction between the road and the vehicle is the brake. Anything that can be attached to a DCD attached to a rope can slide the rope in a controlled fashion, given an adjustable amount of tension in the linear rope path. Pulling down on the rope from below the DCD increases friction dramatically in the DCD by attempting to straighten the rope (Figure 5–2). Almost any load can safely and easily rappel or be rappeled, or lowered, from the top of a rope to the ground by a person

FIGURE 5–2

Any downward tension on the rope increases friction in the DCD and slows descent.

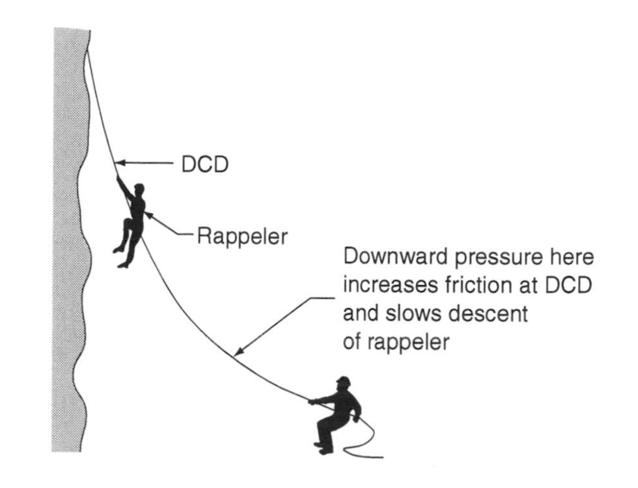

on bottom belay adjusting the tension in the rope appropriately. Therefore, almost any load can rappel or effectively be rappeled (lowered) from the top of the rope to the bottom even with hands off the rope (Figure 5–3A).

▶ **bottom belay**
Pulling down on a rope, when required or requested, by a person positioned below a rappeler.

The **bottom belay** is performed by a vigilant person below the rappeler who pulls down on the rope when required or requested. A bottom belay would be required if the rappeler lost descent control and began falling. It also can be requested by a rappeler to assist in braking after picking off another person or just to relieve a tired braking hand.

The bottom belayer should be positioned in a ready state, with feet shoulder width apart, knees slightly bent, and shoulders and head positioned to constantly watch the rappeler. Probably the most difficult, but most important job of the bottom belayer is to be ever observant and ready to spring into the BELAY ON position should an unexpected descent occur. The bottom belay position can be boring, and the neck muscles are strained by constantly looking up. When possible, do not position yourself directly under the rappeler. This has two distinct advantages. First, you will not be the target of an incoming human missile or dropped equipment, and second, your neck will not be as strained if you can watch the rappeler without looking straight up. You will be able to concentrate more on the person whose safety you are providing. Avoid distractions while on bottom belay and keep some movement in the back and neck to keep them from stiffening up. Always assume the rappeler is going to fall at any second and be ready to snap into the ON BELAY position.

The bottom belayer should be positioned so that the rope going up to the rappeler's DCD is controlled by the dominant hand. The remaining rope wraps around the lower hips and comes out into the nondominant hand for friction control. The belay occurs by dramatically increasing friction in the rappeler's DCD. A right-handed belayer will simultaneously:

1. back up a couple of steps
2. drop down into a linebacker position to take tension in the rope

FIGURE 5–3

(A) A rappeler can be safely lowered to the ground even with hands off the rope by a bottom belayer. (B) The angle of approach the rope makes from the anchor to the edge is critical to the ease of loading for a rappeling person.

(A)

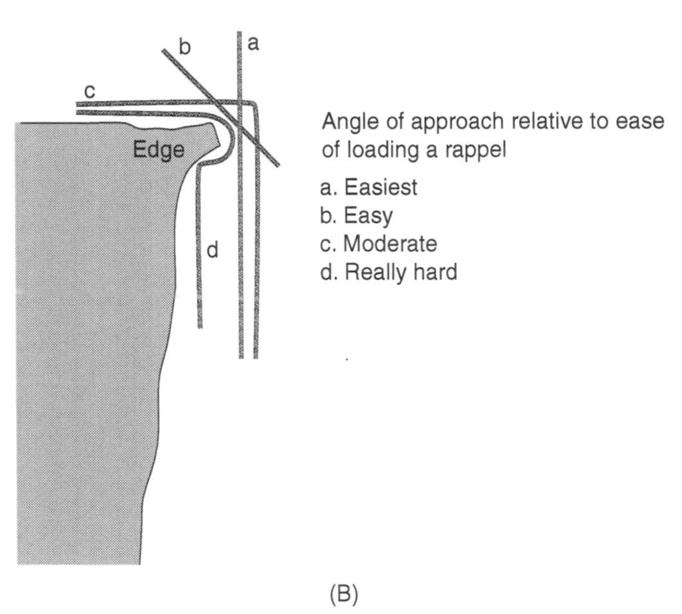

(B)

3. pull down with the right hand
4. drop the running end of the rope across the left thigh with the left hand

Practice snapping into the ON BELAY position quickly and with finesse. Also, practice performing real-time bottom belays with teammates until you develop muscle memory, or a kind of automatic belay response.

Rappeling with finesse, ease, and a smile is dependent on many variables that can be classified into four categories: the rig, the person, the situation, and the environment.

1. The rig can make rappeling easy and fun or very difficult. Engineering an easy, fun rappel rig should be the goal. For everyone, regardless of experience level, loading the edge is somewhat worrisome. It is unsettling to wonder whether the rope system will really hold you, and whether the DCD will

FIGURE 5–4

(A) Special considerations should be made when rigging the rappel rope to ensure the highest possible angle of approach. (B) Very hard rappel load. (C) Team members can help each other load a tricky edge. (D) Loading the rappel rope is not always as easy as just walking backward. (*C, Courtesy Martin Grube; D, Courtesy Chase N. Sargent*)

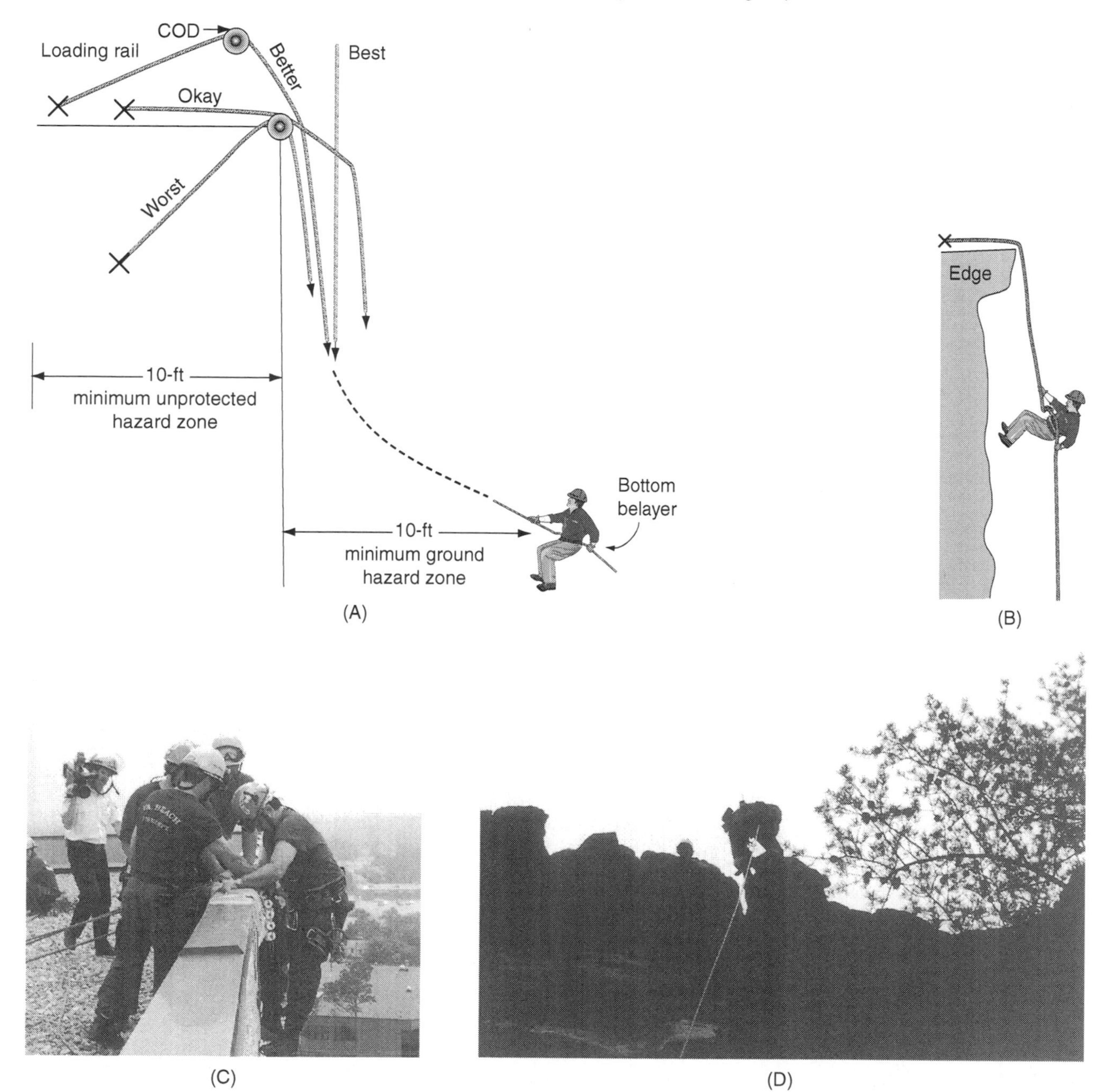

allow you to stop if necessary. Controlling that feeling of anxiety comes with experience, but a tested, simple, and thoughtfully engineered rig is the best medicine.

► **angle of approach**

The angle of the rope from the anchor to the edge relative to the ground. The higher the angle of approach, the easier it is to load the rope.

The **angle of approach** on the leading edge (Figure 5–3B) is critical. Whenever possible, the angle of approach of the rope should be parallel with the wall and perpendicular to the edge (Figure 5–4A). A rope rigged vertically over the head of the rappeler is easy to load. Loading ease diminishes as the approach angle decreases. The worst possible loading situation is when the rope is anchored below the grade level of the edge. Additionally, the rope will, unless otherwise directed, seek a perpendicular position relative to the edge. Anticipating these factors will come with experience and will pay great dividends in terms of team efficiency and the confidence you project to your customers.

Loading a below-grade anchored rappel rope, while awkward, can be achieved with finesse if you use some simple tricks when rigging. Again, a COD anchored above the head helps, but it is not always available. Loading bars are used on most training towers to jack the approach angle into the finesse range of 45 degrees to 90 degrees. Your team can help considerably by easing you down into your edge protection as you load the rope (Figure 5–4B). Finally, you can crawl over the edge sideways and, with a spin and knee maneuver, end up in a semirappeling position on the wall. Do not let go of your grip on the rope while loading, and make sure the rope gets into the edge protection.

A good rope curriculum reinforces familiarity and confidence in the equipment, instructors, and technique. Sometimes knowing that they have a bottom and/or top belay and that they can be completely unconscious and still achieve a successful rappel is comforting. Practice loading from a small wall or loading dock. Use traditional training methods by starting low and easy and working toward more complex and higher loads. Also, team encouragement and a dose of peer pressure can help motivate nervous first rappelers. The appropriate and prudent use of humor can relax those first anxious moments. For example, "Of course you're not going anywhere. Your Vulcan death grip is squeezing the juice out of the rope. Lighten up!" or some similar nonaccusatory comment often works.

2. The *person* has much to do with a successful rappel. Good physical condition and agility is important, specifically upper body and grip strength. Generally, a more experienced rappeler rappels with greater finesse and confidence. A training program should include repetitive rappels in all different rope manners of approach and terrain. Additionally, some people seem to be natural on rope and have an athletic prowess that is hard for those without it to understand. A smart team leader recognizes or even scouts naturals and works them into critical on-rope positions.

   Another consideration is the physical condition of rappelers. If they are exhausted, injured, or in an altered state of consciousness, rappeling is not an option. Be prepared to change to a lowering system when necessary. Finally, remember that alcohol, gravity, and the ground do not mix!

3. The *situation* plays a big role in how the rappel is carried out. The **risk-benefit analysis** helps a team discern its

► **risk-benefit analysis**

An incident management technique where the factors of an emergency situation are studied to determine the risks and compare them to the benefits of attempting a particular action. For example, the benefits of committing live rescuers to extremely hazardous environments for the benefit of recovering dead bodies is negligible.

immediate rescue goals. Whether it is a live rescue as opposed to a body recovery affects the tactics a team chooses to use and the time factors it chooses to follow. Extreme care must be taken when committing live resources to body recoveries. Time is not an issue.

Bail out or emergency egress rappels (see Chapter 6) are a completely different rescue discipline, but they still rely on the art of the rappel. Emergency egress rappels are roll outs under the worst conditions, yet they are remarkably survivable. Rescue rappels usually have two contacts with the anchor(s) and factored time for system redundancy.

Military tactical, SpecWar, SWAT, and police reconnaissance operations are offensive in nature and tend to bring an entirely different emotional focus to the rappel. Timing, muscle memory, and rote training through repeated movements are unique to tactical rope operations.

Rescue and recreation are at opposite ends of the uses of life safety rope and, in many cases, foreign to one another. One of the most common recreational hazards involves rescuers who step out of their rescue environment into a recreational rappel environment. This can be hazardous to the off-duty rescuer and to others in the general vicinity. It seems to create a kind of reality shift from rules to recreation, from helmets to baseball caps, from long pants to Bermudas, and from focus to relaxation. This has been the cause of many slips, trips, and falls from rigging the recreational rappel system with less gear, fewer knots, less redundancy, and maybe less safety margin than the rescue rappel system. Perhaps long hair that would normally be tied up or tucked into a helmet is let down and becomes tangled in the rappel device. Treat gravity and your general proximity to it with the utmost respect at *all* times.

4. The *environment* blurs the line on all rope rescue work, especially rappeling. The physical object from which you are rappeling can be almost any shape and any consistency. Urban and wilderness rappel (Figure 5–4C) sites are as different as wall-walking rappels and free-hanging rappels (Figure 5–4D). Buildings, bridges, antennas, towers, and dishes all have unique hazards. Have you ever lost a tooth filling by standing in front of a microwave dish while someone was calling long distance to Tokyo? Have you ever temporarily forgotten how to tie a figure eight on a bight while someone was speaking on mobile satellite (MSAT)? Cliffs, waterfalls, trees, and caves offer their own set of peculiarities. Did you ever rappel into a 600-foot deep pit on a 600-foot long rope, only to have the rope contract while you were exploring and end up out of reach above your head? Did you ever set your canvas edge protection on fire with the flame from your carbide lamp? Just when you think you have calculated all the possible problems, nature or gravity steps in with another surprise.

Rappeling: Step by Step

This step by step process assumes that the ropes have been properly rigged and that the person has the appropriate personal protective gear on and is receiving expert personal instruction.

1. You are on a safety tether that will have to be removed prior to rappeling.
2. As soon as you touch the rappel rope, shout ON ROPE to the bottom belayer. If a bottom belayer is not being used, then shout ON ROPE to the person operating your top belay.
3. Place your hands on your helmet, silently indicating to your safety person, "please safety check me." The safety person touch checks your harness and all attachments for security and proper attachment.
4. Standing with your back to the edge, correctly attach the rope to the DCD. Make sure the running end of the rope exits the DCD on your control hand side—right side for right-handers and left side for left-handers. *Never let go of the rope with your control hand, unless the DCD is safely locked off. You may use your control hand as a balancing nub, and you may use your nondominant hand to grip the edge, BUT DO NOT LET GO OF THE ROPE WITH YOUR CONTROL HAND!*
5. Attach the top belay rope, when using double rope techniques (DRTs), to your harness.
6. Ask for a final safety check from your safety person.
7. Announce ON RAPPEL to your belayers; confirm BELAY ON back from your belayer(s).
8. Remove the safety tether.
9. Begin to back over the edge by walking backwards and allowing the rope to slide through the DCD.
10. Finesse your weight transfer from where you are standing onto the rope. As stated earlier, much depends on the approach angle of the rope and practice.
11. Once loaded, get a feel for how much grip is required to maintain your position on the rope. If your DCD is a rack, you should only need an OK sign made with your thumb and index finger to retain the rope and direct the angle of the rope into the bottom active bars. If you have to grip the rope to stop, close the angle somewhat to compress the bars or add another bar. If you find you have to jack yourself down the rope, open the angle of the rope to the bottom active bar, or take off a bar. If your DCD is an eight-plate, all you should need to control your descent is a normal handgrip, like a good firm handshake, with your hand in position on your hip. You should be able to completely stop your descent by gripping with your hardest handshake. If you cannot, you can always add your nondominant hand to the rope below the eight-plate as an emergency brake. This two-handed braking technique comes in handy if you pick up another person. Do not forget that you can always call for help from the bottom or top belay person if you cannot stop on your own.

NOTE: Free-hanging rappels are known for making the load on a rope(s) spin. This is caused by spiraled core manufacturing techniques in some ropes, by laid ropes, and/or by sliding on highly used ropes that have been used with certain DCDs, such as eight-plates. Spinning is particularly dangerous when using two ropes, because the ropes can become entangled, preventing movement up or down. Spinning can also cause nausea in the rappeler. Rigging the ropes several feet apart can help, since the angle created between the two ropes can exert enough counterforce to stop the rotation. Better still, rappel with a rack on reliable R/Q rope.

12. Good rappeling position (Figure 5–5) depends on whether you are walking a wall or free hanging. Walking a wall is easier than free hanging, because the wall acts as a tool to help you align and position your body.

    Free-handing rappels require a bit more abdominal exertion to balance out the upper-body weight versus the legs. Therefore, a Class III harness or at the least some type of chest rig to the rope, is highly recommended when performing free-hanging rappels.

    Wall-walking rappels are just that—walks. It is not necessary to jump away from the wall repeatedly while rappeling. Simply take the most comfortable semisitting position with feet about shoulder width apart, knees slightly bent, and the balls of your feet maneuvering on the wall (Figure 5–5). Assuming you are right-handed, your right hand is your control hand. It should be gripping the rope with about the same pressure as if you were reaching down to pull on the emergency brake lever in a small car. Position your hand against your right hip to give your arm stability while on rappel. Your left hand should be positioned wherever it feels natural. That will probably be on the rope above the DCD. Remember, there is almost no descent control *above* the DCD, so that hand is just there for balance as you rappel.
13. Try to adjust your speed to a comfortable level, without stopping repeatedly or causing jerky shock loads on the rope and the rig. Avoid free-falling rappels that can lead to out-of-control falls and injuries. Remember, it is a balance of friction and heat dissipation. A very long, fast descent and a sudden stop can generate enough heat in the DCD to soften or even completely separate the rope.

FIGURE 5–5

Good rappeling position is relaxed and comfortable.

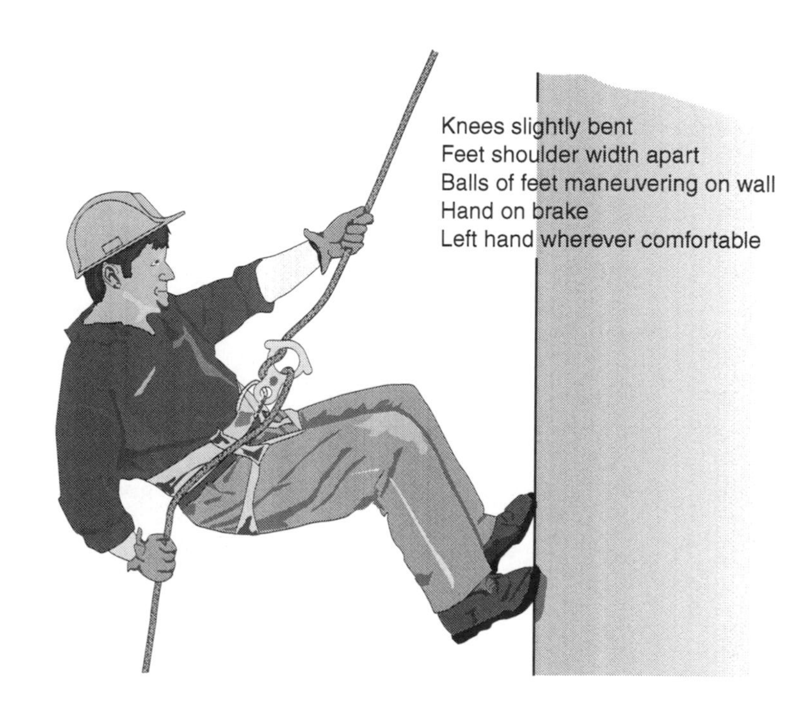

FIGURE 5–6 Locking Off the Eight-Plate While on Rappel

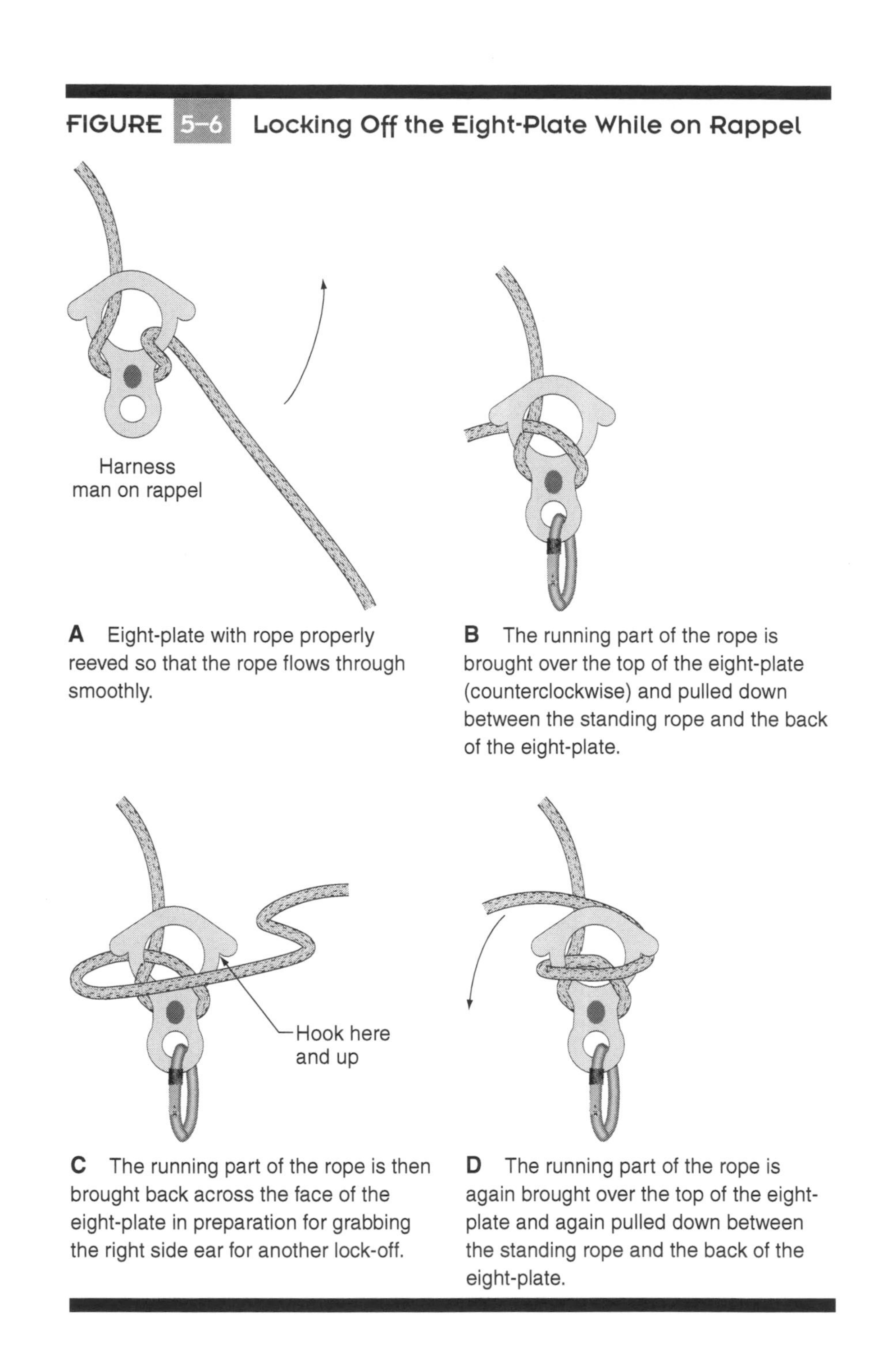

A Eight-plate with rope properly reeved so that the rope flows through smoothly.

B The running part of the rope is brought over the top of the eight-plate (counterclockwise) and pulled down between the standing rope and the back of the eight-plate.

C The running part of the rope is then brought back across the face of the eight-plate in preparation for grabbing the right side ear for another lock-off.

D The running part of the rope is again brought over the top of the eight-plate and again pulled down between the standing rope and the back of the eight-plate.

Locking off the eight-plate (Figure 5–6) to allow hands-free operation is fast and simple. For simplicity, assume the rappeler is right-handed:

1. Bring the rappel to a stop and notify the belayer(s) of your intention to lock off the eight-plate.
2. Change the braking grip of the right hand by turning it over so the palm is up and the thumb is pointing away from the body. To do this, take a temporary brake with the left hand so

the right hand can let go of the rope and turn over. Sometimes people forget to use the left hand as a temporary brake and simply let go with their brake hand. They then attempt to catch the rope quickly with the right hand in the palm up, thumb out position. This technique is not good and will almost always cause a fall.

3. When the right hand is in the palm up, thumb out position, take brake with the right hand and remove the left hand from the rope. The left hand can be placed on the back of the eight-plate to help align it for the lock off.
4. While braking with the right hand, rotate the rope and the right hand counterclockwise up and over the top of the eight-plate. The object is to cause the rope to lodge between the standing, tensioned and anchored, part of the rope, and the top ring of the eight-plate, effectively increasing friction enough to prevent further descent. Eight-plates with ears, or rescue eights, help to guide the rope into position. Once the rope is laid in position on top of the eight, grab the rope with both hands and pull it down into position between the eight-plate and the standing rope. This is relatively easy after some practice, but if you have trouble getting the rope into place, grip the rope tightly between your chest and the eight and throw the shoulders back and away from the eight. This will pull the eight away from the standing part of the rope and will increase the downward pull on the rope to help force it into place.
5. Once the rope is lodged between the standing rope and the eight-plate, you should be stopped. However, the rope can still creep somewhat. Since there is plenty of room for another wrap, repeat step four by bringing the running rope that is between your chest and the eight-plate back to your right. Bring the rope up and over the eight-plate again in a counterclockwise direction. Lay the rope over again softly, position both hands on the rope, and pull it forcefully down between the eight-plate and the standing rope. The eight-plate DCD is now locked and any descent has been terminated.
6. Because you need to be prepared to work with hands free for rescue operations, it is imperative to tie off the eight-plate DCD (Figure 5–7). Tying off the DCD allows you to work and maneuver without worrying that the wraps securing the eight-plate will come off. Begin by forming a bight in the running rope and doubling the rope back so about an arms length, 24 to 30 inches, is made available to tie off the eight-plate. Bring the doubled section of the rope back to the eight-plate with your left hand. The bight should be held out at arms length with your right hand.
7. Bring the bight around the right side of the eight-plate and pass it through, toward your face, the remaining space in the upper ring of the plate.

FIGURE 5–7 Tying Off the Eight-Plate for Hands-Free Operation While on Rappel

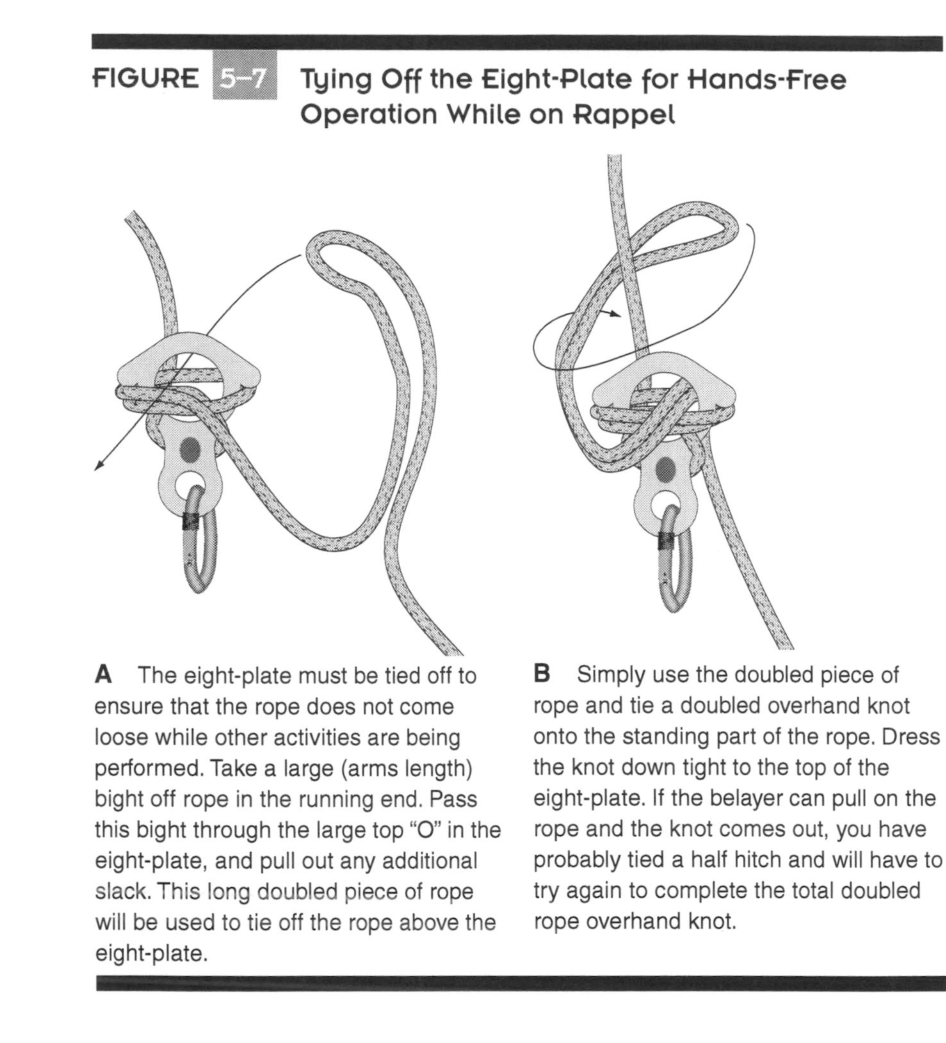

A The eight-plate must be tied off to ensure that the rope does not come loose while other activities are being performed. Take a large (arms length) bight off rope in the running end. Pass this bight through the large top "O" in the eight-plate, and pull out any additional slack. This long doubled piece of rope will be used to tie off the rope above the eight-plate.

B Simply use the doubled piece of rope and tie a doubled overhand knot onto the standing part of the rope. Dress the knot down tight to the top of the eight-plate. If the belayer can pull on the rope and the knot comes out, you have probably tied a half hitch and will have to try again to complete the total doubled rope overhand knot.

8. With the remaining doubled section of the rope and the bight, tie a doubled overhand knot onto the rope above the eight-plate. Twist the overhand knot down tight against the top of the eight-plate for security.

   With practice you should be able to double lock and tie off an eight-plate, while on rope, in less than 30 seconds.

Rappeling Using the Open-End Brake Bar Rack

Rack rappels (Figure 5–8A) are vastly preferred to eight-plate rappels for rescue work. The rope does not twist below the rappeler, the rack is attached semipermanently to the harness (not taken off to be reeved as in most eight-plate applications), and an infinite amount of friction can be added quickly and easily as the rope below the rescuer gets shorter closer to the ground, and a victim's weight can easily be added to the rescuer for evacuation.

Once the eight-plate rappel has been mastered, it is a simple matter to learn the rack (Figure 5–8B). The main concern for

FIGURE 5–8 Rappeling

A Rappeling with an open-end brake bar rack.

B Rack rappeling.

C Rack tied off for hands-free operation.

D Close-up of tied off rack.

rescuers who are learning to rack rappel is that the rack seems to have a lot of moving parts—it clatters and appears ungainly compared to the simple eight-plate. Rest assured, once you get used to the rack, you will find more and more occasions to leave your eight-plate in your gear bag.

To reeve the rack for rappel, identify where the top bar of the rack will lay once you have loaded the rope. Pinch this section of rope between your fingers and lay it on the top bar of the rack—in the training groove if your rack has one. Next add the safety (free-swinging, straight slot) bar and alternately add bars. The number of bars is equal to the amount of friction needed to slide the rope. Four or five bars are usually about right for most people on rappels under 100 feet. The longer the rope is under the rappeler, the more automatic belay is going to occur. The weight of the rope itself adds a lot of friction to the rack. Never rappel with less than three bars. It is always a good idea to start with *more* bars until you get used to the rack.

As you rappel, make your brake hand into an OK sign around the running end of the rope. You should not have to squeeze the rope hard, as you might with an eight-plate. Allow the rack to do the work. All of the descent control is developed by the angle of approach the running rope has on the bottom active bar. Your hand should be in front of your chest (Figure 5–8A) and not on your hip as in rappeling with an eight-plate. Your nonbrake hand should be on the rack around the upper bars and the rack frame. It can gently control the attitude of the rack, give you some fine-tuning of the active bars, and give you some upper-body balance.

Stopping descent while on rappel is as simple as directing the rope *up,* causing the bars to mash together and dramatically increasing friction. You can then add an additional bar or two to increase friction or pick up another person. It should be noted that while a bottom belay will work with rack rappels, the action of stopping is in different directions for the rappeler and the belayer. The rappeler adjusts the running rope up to activate the bottom bar, squashing the bars together, while the bottom belayer tries to pull down to arrest descent. In reality, if the bottom belayer sees that a rappeler is descending too rapidly, it is easy to overpower the arm of the rappeler and bring him to a stop by pulling down hard on the rope.

Locking and tying off your open-end brake bar rack DCD (Figure 5–8C) is another simple and necessary skill. It involves winding the running part of the rope around the bars and securing it with a doubled overhand knot.

1. While on rappel, come to a complete stop at the desired location. Notify belayer(s) of your intentions to lock off.
2. Add all of the remaining bars. This will ensure maximum friction when you prepare to rappel in case you have to bring another person's weight onto your rack, as in a pick-off. Most people find also that brake bars are a little easier to take off than to add while on rappel.
3. With the brake hand in position, raise the running rope to compress the bars. Lay the rope over the top of the rack gently.

Do not attempt to force the rope between the standing rope and the top bar of the rack. You will almost certainly get stuck and require self rescue or, even worse, rescue by your own team.

4. Bring the rope behind the rack and down, passing the plane of the rack frame. Again, encompass all of the bars by bringing the rope up between your chest and the rack.
5. Create a doubled length of rope by forming a bight in the running part of the rope. Use the bight and doubled length of rope to tie off the rope with a doubled overhand knot onto the standing part of the rope above the rack.
6. Twist the overhand knot down next to the rack for security (Figure 5–8D).

The rack is even easier and faster to lock off than an eight-plate. With practice it should be accomplished in much less than 30 seconds.

Emergency Rappel Using a Münter Hitch

▶ **münter hitch**
A friction hitch that can be used for rappeling in a pinch.

The **münter hitch** (Figure 5–9A) is a friction hitch that can be used for rappeling in a pinch. Two instances where this might occur are emergency egress (Chapter 6), or when you absolutely must descend and only have a rope, a harness or belt, and a carabiner. *It must be understood that rappeling with a friction hitch is risky business, with no system redundancy. It is easy also to tie a girth hitch or a clove hitch by accident and become stuck on the rope when you really need to be sliding down it. All other means of completing the task must be considered before rappeling on a münter hitch.*

Assuming you have performed all of the preparatory tasks, such as anchoring the rope, protecting the edge, obtaining a belay, and so forth correctly:

1. Make a münter hitch in the rope and attach it to the carabiner (Figure 5–9B).
2. Connect the carabiner to your belt or harness and make sure it is locked. Assure that the running rope does not roll over the gate lock; it should roll over the spine side of the carabiner instead.
3. Work your way backwards over the edge.
4. The braking of a münter hitch is more similar to the braking of a rack than of an eight-plate. The münter hitch must be closed to achieve maximum friction, meaning your right hand should be in a position to direct the rope up, like on a rack, rather than down toward your hip, like on an eight-plate. This is important to remember, especially for your bottom belayer who will pull down hard if you begin to fall. While the bottom belay will open the münter hitch, sufficient friction will be available to substantially slow any uncontrolled descent.
5. Gather a feel for the required friction as you load the edge. Raise your brake hand to slow down, and lower it to go faster.

FIGURE 5–9 Münter Hitch

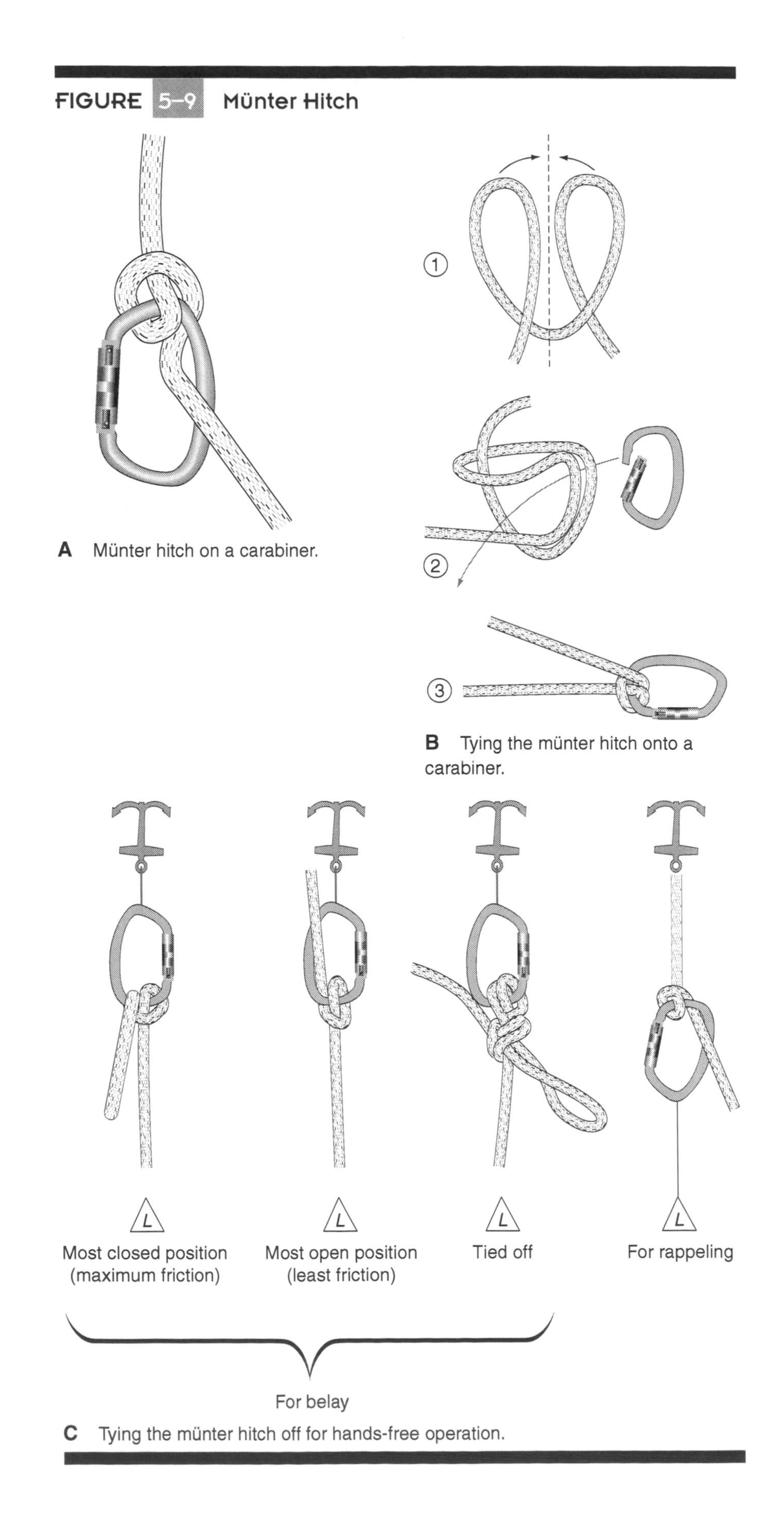

A Münter hitch on a carabiner.

B Tying the münter hitch onto a carabiner.

C Tying the münter hitch off for hands-free operation.

Once loaded safely in your edge protection, assume a comfortable rappeling position and work your way down the rope.

The münter hitch can be tied off for hands-free operation or just to take a rest (Figure 5–9C).

1. Close the hitch by bringing up the running part of the rope. Hold the running part of the rope next to the standing part of the rope with your hands.
2. Create a bight in the rope long enough, 20 to 24 inches, to tie a doubled overhand knot above the carabiner/descender.
3. Dress and secure the doubled overhand knot next to the carabiner/descender.

PERFORMING A SELF RESCUE

Rappeling positions the DCD somewhere in the upper torso area, and racks may be a bit closer to the face. There have been numerous occasions where something gets caught in the DCD and prevents further descent. It might be a T-shirt, hair, a loose piece of harness webbing, or a helmet strap. Even breasts and chest hair have been caught in DCDs, drawn there by a loose shirt. The immediate problem is, at best, embarrassing and, at worst, disfiguring. The intermediate problem could be fatal if the jammed DCD is not freed up.

The jammed DCD issue is rare but not unheard of in rescue teams, primarily because rescuers usually train with a level of professional discipline that requires appropriate uniforms and personal protective equipment. People, sometimes including off-duty rescuers, let their hair down in less-disciplined and recreational environments, which sometimes leads to trouble.

There are several ways to manage this potential problem. The first is simple prevention. Educate all rappelers about the danger of a jammed DCD, and always train to fundamental correctness. Make sure loose clothing is tucked or simply traded for tighter clothing. Secure long hair under a helmet, and make sure all loose straps from personal protective equipment are secured and not flopping around while on rappel. The second way to manage a jammed DCD is to lower the rappeler to the ground or raise the rappeler back to the top where other team members can help disentangle the device. The lower and raise option works only if the rappel system is rigged dynamically and there are people prepared to activate a raise or a lower.

▶ **self rescue**

Ability of a rescuer to escape unaided from a given situation.

Self rescue is the ability of a rescuer to escape unaided from a given situation. Self rescue should be one of the first skills rappelers learn, especially rescuers that may rappel. Self rescue, unloading the DCD while on rappel, can be practiced and is performed easily by using a personal self-rescue rope grab.

1. *NEVER, NEVER* get on a rope without your own, dedicated personal rope grab, a prusik loop, or one-handed, quick-on, quick-off, mechanical ascender with leg sling. The prusik loop should be long enough, for most people about a 9-foot single length before making the three-wrap double

fisherman's loop, so when it is attached to the rope above the jammed DCD, it is easy to place your foot into the bottom bight to step up and free yourself. A mechanical rope grab should have a sling attached that is long enough to do the same.

2. Find out what is caught in the DCD that is preventing descent. Sometimes it can be yanked free or torn away and descent continued. *Do not attempt to use a knife or scissors on a single rope entrapment. The proximity of the rope to the entangling item is too close for comfort. Only in a dire emergency, such as when someone's airway is compromised or a fatal injury is about to occur, should an attempt be made to free the entangled item with a sharp object. Even then, it should only be used when there is some means to back up the original line.*
3. Secure yourself to the line by locking off the DCD, as discussed earlier in this chapter. Sometimes the item is so entangled in the DCD that routine lock off is impossible. If this occurs, use a doubled arm's length of rope from below the DCD to create a bight about 24 inches long. Use the bight to secure a doubled overhand knot onto the standing line above the DCD. This will prevent the DCD from traveling while the entangling object is freed.
4. Attach the self-rescue rope grab to the standing rope above the DCD. If using a prusik loop, two wraps is sufficient to hold your weight, and it is easier to untie than three wraps.
5. Since the object of the self rescue is to release the entanglement from the DCD, the weight of your body must be transferred away from the DCD and onto the standing line *above* the DCD. Slip one foot into the bight of the prusik loop or the leg sling of the mechanical ascender and stand up. Your weight will transfer off the DCD and onto the rope grab. Lock your leg so you do not fatigue your leg muscles. Use your nondominant hand and arm to remain upright on the standing line. Use your dominant hand to free the entangled DCD. Untie the lock-off and tie-off on the DCD. Make sure your brake hand is back on the running part of the rope below the DCD in the rappeling position.
6. Do not get the DCD jammed again!

PERFORMING THE PICK-OFF

The final introductory rope rescue skill that should be mastered is the ability to pick off someone that has become stuck at an elevated point on a rope, swinging stage, rock outcropping, or somewhere else (Figure 5–10). *The pick-off is designed for use where the victim is relatively ambulatory and injuries are not considered life threatening.* The object of the pick-off is to draw the victim's weight onto your DCD, preferably, or if necessary onto an approved Class II or Class III harness or even the connection point on a lowered rope,

FIGURE 5–10
A rescuer rappeling an ambulatory patient during a pick-off maneuver. (*Courtesy Chase N. Sargent*)

and then descend to a safer location. It is always best to have the victim's weight placed on the DCD or the lowering rope connection point, if a lowering rope is used instead of rappeling, as opposed to your harness. However, quality NFPA third-party certified Class I, II, and III harnesses are designed to accept two-person loads for rescue work and are sometimes the tool of choice for the pick-off.

1. Locate your target, victim or patient, and determine what rope rescue system to engineer to make access quick and simple. Check your bombproof anchors and change of direction points, if necessary, to bring your rope(s) just to the side of the area directly above your target. Decide whether someone is going to be lowered to the target for the pick-off, which is preferred, or if you are going to rappel on a single line due to equipment and/or time constraints. If you are going to rappel using an eight-plate DCD, consider double wrapping it for additional friction when the victim's weight is added to your own.
2. Rappel or be lowered to a point that is level with the target. Be sure you are secured in that position, that the eight-plate is locked and tied or that signals from above indicate that your lowering system has been secured.
3. Do not overlook reasonable medical considerations involving your patient, who is probably also very nervous. Some encouraging words delivered in a calm manner can go a long way toward achieving cooperation with the victim.
4. If the victim has on a reasonable harness and simply needs to be extracted from the predicament, simply attach a prusik cheater or pick-off strap to the victim's harness. Make sure it is secure, remove the victim from the position or disentangle the victim from the rope, and lower the victim's weight onto your DCD, lower rope connection point, or harness. Cradle the victim between your legs to prevent banging on the wall as you rappel.

Caution
The hasty hitch is a quickly improvised webbing hitch that is not particularly comfortable, nor is it redundant in any way. It should be considered an emergency egress hitch to be used only after all other means of victim attachment have been considered and eliminated. It should be applied only to victims that are ambulatory, conscious, and cooperative. Make sure the victim's arms are kept down *during the lowering or raising process to help prevent the victim from sliding out of the hitch.*

5. If the victim has no harness, one will need to be applied. If the situation is entirely nonemergency and the victim is in a relatively safe zone, you can take a presewn harness to the victim and have the victim don it before removal. If time is critical, such as when handing under a helicopter in bad weather, the hasty hitch may be the best choice. Attach the hasty hitch to the victim. Be sure it is not wrapped around the victim's rope, safety cables, or fall protection. Snug the hasty hitch tight to the body and connect the hitch's carabiner to your DCD, lowering rope, or Class II or III harness. The victim is ready to be raised or lowered as appropriate.

SUMMARY

Rappeling is a relatively easy method of transportation that involves sliding down a rope. It is a great way to quickly access victims for medical assessment or even for simple evacuations. There are many associated skills that a *rescue* rappeler must master before attempting to rescue someone while on rappel. Comfortable control of the DCD and good personal attitude on the rope should not be taken for granted. The ability to self rescue and handle one's own problems is essential. Professional rescuers distinguish themselves from amateurs by spending literally hundreds of hours hanging on ropes and working out solutions to progressively more complex on-rope problems.

KEY TERMS

Rope calls
Bottom belay
Angle of approach
Risk-benefit analysis
Münter hitch
Self rescue

REVIEW QUESTIONS

1. Encourage the use of the word(s) _____ for anyone on the team to use at any time.
   a. on belay
   b. stop
   c. on rope
   d. safety
2. _____ and _____ are the best choice of words for movement in a rope system.
   a. haul, lower
   b. raise, lower
   c. pull, release
   d. tension, slack
3. It is estimated that 80% of all communication failures are a result of inadequate _____.
   a. language (or coding) skills
   b. feedback
   c. instructions from the sender
   d. medium

4. The _____ on the leading edge is critical to anxiety control.
   a. angle of approach
   b. edge protection
   c. safety tether
   d. safety person
5. When rappeling, never let go of the rope with your _____, unless your DCD is safely locked off.
   a. safety prusik
   b. belay stance
   c. control hand
   d. safety tether
6. The _____ is a friction hitch that can be used for rappeling as a last resort.
   a. münter hitch
   b. hasty hitch
   c. clove hitch
   d. kanoodler hitch
7. The ability of a rescuer to escape unaided from a given situation is called _____.
   a. free rappeling
   b. personal escape
   c. self rescue
   d. emergency egress
8. The pick-off is designed for use when the victim is relatively ambulatory and injuries are not considered _____.
   a. life threatening
   b. due to time constraints
   c. to be in the realm of rope rescue work
   d. under the authority having jurisdiction
9. When performing the pick-off, it is always preferable to have the victim's weight placed on your _____ instead of your _____.
   a. harness, carabiner
   b. descent control device, harness
   c. harness, descent control device
   d. rope, harness
10. When performing a pick-off rescue when time is critical, such as hanging under a helicopter in bad weather, _____ may be the best choice.
   a. the hokie hitch
   b. the modified Swiss seat
   c. a Class III harness
   d. the hasty hitch

CHAPTER

6 Personal Emergency Escape Rope Systems (PEERS)

OBJECTIVES

Upon completion of this chapter, you should be able to:

- define the six primary equipment components of a PEERS system.
- define the margin of safety as listed by NFPA 1983 for personal emergency rope.
- demonstrate the proper method of reeving personal escape rope to three different DCDs.
- identify dangerously reeved personal escape rope systems.
- identify and define at least five actions to consider before exiting using the PEERS equipment.
- identify accepted rope inspection techniques.
- define a pseudo-anchor.
- identify at least five safety rules for practical training evolutions involving rope rescue.

Note: Emergency workers who have been in the unfortunate position of bailing out of a location that has become untenable have always had several miserable options: Wait for co-workers to find and remove them, which is not likely; wait for the fire to go out, which is also unlikely; yell for someone to bring them a ladder, which might happen; and when all else failed, jump and hope for the best. When you are on fire, it does not matter how high it is. When you are on fire, you will jump! That very brief cooling off period as you are falling somehow makes that stop at the bottom a little more appealing.

A DAY ON THE ROPES

You and your partner are searching deep in the middle of a burning three-story duplex house fire. The whimpering of a tiny child draws you and your partner past the limits where you know it is safe. Still you search. You are drawn into a room filled with smoke so thick there is absolutely no sight of the fire, yet the heat is searing. You know your turnout gear is close to failure, and that your breathing apparatus must be nearing the 25% remaining air level that triggers your low-air alarm. Your partner, ever controlled, tells you in an uncharacteristic voice, "Let's roll out of here man; she's going to go." At that moment you stumble across a little body, almost crushing it with your knee. Simultaneously your partner says, "Man, we're losing it; the hallway's fully involved and I don't hear any cavalry coming!" You spot the faint flash of one of the pumper's red strobes through a heat cracked window. Instinctively, you grab the baby and head for the only way out of the increasing inferno. Your partner cleans the glass and framework out of the window with his Halligan Tool. He lays over the window ledge to look for a safe way to the ground. The only truck company on the scene is involved with over ladder rescues on the C/D corner, and the 24-foot extension ladder thrown on the A side is at least 12 feet east and well below your window. Your partner has caught the attention of a pumper operator who has notified a newly arriving engine company crew. The freshly aspirated air in the room has intensified the fire moving down the hallway. Flames are becoming inseparable from the dense smoke. The room temperature is soaring.

Your low-air alarms are going crazy now, the ceiling temperature is approaching 1200°F, and for you and your partner, all else has failed! There are no other options than the window, and you have maybe sixty seconds before you, your partner, and the baby are toast.

Fortunately, your department has given you the equipment and training necessary to do that once in a lifetime, low-incident/high-risk emergency escape.

You hand your little victim to your partner, who drops the baby gently to the waiting hands of the engine crew on the ground. It is about a 20-foot drop. The baby lands safely and is rushed by one of the crew to the medical sector station. You and your partner have only three choices now: rope escape, jump, or die.

Your partner grabs the terminal end carabiner out of his PEERS kit and snaps it onto his Halligan Tool. An arm's length away, he four-wraps the carabiner on his truckee belt. Two seconds later, carefully holding tension on his escape rope and pseudo-anchor, he is out the window in an ugly but soothing rappel to safety. You have found a cast iron radiator and thrown a round turn around the piping. You five-wrap the carabiner on your truckee belt, straddle the ledge, and roll out the window. Seconds later you are safely on the ground and a full-volume firestorm rolls out of the window you just vacated.

HAZARDOUS ENVIRONMENTS

Increasingly, fire and rescue workers are asked to operate in potentially hazardous environments, with little or no means of egress. Searching the twentieth floor of a burning condominium for sleeping or trapped people when the fire is raging below you is similar to entering a confined space without a retrieval line. The environment is Immediately Dangerous to Life and Health (IDLH), and it has a limited means of egress and the potential for engulfment. Fire and Rescue departments go out of their way to prepare for all possible emergencies when sending teams into confined spaces and,

FIGURE 6–1

Personal Emergency Escape Rope Systems offer an alternative to jumping if firefighters become trapped in structure fires. (*Courtesy Martin Grube*)

yet, somehow tend to overlook the retrieval means for firefighters and rescuers in mid- and high-rise structures.

NFPA 1983

Standard on Fire Service Life Safety Rope and System Components

Now that NFPA 1983, *Standard on Fire Service Life Safety Rope and System Components 1995 Edition,* has gained some tenure and the 2000 edition is repeating the escape trends, fire departments are beginning to take a legitimate look at emergency escape rope systems. Traditionally, fire departments have relied entirely on ground ladders and aerial apparatus to make safe personnel who have only one way out of an untenable situation. Some departments with rope rescue or technical rescue teams have developed as an offshoot, **emergency egress** rappeling systems and techniques based on long 1/2-inch rescue ropes (Figure 6–1), steel hardware, and either pre-sewn Class II or improvised webbing harnesses. Both means of last ditch escape have proven life-saving over the years, but they also have limitations that can, in some situations, make them completely useless.

► **emergency egress**

Any means of escaping a structure, container, or vessel under emergency conditions.

Aerial apparatus are limited primarily by their length, capacity, and horizontal approach to a building. Most 100-foot-long aerial apparatus extend only to the sixth or seventh floor (Figure 6–2) and considerably lower if there is a parking garage or other structural element jutting out the side of the building (Figure 6–3). Ground ladders, again severely limited by length, also have people power considerations. It takes at least two and sometimes four or five people to raise a 35- to 50-foot ground or Bangor ladder (Figure 6–4).

Long half-inch rope rappel systems with Class II pre-sewn or improvised webbing harnesses, a DCD, and a couple of carabiners provide a reasonable margin of emergency escape capability. They are of little value, though, unless they are carried with the crew. Unless the equipment is literally attached to the firefighters or rescuers themselves, there is little hope that it can be found, donned, and rigged when last-ditch egress becomes necessary. Usually, the 22.5-pound,

FIGURE 6–2

Even modern ultralong aerial ladders are too short to reach many areas where firefighters can get trapped. (*Courtesy Martin Grube*)

FIGURE 6–3

Protruding building elements, such as parking garages, can limit access. (*Courtesy Martin Grube*)

300-foot 1/2-inch diameter rope and bag is left in the stairwell with the standpipe connections, and donning a Class II harness or tying on an improvised webbing harness, such as a Swiss seat or other adaptations, in a life-threatening emergency is complicated. This is all compounded by the fact that emergency egress is maybe a once in a lifetime event, and firefighters often do not keep their skills highly polished on *maybe once in a lifetime* events. In addition, because

FIGURE 

Fully extended aerial ladders are still a long way away on most high-rise buildings. (*Courtesy Martin Grube*)

firefighters get very good at running fires, since they do them frequently, management often does not put a high emphasis on emergency escape practices due to insignificant statistical support. Therefore, they often do not mandate routine practice. Obviously, firefighters are good at the tasks they perform frequently through actual responses or mandated training, and they are not as good as those they perform infrequently. Ultimately, firefighters' survival profile is again compromised by the hazard curve paradox.

PERSONAL EMERGENCY ESCAPE

► **Personal emergency escape rope system (PEERS)**
A combination of equipment and training that gives trapped firefighters a last ditch means of egress after all other means have been considered.

A **personal emergency escape rope system (PEERS)** has to follow three rules in order to be successful. These rules will be refered to throughout this chapter. First, there must be a commitment by management to provide for the safety of personnel assigned to high-hazard operations by providing continuous training and every possible means of escape. Second, a PEERS equipment inventory must be purchased and maintained. Third, we must put teeth into policy that mandates that PEERS equipment become part of the personal protective ensemble for personnel operating in structures exceeding 30 feet in elevation.

General Strategy

Keep emergency escape and rope rescue concepts mentally separate. They are two different kinds of tools used for two different kinds of jobs. The idea seems simple enough—"I'm trapped, and I'm going to rappel out of this window." With the PEER system, the idea is to rappel just as far as necessary to escape harm. The PEER equipment will allow you to go down only a floor or two, or sometimes a bit more when using pseudo-anchors, to escape harm in mid- and high-rise structures. In house fires, it may allow you to safely descend to the ground, depending on the availability of anchors inside the house.

FIGURE 6–5

(A) A PEERS Pak consists of carabiners, escape cordage, a descent control device, and deployment bags. (B) Personal Escape Devices (PEDs) are being manufactured in many different configurations. They are light, small, and strong enough to handle rigorous fire service needs. (C) The Christmas tree PED has several configurations that vary friction. (D) A carabiner can be wrapped for emergency egress. (E) A ladder truck belt and hook can be used for emergency egress. (*A–C, Courtesy Martin Grube; D–E, Courtesy Kurt A. Southall*)

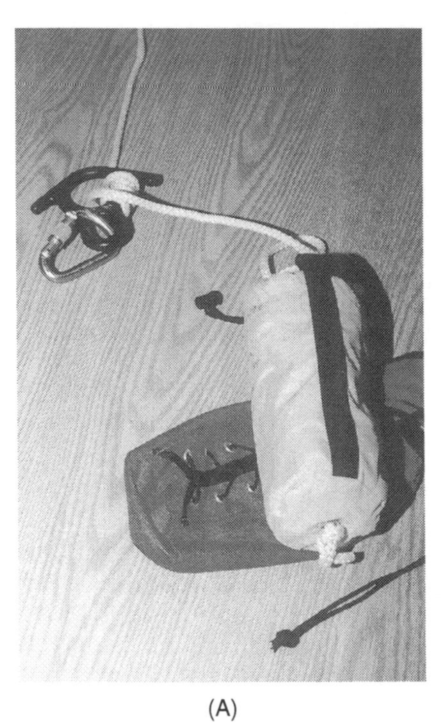

(A)

(C)

(D)

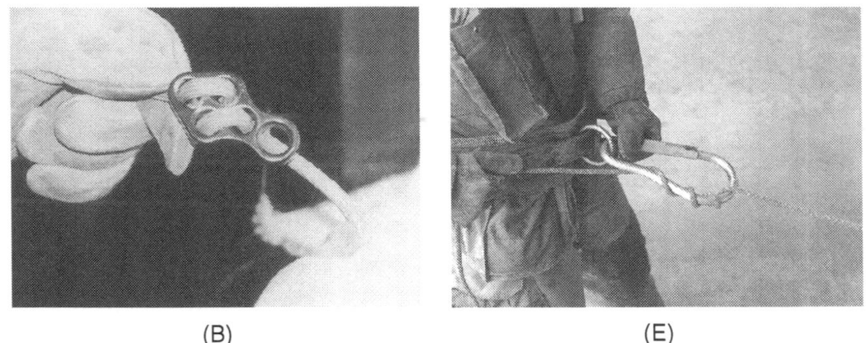

(B) (E)

System Components

There are several personal emergency escape concepts that meet the above criteria and satisfy the requirements of the policy example listed at the end of this chapter. The PEER system example in this book has been extensively researched and tested to be easy to learn, easy to teach, relatively inexpensive, and reliably safe.

PEERS components (Figure 6–5A) should include the following items:

1. 50- or 75-foot length of personal emergency escape rope with a minimum breaking strength of 3,000 pounds
2. Life safety harness, ladder/escape belt, or escape belt
3. Two large aluminum carabiners
4. A DCD: an eight-plate, a manufactured personal escape device, a descent control carabiner, or a life safety belt hook
5. PEERS protective storage bags (bag in a bag)
6. 18-inch tubular webbing edge protection

FIGURE 6–5 continued

(F) A properly wrapped carabiner causes the rope to travel on the spine side of the carabiner away from the gate. (G) An improperly wrapped carabiner can unlock and open while on rappel. (H) All improvised webbing harnesses must be preloaded and tested in a safe place before loading. (I) A firefighter's uniform or last chance emergency egress belt. (*F–I, Courtesy Kurt A. Southall*)

(F)

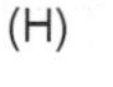

(H)

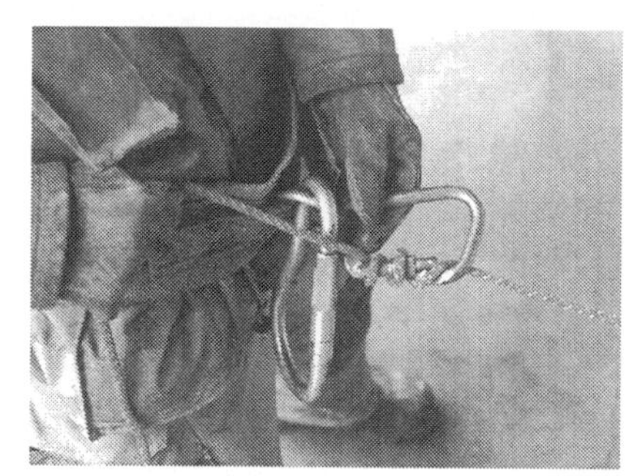

(G)

(I)

NFPA 1983

Standard on Fire Service Life Safety Rope and System Components

The PEER system needs some mode of transportation. Rope rescuers use nylon or polyester rope as their highway to safety.

In order to transport enough rope on your person to always have it available, it needs to be light. That means it must be shorter and narrower than rope rescue system components. NFPA 1983-95 edition (Section 4-2.1) states that a "rope designated as a personal escape rope shall be designed to have a maximum working load of at least 300 pounds force (1.34 kN)," and Section 4-2.1.1 allows for a 10:1 maximum working load. Translated, it means that cordage, or rope, used for personal emergency escape must have a minimum breaking strength of at least 3,000 lbf. Some manufacturers are now listing their 8-mm, or 5/16-inch, diameter cordage at a little more than 3,000 lbf minimum breaking strength. Most 8-mm cordage on the market today is a little thick, with maybe an extra strand of nylon or polyester in the core, making it really about 8.2 or 8.3 mm in diameter. It is safer to use 9-mm, a little less than 3/8-inch, diameter cordage and appreciate the little bit of extra nylon or polyester and breaking strength. Convincing people of the safety of 8- or 9-mm rope for practice or use might prove to be quite a feat. Expert example, patience, and an ever-present top belay of thick (1/2-inch diameter) life safety rope will help considerably during training.

Rope length is variable, with 50 feet being the minimum and 100 feet being a little long and heavy. Seventy-five feet seems to be a good compromise for both weight and function. Seventy-five feet of 9-mm cordage weighs 3.88 pounds and packs nicely into a bag about the size of a large salami. It is long enough to attach to almost any improvised anchor in the average room and still have some rope left for the escape rappel.

Next, an emergency escape rope system needs a method of generating friction with the rope in an amount equal to or greater than the force generated by a body falling to the earth. There are several dedicated personal emergency escape DCDs available that are very light and strong and work well when used in accordance with manufacturer's instructions (Figure 6–5B and Figure 6–5C). Two of the most common and proven descent control tools are the eight-plate and the carabiner, or life safety belt hook, wrap. The eight-plate DCD was discussed in detail in Chapter 3. As a reminder, with 8- or 9-mm cordage rappels there is much less friction, or surface area of rope, contacting the DCD. One way of increasing the friction and preventing the cordage from coming off the eight-plate is to wrap it in a nontraditional manner. That is, make several round turns around the top O-ring part of the eight instead of reeving it as if rappeling on 1/2-inch rope. Use three or four wraps, rather than the one which is used when doing 1/2-inch diameter rope rappels.

The carabiner and belt hook wraps seem like a step backward to the old days before good descenders were readily available and, in a way, they are a step back (Figure 6–5D and Figure 6–5E). Referring to personal emergency escape rule number three, a large locking belt hook should already be a part of every rope rescuer's ensemble. This means a DCD is always present and available. For engine people or rescuers, a locking carabiner is a simple, light, and inexpensive addition to their personal protective equipment.

Properly wrapping a locking carabiner or locking belt hook involves wrapping the spine side so that the standing part of the cordage comes off toward the anchor, away from the locking gate. The running part of the cordage also rolls away from the locking gate, as shown in Figure 6–5F.

Improperly wrapping the hook of the carabiner so the gate becomes loaded can cause the lock to spin to the unlocked position, allowing the gate to open enough for the rope to separate from the DCD (Figure 6–5G).

Long missing in the perfect emergency escape plan has been the means of attaching the DCD to the escapee. While there has been some progress, turnout gear and protective ensemble gear manufacturers have been reluctant to incorporate harnesses. The perfect solution would be a removable Class II harness *inside* the vapor barrier of the pants that could be inspected, cleaned, and replaced as often as necessary. The harness tightening would have to be a function of closing the pants so it was not time-consuming to don the pants every time you went out the door. Unfortunately, this has not yet become reality. Finally, worries about, to date, NFPA compliance and third-party testing costs combined with a generally lethargic market for such a combination and has left an unfortunate and dangerous gap in the emergency escape formula.

This leaves only a couple of viable options: the pre-sewn and improvised harnesses, and the life safety belt or escape belt. The harnesses are described in detail in Chapter 3 for rope rescue operations. Their limitation for emergency escape is simply the time it takes to don them safely. This, in itself, is a violation of rule number three. If it must be donned in an emergency, it is likely to fail. Improvised webbing harnesses are a better choice, since firefighters and rescuers may tend to keep a light piece of 20-foot-long, 2-inch tubular webbing in their pockets. However, firefighters and rescuers then run into proficiency problems due to a lack of practice, and if the webbing comes untied as they begin to rappel, the obvious result is, again, unacceptable. *Never use an improvised webbing harness until it has been properly checked by another person, and it has been pre-loaded in a safe place.* All this is going to be very hard to do when life safety is threatened.

By strictly adhering to the three rules, as listed on page 185, and until someone finally incorporates a good egress harness into the bunker gear, the most promising attachment device for the escapee is the ladder/escape belt. It is defined by NFPA 1983 (Section 1-3) as, *A belt that is certified as compliant with the applicable requirements of this standard for both a ladder belt and an escape belt, and that is intended as a positioning device for a person on a ladder as well as for use only by the wearer as an emergency self-rescue device.*

NFPA 1983

Standard on Fire Service Life Safety Rope and System Components

One consideration is the uniform last chance belt that has been marketed since at least 1980. This uniform belt (Figure 6–6) is worn as part of the everyday station uniform and therefore is always on, except at night where getting into station pants before donning bunkers would have to become common practice. Manufacturers are already making and having uniform station belts certified as NFPA 1983-95- and 2000-edition compliant. This might be the only economical alternative at about $20 each, but the department would have to buy one for every operational person, for example, a department with 300 suppression personnel. If a department were to purchase a complete ladder/escape belt for each entry position on each apparatus, it would only need to buy 100 if there are three shifts. If management and budget constraints simply prohibit the expense of good ladder/escape belts, and the PEERS program dies because of it, using the relatively inexpensive but NFPA-compliant uniform escape belt may salvage efforts, and save some lives.

The uniform escape belt contains a D-ring attachment to which the PEERS DCD connects. Rappeling in it is grossly uncomfortable, and an unhealthy shock load can cause serious damage to internal organs. But, of course, the alternative is much worse.

Several manufacturers have designed good NFPA-compliant ladder/escape belts that are worn as part of the protective ensemble (Figure 6–7 and Figure 6–8). They wear nicely with breathing apparatus, double as a belt that holds multiple tools, and provide back

FIGURE 6–6

A firefighter's uniform (last chance) belt can be accessed through turnout gear to make connections for an emergency escape.

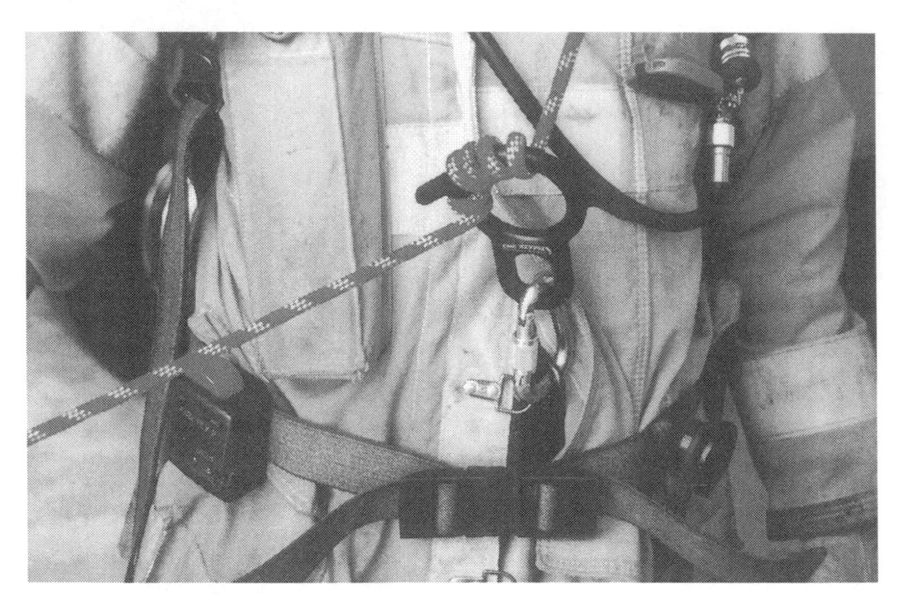

support. This is by far the most realistic attachment device, due mostly to its versatility. The belt is used every day and, therefore, the wearer develops proficiency and confidence. Most NFPA-compliant ladder/escape belts rappel in relative comfort. Some departments are issuing the less expensive uniform escape belt to a majority of their operational personnel and issuing the more expensive NFPA-compliant ladder/escape belts to the truckee positions on the ladder trucks (Figure 6–9).

Performing Personal Emergency Escape

Bailing out of a burning building in an emergency must become a matter of muscle memory—completely automatic, with no chance of forgetting the procedure.

The PEERS equipment and technique are designed to be the final alternative as an escape method after all other means of escape have been analyzed and eliminated. Emergency escape actions include but are not limited to the following:

Notify incident command and other firefighters of your condition and location.

Consider staying in place as long as possible, based on fire conditions, air supply, and structural conditions.

Consider exiting the way you came in.

Consider climbing up and away from untenable conditions.

Consider alternative stairways and fire escapes.

Consider ground ladder, Squrt™ boom, or aerial ladder escape.

Consider using 1/2-inch diameter rope as an alternative rappel rope to the emergency escape rope.

After all other means of escape have been considered and eliminated, proceed with personal emergency escape steps (Figure 6–10):

1. Again, in order for the PEER system to work effectively, the harness or ladder/escape belt must be in place on the

FIGURE 6–7

The Yates truckee belt (emergency escape harness) is a great tool belt and emergency escape belt. This model is convertible to a Class II harness with legs that are kept in the belly pouch until needed.

FIGURE 6–8

The Yates truckee belt with legs extended into the Class II position.

firefighter's body when it is needed. This facilitates staying as long as possible awaiting rescue, and it helps ensure that the harness or ladder/escape belt is on correctly, as opposed to attempting to put it on under less-than-desirable conditions. Don the ladder/escape belt or life safety harness before entering the structure.

2. The **PEERS Pak** containing the rope should be attached to a turnout gear pocket or the bottom of the back of the SCBA pack frame for engine company personnel not wearing ladder/escape belts. For personnel wearing a ladder/escape belt, the PEERS Pak may be attached to the bottom of the ladder/escape belt or the bottom of the self-contained breathing apparatus (SCBA).

► **PEERS Pak**

A double bag system containing the equipment required for emergency escape: escape cordage, DCD, two carabiners, a harness (if one is not a part of the protective ensemble), and edge protection.

FIGURE 6-9

(A) Personal escape harness and carabiner attachment. (B) "A" Arm Truckee belt rigged for rappel. (*Courtesy Martin Grube*)

(A)

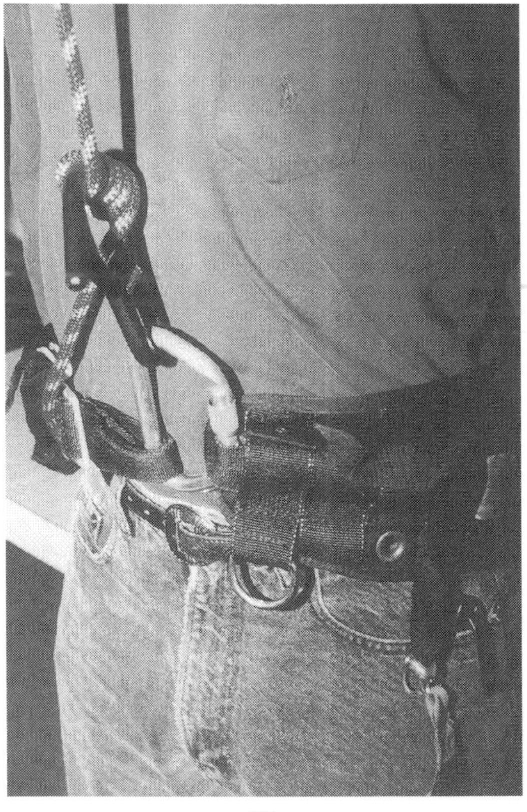

(B)

3. Locate a suitable egress location, such as a window or balcony. Clear any obstructions and smooth out any sharp edges, aluminum or steel window tracking, and glass. Make sure you have a safe floor, or landing zone, below the egress location.
4. The standing part, or anchor side, of the emergency escape rope has a permanent figure eight on a bight knot tied in the end and is attached to a carabiner by several round turns in

FIGURE 6–10

(A) PEERS sequence: Choose the anchor, deploy the rope, pad the edge. (B) Roll out, making sure not to shock load the rope. (C) Maintain the roll out until clear of the window. (D) Continue or smooth roll out. (E) Orient yourself as time allows. (F) Rappel to safety. (*Courtesy Martin Grube*)

(A) (D)

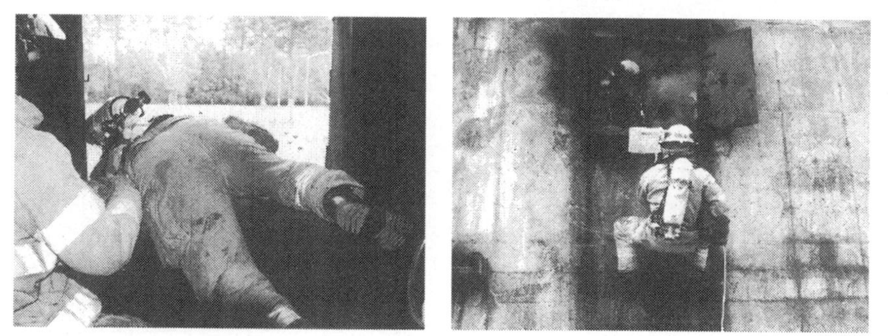

(B) (E)

(C) (F)

the bight (Figure 6–11A). This carabiner will stay in the bight, so the rope can be wrapped around a suitable anchor and connected back to the rope or connected directly to an anchor via the carabiner. Locate a suitable anchor as close to the egress site as possible. The most likely close anchor point will be a structural component (Figure 6–11B). A large enough hole can be made in most walls to wrap structural components (Figure 6–11C). *Pad all edges with linens, pillows, or any other soft material.* Standpipes and iron sprinkler piping may also be good anchor sites. If time permits and the equipment is available, consider using 1/2-inch diameter R/Q rope as a safer alternative to the emergency escape rope. The longer rope can be attached to a bombproof anchor in the stairwell or other remote site. If necessary, the emergency escape rope can then be attached to the 1/2-inch diameter R/Q rope.

FIGURE 6–11

(A) Round turns on anchor carabiner help keep the cordage in place. (B) Anchor site to bombproof structural component. (C) Holes in walls can make good anchor sites. (*Courtesy Martin Grube*)

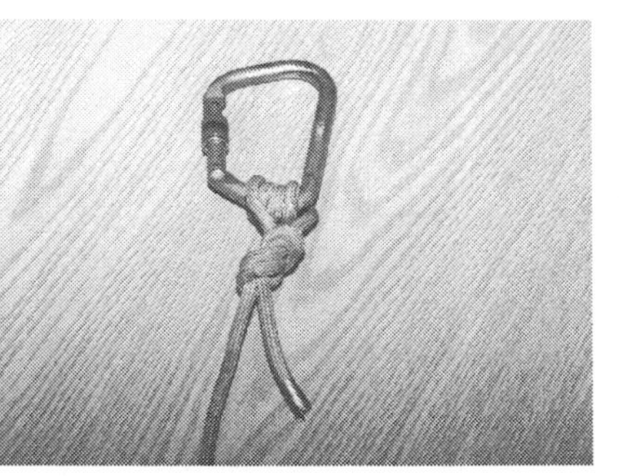

(A)

(B)

(C)

FIGURE 6–12

(A) Halligan tool used as a pseudo-anchor. Note half-inch R/Q lifeline used as safety belay. (B) Care must be taken to hold the pseudo-anchor in place until the full load is placed on it. (C) Pseudo-anchor holding rappeler. Note safety person spotting halligan tool and belay rope. (*Courtesy Martin Grube*)

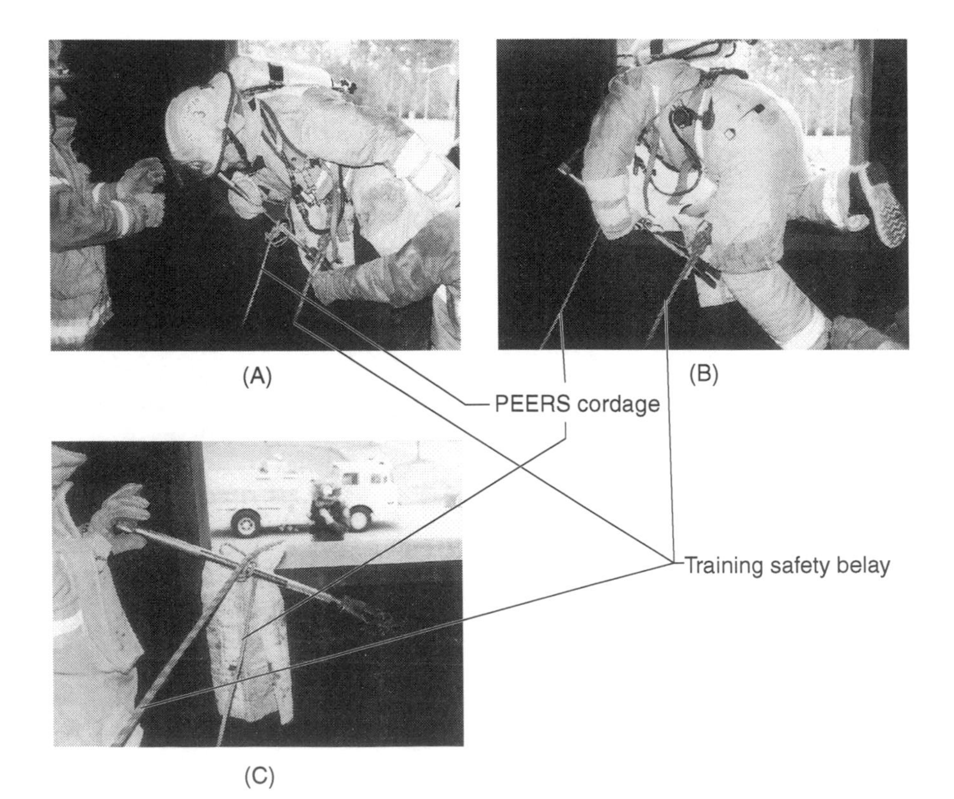

3a. Under extreme conditions, pseudo-anchors may be the anchors of choice (Figure 6–12A). Attach the end of the PEERS rope to the pseudo-anchor by connecting to it directly with the carabiner. Or, wrap several round turns of the PEERS rope around the pseudo-anchor and connect the anchor carabiner back onto the PEERS rope itself.

4. Make sure the DCD is properly reeved to the rope and that the DCD is attached to the life safety harness or ladder/escape belt using the provided carabiner.
5. Load the rope as slowly as possible and rappel to the next floor or the next safe, accessible place. If a pseudo-anchor is being used, make sure tension is applied continuously throughout the loading process to hold the anchor in place (Figure 6–12B and Figure 6–12C). Be sure the DCD clears the edge to prevent it from hanging up and preventing safe descent.
6. There is a permanent figure eight on a bight and carabiner tied into the end of the rope to prevent you from sliding off the end if the rope does not make it to a safe place. The carabiner can be locked onto the harness or belt while awaiting rescue. If a safe place is made, off-load the rope and proceed to safety. Notify command of your situation as soon as possible.

NFPA 1983

Standard on Fire Service Life Safety Rope and System Components

Mayberry Fire Department
Sample Policy

Personal Emergency Escape Rope System (PEERS)

Purpose

To provide guidelines for the use, maintenance, and storage of the personal emergency escape rope system components.

Scope

Applies to all uniform personnel.

Definitions

Personal Escape Rope—A single-purpose, one-person, one-time-use emergency self-escape (self-rescue) rope; not classified as a life safety rope.

Category I Personal Emergency Escape Rope—Unused personal escape rope, less than two years old, with an updated inspection history card on file, passing inspection according to manufacturer's recommended procedures and otherwise conforming to all requirements of this standard and NFPA 1983 (current edition).

Category II Personal Emergency Escape Rope—Personal emergency escape rope, less than five years old with an updated inspection history card on file, passing inspection according to manufacturer's recommended procedures and otherwise conforming to all requirements of this standard and NFPA 1983 (current edition). Any given pic of the sheath along its length must have 50% or less separation of fibers (Penberthy inspection technique). May be used for the purposes of emergency escape training.

Category III Escape Rope—Emergency escape rope, greater than five years old or otherwise fails to meet the requirements of Category I or Category II Personal Emergency Escape Rope. Category III Escape Rope shall have its inspection history card attached to it and be sent to Support Services for accountability purposes.

Descent Control Device (DCD)—A steel or aluminum device that attaches to a load (rappeler's harness or belt) reeved to a fixed rope, whose interface causes enough friction to control the descent of the load.

Ladder/Escape Belt—A fabric configured device that fastens around the waist and is intended for use both as a positioning device for a person on a ladder and as an emergency escape device for the wearer only.

Life Safety Harness—An arrangement of materials secured around the body used to support a person during rescue or emergency escape situations.

Penberthy Inspection Technique—A conservative rope inspection method where every individual pic of the sheath of the rope is examined for the separation of rope fibers. A pic is a single square sample of the weave of the sheath of a rope. On a new rope, 100% of all pic fibers are connected. Wear on the rope causes increased separation of pic fibers, which is an excellent indicator of rope performance characteristics.

Pseudo-anchor—Anchor sites that should be considered last-chance anchors. Pseudo-anchors are not bombproof structural components, but instead are tools and other sturdy devices that can be wedged or otherwise jammed into position to allow for temporary, one-use-only emergency escape rappels. Pseudo-anchors have the advantage of being movable, allowing for re-placement as fire conditions dictate. Pseudo-anchors allow more rope to be extended out the window, and they keep anchored rope farther from the effects of fire.

Inspection and Care

Software

1. All personal escape rope shall be inspected in accordance with the manufacturer's recommendations. An identification marker shall be attached to every life safety and personal escape rope. The marker will reference an inspection and documentation card (usage and history reference) that is stored in the office where the rope is issued. All life safety and personal escape rope should be stored in the bag provided, in a chemical-free, rust-free, well-ventilated compartment. Keep all nylon products out of direct sunlight and avoid extended exposure to fluorescent lights.
2. All personal emergency escape rope shall be physically and visually inspected on at least a quarterly basis (once every three months). The inspection and any observations shall be recorded on the documentation card. Upon inspection, if a rope fails to meet the higher category inspection criteria, the new category classification of the rope shall be marked on the next available line of the documentation card with a black marker (Figure 6–13). (Example: *This rope has been downgraded to category II.*)
3. Personal escape rope shall be physically and visually inspected on at least a monthly basis. The inspection and any observations shall be recorded on the documentation card. Upon inspection, if a personal emergency escape rope fails to meet the manufacturer's inspection criteria or has any sheath pic

FIGURE 6-13

Rope inspection card. *(Courtesy Kurt A. Southall)*

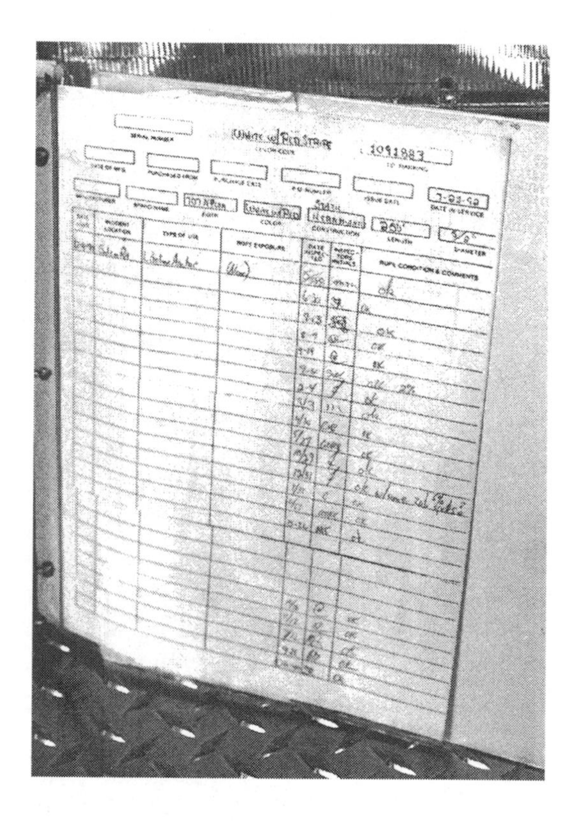

exceeding 50% wear, or there is any question as to the integrity of the rope, it shall be removed from service. The documentation card shall be appropriately marked and attached to the rope and sent immediately to Resource Management.

4. All life safety personal escape rope, harnesses, belts, and storage bags shall be kept clean and free from dust, dirt, rust, chemicals, or other contaminants that could damage the synthetic fibers. Most dirt can be brushed away from nylon; however, extremely dirty software can be cleaned by flushing it with water and air drying it out of direct contact with sunlight. Software that has suffered severe exposure to contaminants shall be tagged with the appropriate information, documented on the inspection history card, and sent to Resource Management for inspection and evaluation.

Hardware

Carabiners, eight-plates, racks, pulleys, rigging plates, swivels, and other hardware issued by the Mayberry Fire Department will be marked by Resource Management prior to issue. Each piece of equipment will have an inventory number and shall become part of the apparatus inventory. Station color shall be marked in a nonabraded, nonmoving location (i.e., do not paint the carabiner threads or pivot pin).

Hardware dropped more than 20 feet or otherwise damaged must be sent to Resource Management for evaluation. All hardware must be inspected and cleaned as necessary on a quarterly basis.

NOTE: PEERS certification is highly recommended where firefighting activities may occur outside the reach of ground or aerial ladders.

PEERS

Any firefighter entering any building, ship, or other structure that exceeds three stories or thirty feet in height or otherwise working outside the reach of ground or aerial ladders shall don and carry the PEERS

NOTE: The PEERS Pak includes the rope, protective bags, carabiners, and DCDs. The PEERS equipment includes the addition of the life safety harness or ladder/escape belt.

equipment. There will be one complete PEER System for each firefighter position on each engine and ladder truck.

PEERS Components:

1. Seventy-five-foot length of personal emergency escape rope, with a minimum tensile strength of 3,000 pounds
2. Life safety harness, ladder/escape belt, or escape belt
3. Two large aluminum carabiners
4. Descent Control Device (DCD): an eight-plate, a manufactured personal escape device, a descent control carabiner, or a life safety belt hook
5. PEERS protective storage bags (bag in a bag)

Procedures for using the PEERS equipment:
The PEERS equipment is designed to be the final alternative escape method after all other means of escape have been analyzed. Emergency escape actions include but are not limited to the following:

Notify incident command and other firefighters of your condition and location.

Consider staying in place as long as possible, based on fire conditions, air supply, and structural conditions.

Consider exiting the way you came in.

Consider climbing up and away from untenable conditions.

Consider alternative stairways and fire escapes.

Consider ground ladder, Squrt™ boom or aerial ladder escape.

Consider using 1/2-inch diameter rope as an alternative rappel rope to the emergency escape rope.

Once all other possible means of escape have been eliminated, consider escape using the PEERS equipment. *The PEERS equipment is designed to allow you to rappel ONE OR TWO FLOORS ONLY to escape absolutely untenable conditions.*

1. In order for the PEER system to work effectively, the harness or ladder/escape belt must be in place on the firefighter's body when it is needed. This facilitates staying as long as possible awaiting rescue, and it helps ensure that the harness or ladder/escape belt is on correctly, as opposed to attempting to put it on under less-than-desirable conditions. Don the ladder/escape belt or life safety harness before entering the structure.
2. The PEERS Pak shall be stored in a turn-out gear pocket or attached to the bottom of the SCBA for engine company personnel not wearing ladder/escape belt. For personnel wearing a ladder/escape belt, the PEERS Pak may be attached to the bottom of the ladder/escape belt or the bottom of the SCBA.
3. Locate a suitable egress location (window or balcony). Clear any obstructions and smooth out any sharp edges, aluminum or steel window tracking, and glass. Make sure you have a safe floor, or landing zone, below the egress location.
4. The standing part of the emergency escape rope has a permanent figure eight on a bight tied in the end and is attached to a carabiner by several round turns in the bight. This carabiner will stay in the bight, so the rope can be wrapped around a

suitable anchor and connected back to the rope, or connected directly to an anchor via the carabiner. Locate a suitable anchor as close to the egress site as possible. The most likely close anchor point will be a structural component. A large enough hole can be made in most walls to wrap structural components. Pad sharp edges with linens, pillows, or any other soft material. Standpipes and iron sprinkler piping may also be good anchor sites. If time permits and the equipment is available, consider using 1/2-inch diameter R/Q rope as a safer alternative to the emergency escape rope. The longer rope can be attached to a bombproof anchor in the stairwell. If necessary, the emergency escape rope can then be attached to the 1/2-inch diameter R/Q rope.

5. Under extreme conditions, pseudo-anchors may be the anchors of choice. Attach the end of the PEERS rope to the pseudo-anchor by connecting to it directly with the carabiner, or, wrap several round turns of the PEERS rope around the pseudo-anchor and connect the anchor carabiner back onto the PEERS rope itself.
6. Make sure the DCD is reeved to the rope and that the DCD is attached to the life safety harness or ladder/escape belt using the provided carabiner.
7. Load the rope as slowly as possible and rappel to the next floor or the next safe, accessible place. If a pseudo-anchor is being used, make sure tension is applied continuously throughout the loading process to hold the anchor in place. Be sure the DCD clears the edge to prevent it from hanging up and preventing safe descent.
8. There is a permanent figure eight on a bight and carabiner tied into the end of the rope to prevent you from sliding off the end if the rope does not make it to a safe place. The carabiner can be locked onto the harness or belt while awaiting rescue. If a safe place is made, off-load the rope and proceed to safety. Notify command of your situation as soon as possible.

Use of PEERS Equipment

The operational, emergency egress, or training use of the Personal Emergency Escape Rope System, with the exception of the Mayberry Fire Department Technical Rescue Team, shall be limited to the technique and procedure as outlined in the authority having jurisdiction (AHJ) and Mayberry Fire Department Rope Rescue I, II, and III manual or as outlined in this policy. Training or research with techniques or equipment other than as defined in the AHJ and Mayberry Fire Department Rope Rescue I, II, and III manual or as outlined in this policy must be requested in writing and written permission must be granted by the Chief of Special Operations and the Chief of Training.

Use of associated equipment at the technical level (Technical Rescue Team) is outlined in the Special Operations, Rope Rescue Policy.

Equipment Complements

Engines—Each engine shall be equipped with one complete PEER system for each position on the truck.

Ladder Trucks—Each ladder shall be equipped with one complete PEER system for each position on the truck.

Battalion Office—Each Battalion Office will have, for the purposes of emergency escape rope training, (4) Category II PEERS sets. Documentation and maintenance cards will be kept up to date on each piece of software and kept on file in the Battalion Office.

Use of Personal Equipment

The use of personal rope equipment is allowed providing that all hardware meets the same manufacturers' specifications as Mayberry Fire Department issue equipment, all software meets the same manufacturer's specifications as Mayberry Fire Department issue, and a copy of the history and inspection card is kept updated and made available as requested. Any personal equipment that does not meet the same manufacturer's specifications as Mayberry Fire Department issue equipment shall be prohibited from use unless written permission has been requested from and granted by the Chief of Special Operations and the Chief of Training.

Safety Rules for Practical Training Evolutions and Demonstrations

All operational personnel shall perform and document actual emergency escape drills at least six times per year. At every drill, at least one escape shall be performed with a pseudo-anchor (including belay line), and one escape shall be performed with a bombproof anchor.

1. All practical training evolutions shall fall within the parameters of the AHJ and Mayberry Fire Department Rope Rescue I, II, and III manuals and the information included in this policy.
2. One person will be appointed lead instructor of the drill, preferably the person with the highest AHJ rope rescue certification. The lead instructor *and* ranking participating fire officer shall assume responsibility for a final GO or NO GO for every activity. The lead instructor and ranking participating fire officer shall assume the responsibility for the safety of all participants and observers and for the safety of any members of the public.
3. A pre-drill briefing will occur with *all* participants describing the goal of the evolution and the steps that are planned to accomplish the evolution. All safety procedures will be stated by the lead instructor and understood by all participants.
4. All participants will be wearing appropriate personal safety gear as specified by the lead instructor.
5. Before anyone is committed to any rope system, the lead instructor will personally inspect all components of the system, a representative load will be used to exercise the full length of the system before a human load is committed, and the lead instructor will vocally certify that the system is ready for operation and that assignments have been made by qualified individuals.
6. All PEERS training using pseudo-anchors shall have a pre-rigged top rope belay system using 1/2-inch rope in the event of anchor failure or slippage. All pseudo-anchors shall be safety tied to prevent them from falling over the edge.
7. The lead instructor will ensure the proper maintenance and record keeping of all of the equipment used for the training and the appropriate documentation for the training report.

■ REVIEW QUESTIONS

1. The NFPA requires a _____ safety to strength ratio for personal emergency escape rope.
   a. 5:1
   b. 10:1
   c. 15:1
   d. 20:1
2. Aerial apparatus are limited primarily by their _____ approach to the building.
   a. length, capacity, and horizontal
   b. weight, expense, and reach
   c. weight, manpower, and vertical
   d. hydraulics, weight, and vertical
3. Ground ladders, severely limited by length, also have _____ considerations.
   a. rated capacity
   b. heat exposure
   c. mechanical
   d. people power
4. The three rules for successful personal emergency escape systems are commitment by management to training and providing means of escape, purchase and maintenance of PEERS equipment, and _____ .
   a. equipment of all emergency response suppression personnel with NFPA-approved turn-out gear.
   b. enforcement of policy that mandates the PEERS equipment become part of the personal protective ensemble for personnel operating in structures exceeding thirty feet in elevation.
   c. use of every available means to purchase the largest and most technologically advanced aerial apparatus available.
   d. teaching of rappeling to all response personnel.
5. PEERS rope should be at least _____ long and have a minimum breaking strength of _____ pounds.
   a. 100 feet, 4,500
   b. 100 feet, 3,000
   c. 50 feet, 4,500
   d. 50 feet, 3,000
6. Descent control devices used for the PEER system may be an eight-plate, _____ , a _____ , or a life safety belt hook.
   a. a manufactured personal escape device, descent control carabiner
   b. a rappel rack, bobbin style descender
   c. a münter hitch, rappel rack
   d. a manufactured personal escape device, rappel rack

7. _____ feet of emergency escape rope is the best compromise for both function and weight.
   a. 50
   b. 75
   c. 100
   d. 250
8. Properly wrapping a locking carabiner or life safety belt hook involves wrapping the _____ so the rope comes off toward the anchor, away from the locking gate.
   a. harness
   b. anchor
   c. maximum axis
   d. spine side
9. Two problems with issuing uniform last-chance belts to personnel are that they are grossly uncomfortable to rappel on, and at night _____ .
   a. they are very hard to see
   b. are almost impossible to buckle without a hand light
   c. you may leave them in your uniform pants
   d. the DCD can get disconnected from the connecting carabiner
10. Firefighters are good at the actions they perform frequently through actual responses or mandated training, and they are not good at those actions they perform infrequently. Ultimately, their survival profile is compromised by the _____.
    a. lack of good equipment
    b. Hazard Curve Paradox
    c. hazards of their business
    d. bean counters
11. Expert example, patience, and an ever-present _____ of 1/2-inch R/Q rope will help alleviate fears considerably during training.
    a. amount
    b. 200-foot length
    c. spool
    d. top belay
12. The PEERS equipment and techniques are designed to be the _____ after all other means of escape have been analyzed and eliminated.
    a. initial escape method
    b. final alternative escape method
    c. preferred but nontraditional egress method
    d. tools of choice

13. Remember never to use an improvised webbing harness until it has been properly checked by another person and _____ in a safe place.
    a. loaded
    b. rappeled on
    c. inspected
    d. maintained
14. _____ are not bombproof structural components, but instead are tools and other sturdy devices that can be wedged or otherwise jammed into position to allow for temporary, one-use-only emergency escape rappels.
    a. Semi-rigid anchors
    b. Friction devices
    c. Pseudo-anchors
    d. Grenade-proof anchors

CHAPTER

7 Rigging and Operating Lowering Systems and Patient Litters

OBJECTIVES

Upon completion of this chapter, you should be able to:

- define the components of a two-line lowering system.
- demonstrate an understanding of force vectors.
- define the Golden Angle.
- engineer a change-of-direction component for a lowering system.
- engineer a two-line lowering system.
- engineer a hokie hitch load-releasing cordage configuration.
- explain how to pass a knot through a descent control device in a lowering system.
- demonstrate applying the Integrated Harness System (IHS) to a patient in a wire basket type (Stokes) litter.

A DAY ON THE ROPES

Carl was in severe shock. His legs were crushed and he was exhausted. We had finally extracted him from tons of crushing cricker 41, limestone rocks about the size of baseballs. Three hours earlier Carl had been trying to break up a clog in the building-sized shaker that was helping transfer cricker from the collier ship to the storehouse. When the cricker clog broke free, it sucked Carl down with it. The operator quickly shut down the shaker, but Carl was already about 25% consumed. He had been trapped up to his hips. All efforts to remove him failed.

Our team responded after routine efforts to remove Carl proved fruitless. Plan A was to dam the running cricker away from Carl using a barrel and other rounded, self-shoring forms. It failed. The cricker pieces were bigger than sand and slurry, which is where self-shoring devices show their worth. Plan B involved sucking the cricker up in huge vacuum pipes supplied by public utilities. It too, failed. The big SuperVac truck that could have handled the cricker 41 was too heavy to be lifted onto the shaker by the dock crane. The smaller unit did not have the suction required to lift size 41. Plan C worked so well that we all stood there astonished as the cricker ran away from Carl. We had our rescue technicians oxy-acetylene cut a strategically placed triangular hole below and away from Carl. Two 3-foot cuts were made into the angled wall of the shaker's 1/2-inch thick steel wall. The cuts converged to make a 90-degree angle. This was done to allow a controlled release of the cricker from the shaker. Too fast and Carl could have been pulled in two. Too slow and Carl would die from his shock and crush injuries. The angle would, at first, cause a slow release of the contents. As the rate of release was determined, the angle of the cut could be increased to allow a faster run. The two original cuts were connected with an intersecting cut leaving a triangular opening about the size of a breakfast pastry. The run was too slow. Another cut was made making the triangular opening the size of a neatly folded American flag. The cricker then ran smoothly into the Elizabeth River far below the shaker.

Carl was free from the cricker, but he was far from safety and about 120 feet above the waiting ambulance. Carl had floated effortlessly into the jury-rigged bosen's chair we had positioned under him prior to the cricker release. Big Teddy Munden was hanging out in space attached to a safety rope in the back of his Class III harness, stabilizing Carl. The surface supporting Carl and Teddy melted away from under them as the cricker continued to pour into the river. We were able to tag-line them over to a level and stabilized area near the top of the shaker for further medical evaluation and complete patient packaging. Within minutes, Dr. David Cash and a couple of the paramedics had Carl ready for his next ride.

The crane above the shaker had been secured for use as an elevated anchor. A member with a radio was staged inside the operating compartment with the operator to make sure the crane could not be activated. The crane was locked and chocked and secured from any unnecessary movement, which was *any* movement at all. At the boom, a double pulley COD was anchored, and a mainline and a belay line, one red and one blue, were passed down to the Stokes basket, Carl, and the waiting paramedic/rescue technician. From the double pulley, the red main line was reeved onto an anchored rescue rack DCD for the lower. The blue belay line was attached to tandem three-wrap 8-mm prusiks anchored with a hokie hitch load-releasing configuration. A four to one simple pulley system was piggybacked onto the mainline to raise the load up and over the sides of the shaker. A well-packaged and medicated Carl and his attending paramedic were raised up the sides of the shaker with the command of "tension." The belay line took up the slack through the prusiks as the load was raised. At the edge, the command was switched to LOWER. The piggyback was removed; the rack was fine-tuned for the combined weight of Carl, his attendant, and the equipment; and the command, "slack" was issued. The belay line was kept just slightly slack and almost parallel to the mainline. The prusiks stood ready to grab the load if necessary. Carl was on his way. Five minutes later he was being loaded into the back of the ambulance for a ride to the hospital.

▶ **lowering systems**

Rope systems engineered to use the effects of gravity to transport people and equipment from a high point to a low point.

Lowering systems (Figure 7–1) are unusually simple, comparatively fast, and the preferred rope rescue system under most circumstances. Lowering systems have the following advantages over other systems:

1. Lowers generally develop much less system tension than raises. Gravity provides the motivation, and all we do is direct the load toward the safest path to the ground, or at least to a safer place than where it started. Raise systems require much more energy (work), because we must deal with the effects of gravity, while we counteract gravity. We also must deal with the effects of whatever simple machine, or mechanical advantage system component, we have introduced into the system. Therefore, we either need more people for raise systems, or we need fewer people to work harder, which is often more dangerous and slows the whole operation down.
2. Lowers greatly reduce the possibility of shock loading a system and all of its components.
3. Lowers mandate that a majority of the team will be above the moving parts of the system. Equipment, rocks, or whatever are less likely to fall on people's heads.
4. The system and all of its operating components are laid out in a, preferably, clean and mostly horizontal configuration at the top where it can be worked and massaged into a fluid-like motion. Alterations can be made quickly, because more

FIGURE 7–1

Lowering systems take advantage of gravity to move rescuers, patients, and equipment from high places to low places. (*Courtesy Martin Grube*)

people and equipment are staged at the top rather than at the top *and* bottom, or side to side as in a highline.

5. The lower system is safer than having someone rappel to the target. Because we remove all of the transportation worries from the person(s) on the rope, they can concentrate on maneuvering, patient care and packaging.
6. There are fewer components to rig, and the systems are simpler, so people with less experience can more safely rig lowers than raises or highlines.

The basic components of a simple lowering system are the main line, belay line, anchors and attachments, DCD, brakes, edge protection, rescue workers, and a load. More involved systems require COD pulleys, load-releasing hitches, elevated anchors, swivels, and various devices to enable your system to approach fluid motion.

ENGINEERING BASIC LOWER SYSTEMS

1. The lowering operation must begin with the location of your patient. The lower system must be engineered to be at a point directly over the patient's position or over a position where rescuers can bring the patient so main and belay ropes can be attached.
2. Anchors must be located and assessed and must be bombproof and preferably in line with the patient. If they are not in line with the patient, they will need to be directed there with the help of some COD system components. Use at least two separate anchor systems—one for the main line and one for the belay line. Also be sure both anchors have a secondary or a backup.
3. Establish which rope will be used as the main line and which will be used as the belay line. They should be different colors, if possible.
4. Anchor the DCD you have chosen, preferably the R/Q closed end brake bar rack. Remember, the purpose of anchoring the DCD is to cause the anchor to carry most of the load. The small remaining amount of load that is not being carried by the anchor is the control amount that is being managed by an operator. The control amount is the difference between the load moving and the load not moving as determined by the operator.
5. Secure the belay line with tandem three-wrap (t3w) prusiks, or the nondestructing mechanical rope grab of choice. If no mechanical advantage system is being employed, build the hokie hitch load-releasing configuration and place it between the prusiks and their anchor.
6. Elevate the leading edge, if possible, to get both ropes up and off the edge and to help with the ease and safety of loading and unloading the system.

NOTE: The numbers used in the following force vector discussion are theoretical approximations. It is much easier to understand the underlying concepts by first studying the theory. In reality, there are many additional factors that can add to or subtract from theoretical figures. These include but are not limited to edge loading, friction coefficients, wind load, water weight, humidity, equipment condition, human interaction, etc. As your understanding of the basic physical concepts becomes solidified, the many more ambiguous, unseen, and unknown variables are much easier to anticipate, estimate, and calculate.

7. Perform the final safety check and ask The Question. Have someone else certify the system, pre-load the system, and make any necessary changes.

There are three important factors that have a great influence on the effectiveness of lowering systems and rope rescue systems in general. They are force vectors, change of directions (directionals), and the Golden Angle. All of the factors are interrelated and understanding their practical applications makes the difference between good riggers and great riggers.

A rope system engineer should have as accurate an understanding as possible of the forces on all of the components of a system. It is not as simple as thinking, "I have 300 pounds on the rope, so I must have 300 pounds on the anchor." There are myriad factors that contribute to the amount of force being delivered to each component of the system. In short, there are many unseen forces built into your system. Essentially, you can control the forces only if you understand what causes them and what increases and decreases them.

Force Vectors

► force vectors

Any angle or curve in a rope system that is not in a straight line between the anchor and the load. All vectors add to overall system forces.

Force vectors occur anywhere in your system that is nonlinear. Vector comes from the word *vehere* or *vectus,* meaning to carry. Specifically, it is a measurable force or mass that has direction. A force vector can then be thought of as any angle or curve in your system that is not in a direct line with the load and the anchor. All force vectors *add* to your overall system forces.

Ideally your load is simply hanging *straight* down from a fixed anchor like a plumb bob. Rarely will you get an opportunity to perform a plumb-bob-type rescue. More common are component interactions that are angular or curved, that is nonlinear. Literally hundreds of possibilities exist, including the angle created in an anchor sling when it is placed around an object, the angle of the legs in a litter bridle, the angle created when you direct a rope more in line with your target, the catenary curve caused by gravity on a rope that is not vertical and is anchored at both ends, the angles created by the pulley's mechanical advantage systems, the angle in a rope as it goes over the edge, the angles made by the parts of load-sharing anchors, the angle created by the load on a rope tensioned between two buildings, or the angle emerging from your leg straps as they connect to your main harness attachment. The list is almost endless. All of these force vectors alter the effect the load has on the anchor. Force vectors can help you, or they can hinder you.

For practical rope rescue applications, there are two distinctly different instances where force vectors influence your rigging. They are a COD and a load that is shared between two fixed anchors.

Change of Direction

► change of direction (COD)

A system component that alters the direction of the rope.

The **change of direction (COD)** (Figure 7–2A) is one of the most common and helpful tools available to the rig master. Following is a typical example. You are lowering a rescuer to the target who is

FIGURE 7–2

(A) Change-of-direction (COD) system components direct lines to where they will be most useful.
(B) The rope always seeks perpendicular to the edge. CODs can direct ropes over the top of the target and prevent see-saw type abrasion on the rope. (*A, Courtesy Martin Grube*)

(A)

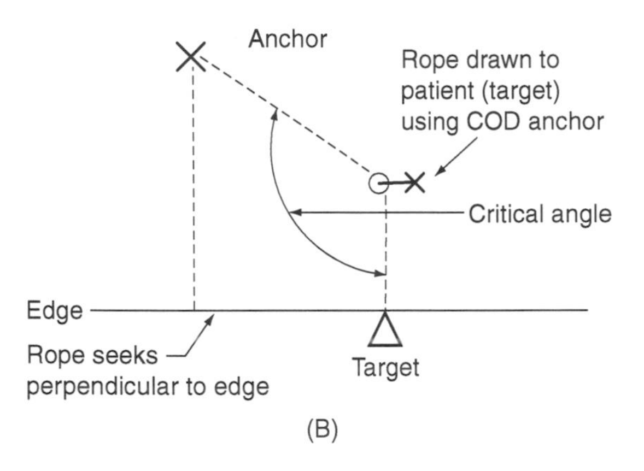

(B)

about 20 feet to one side of the target (Figure 7–2B). You use a COD to pull the rope in line with the target from above. There is an angle created in the line from the point at which the COD carabiner or pulley is attached. The angle can be expressed as a value from 1 degree (extremely narrow) to 179 degrees (very wide—almost a straight line). Analyzing the angle tells you a lot about how much force you have added to your overall system. If the COD vectors the rope 90 degrees, you have added a lot of force to your overall system, maybe even enough to break something or to pull an anchor out of the ground (Figure 7–3A). If the COD vectors the rope very little, you have only influenced the linear intentions of the line a little and have added little additional stress to your overall system (Figure 7–3B). The bottom line is that the COD force vector *always* increases the overall tension in the system and is directly proportional to the angle created (Figure 7–3C).

NOTE: A good rule of thumb for CODs is the wider the angle created by the force vector, the less the force added to the system. Conversely, the narrower the angle created by the force vector, the greater the force that is added to the system.

The other situation in which force vectors must be understood by the rig master is when a rope, or any system component, has been anchored at two ends (Figure 7–4), and a load (force) is directed between the two anchors, creating an angle of force. Not only is the load exhibiting a force on the two anchors, but the two anchors are exhibiting a force on each other. For example, two trees

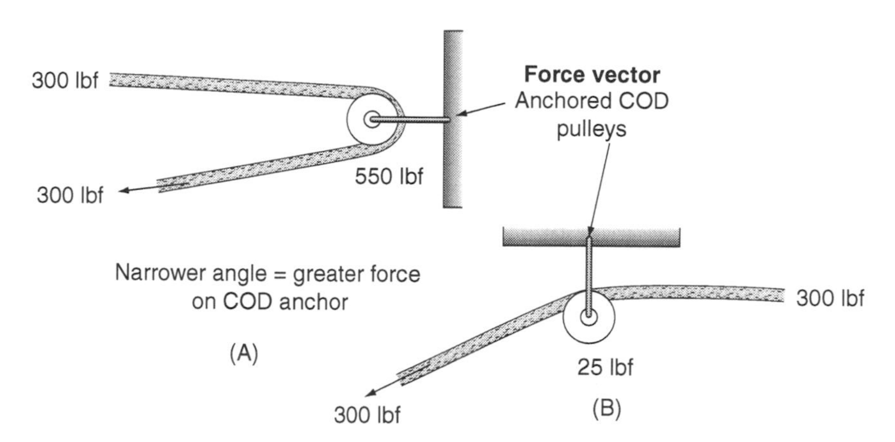

FIGURE 7-3

(A) Narrow angle CODs can double the forces on the COD anchor. (B) Wider angle CODs have less force on them than comparatively narrower angle CODs. (C) A dynamometer can be used to measure COD forces. (*C, Courtesy Martin Grube*)

FIGURE 7-4

System components anchored at two ends develop forces on each anchor that increase exponentially as the angle gets wider.

NOTE: In shared anchor systems, the narrower the angle created by the force vector, the less the force is on the anchors, conversely, the wider the angle created by the force vector, the greater the force on the anchors.

are being used as anchors, but they are not in line with the target. In fact, they are 20 feet apart and a line drawn between them would be parallel to the edge. You would like to tie a section of rigging rope between the trees and attach your main line to it. As tension is applied to the main line, an angle is created in the rigging rope. Analysis of this angle is critically important. Again, the COD force vector *always* increases the overall tension in the system in direct proportion to the angle created. But in this case, the phenomenon is just the opposite of the COD force vector issue.

Golden Angle

Golden Angle

A force vector with the angle of 120 degrees, or one-third of a circle (360 degrees). The Golden Angle effectively triangulates system forces.

Understanding the **Golden Angle** helps to bring the whole force vector issue into focus. Imagine a circle drawn on a piece of paper. Starting with a point (dot) made exactly in the middle of the circle, draw straight lines radiating toward the circle edge (Figure 7–5) to divide the circle into three equal sections. A complete circle has 360 degrees. The three angles created in this circle are 120 degrees each. Some people call this the Golden Angle, and it is very important to architects, mathematicians, engineers, and especially rope rescue rig masters.

Material that is formed into an equilateral 120-degree angle duplicates whatever the load is at the angle and transfers it onto both legs. Instead of the load being shared equally with the legs of the angle each bearing half the load, at 120 degrees the legs *each* carry the same force as the point at the center point of the angle. Each anchor still supports half the load, but they also pull against each other with a force equal to half the load. This means that as rope system engineers build angles into otherwise linear systems, forces are multiplied sometimes dramatically. It is vitally important to recognize the effects that angles, or force vectors, have on our rope systems. The Golden Angle is always a good starting point because ropes, slings, or even strings of carabiners all bear the same force when loaded in the shape of 120-degree equilateral triangles.

Practically speaking, it helps to refer back to the rule of thumb for COD components. The wider the angle created by the force vector, the less the force added to the system. Conversely, the narrower the angle created by the force vector, the greater the force added to the system. With the Golden Angle in mind, assume that you have built a rappel system that is simply a single rope anchored to a tree some distance from the edge. Because this edge has many sharp rocks, you have decided to run the rope up and through a pulley that is attached to a large branch of a tree near the edge, about 12 feet high (Figure 7–6). This arrangement gives you a great leading edge and cleans up any edge protection problems you had. Consider the angle created by the COD pulley in the tree. The pulley would be the center point of the angle with one leg of the angle run-

FIGURE 7–5

A circle can be divided into three equal 120-degree angles.

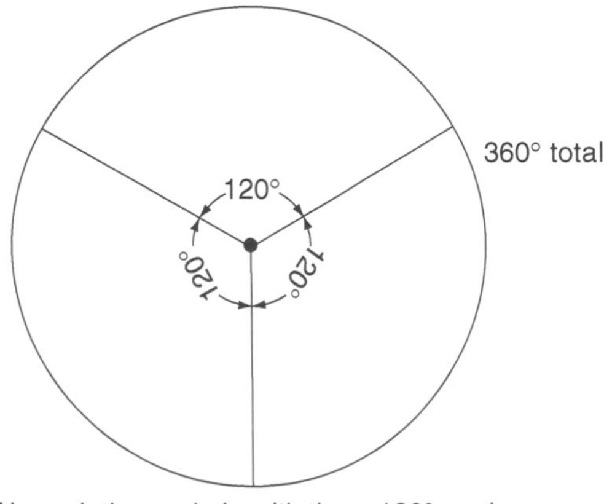

Triangulating a circle with three 120° angles

FIGURE 

The leading edge should be elevated whenever possible. Considerations should be made to keep the COD angle as wide as possible.

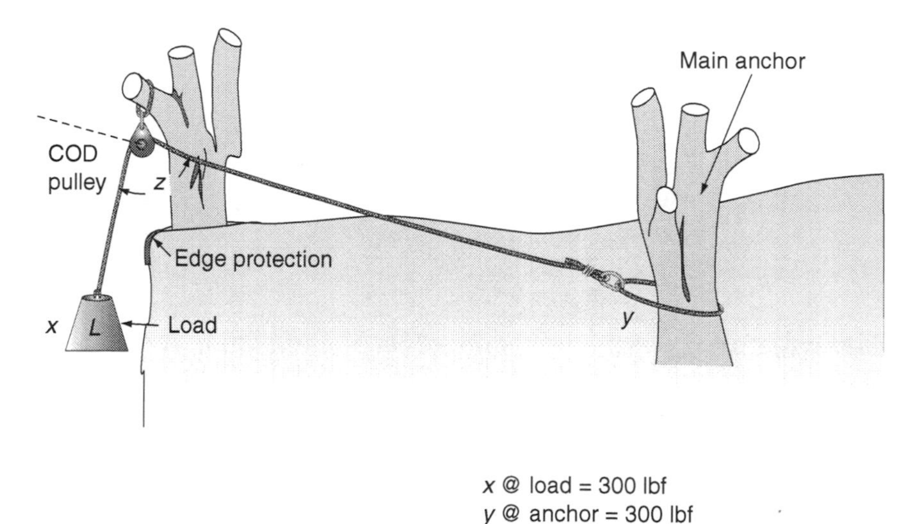

x @ load = 300 lbf
y @ anchor = 300 lbf
z @ COD angle of 85° = 450 lbf

ning to the anchor, and the other leg running down the face of the rock wall. Assume you have placed a 300-pound person on the rope. You would think that you had transfered all 300 pounds to the anchor, but the pulley *must* be bearing some of the weight. It is easy to estimate the load on the tree if you make an accurate reading on the angle in the rope, and you know the weight of the load on the rope. If the angle is 120 degrees, you have equalized the system forces and you have 300 lbf at the load, 300 lbf at the anchor, and 300 lbf at the pulley in the tree. If you have a wider angle, such as 150 degrees, you have 300 lbf at the load, 300 lbf at the anchor and only 150 lbf on the COD pulley. Conversely, if you have a narrower angle than 120 degrees, the load on the COD pulley becomes *greater* than the load itself. For example, if the angle created in the COD pulley is only 85 degrees, the load is 300 lbf, the anchor has 300 lbf, and the force at the COD is almost *450 lbf.* At zero degrees, meaning the ropes must now be parallel, the rope has taken a 180-degree turn into and out of the pulley, and the force is doubled on the COD. So, if the angle is zero or even a couple of degrees, the rope going to the load is almost parallel to the rope going to the anchor. The load exerts 300 lbf, the anchor is receiving 300 lbf, and the COD is receiving *600 lbf.* You have doubled the forces in your system.

NOTE: In single COD force vectors, the most force you can generate at the point of the angle is two times the load. Do not confuse this with shared-anchor force vectors, where the load on the anchors increases exponentially as the angle gets wider and more and more force is created by the counteraction of the competing anchors.

You have made a mechanical *dis*advantage system of sorts and, if you have chosen your tree branch COD haphazardly, catastrophic failure could be the result.

The Golden Angle stays the same for the reverse phenomenon of shared anchor systems. At 120 degrees angle between the anchors, *each* anchor bears half the force of the load and half the force pulling from the other anchor. With 300 lbf between the anchors, each anchor receives 150 lbf from the load and 150 lbf from the other anchor for a total of 300 lbf.

As an example, again imagine that you are using two trees as anchors (Figure 7–7A), and they are not in line with your target. They are 20 feet apart, and a line drawn from tree to tree would be parallel to the edge. You tie a section of rigging rope between the

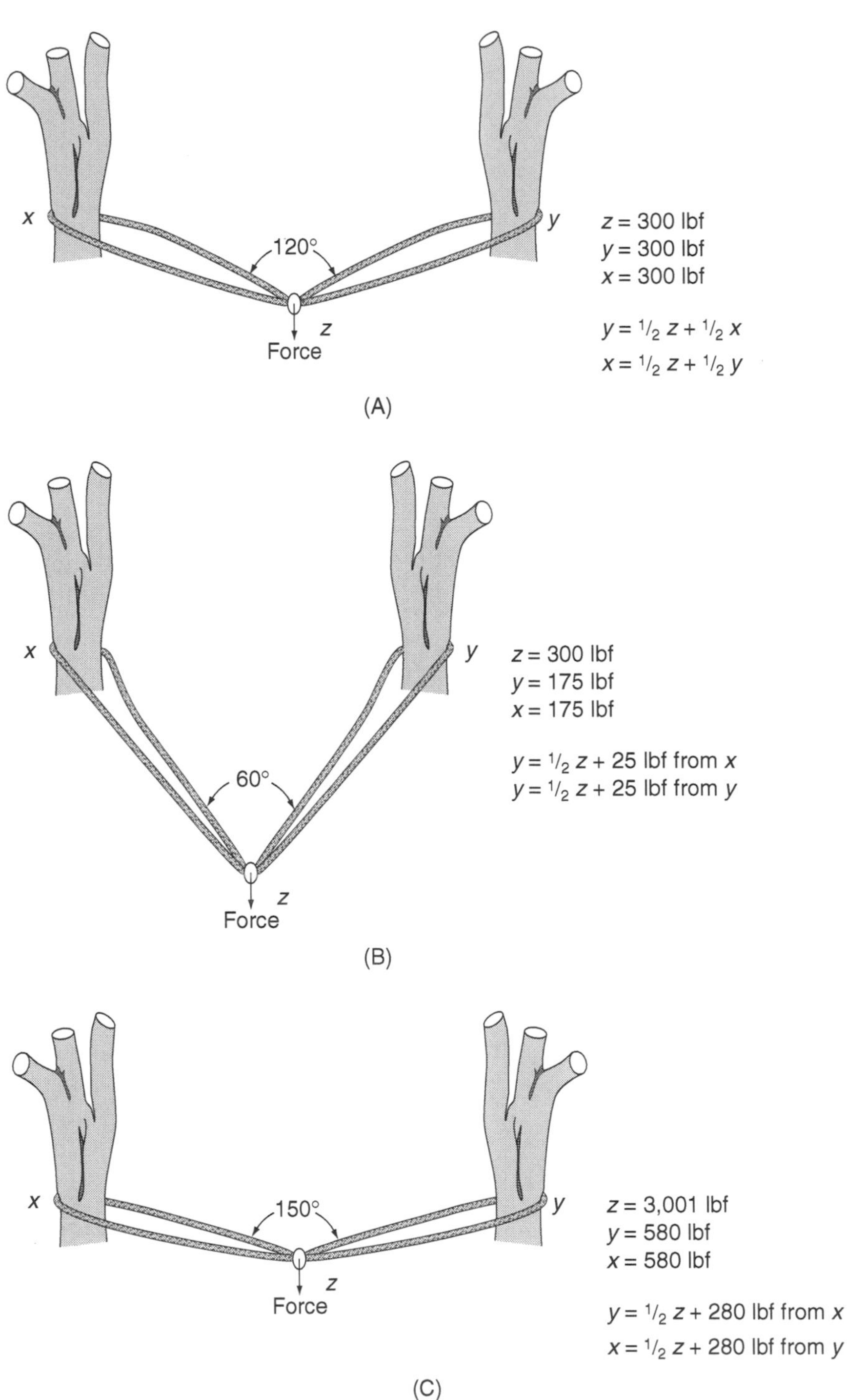

FIGURE 7-7

(A) Load sharing anchors should be as narrow as possible. Each anchor receives force from the load and some force from the opposing anchor, which is dependent on the force vector angle. (B) The narrower the angle, the lower the force vector. Narrower angles exert less force on the adjacent anchor.
(C) The wider the angle, the greater the force vector. Wider angles increase the force exerted on each anchor considerably.

trees and attach your main line to it. As tension is applied to the main line, an angle is created in the rigging rope. If the main line force is 300 lbf, and the angle in the rope is 120 degrees, the force on the trees is 300 lbf *each*. Remember, the rule of thumb for shared anchors states that the narrower the angle created by the force vector, the less the force is on the anchors. Conversely, the wider the angle created by the force vector, the greater the force on the anchors.

If the angle created in the rigging rope is 150 degrees, the force on each anchor would be 580 lbf, which is *almost double the load of the main line, on each anchor* (Figure 7–7B). As the angle of the rigging rope is made wider, approaching 170 degrees, the load on the anchors becomes almost astronomical; 300 lbf would create more than 2,000 lbf on each anchor. The force applied by the load is still only 150 lbf to each anchor. The remaining force is generated by the angle causing the anchors to pull against each other with at least 1,850 lbf.

Conversely, with shared anchor systems, the narrower the angle, the less each anchor pulls against the other anchor, and component forces are greatly reduced (Figure 7–7C). If the angle created by the load is relatively narrow, force on the anchors will also be relatively low. At 60 degrees, the load on each anchor is only 173 lbf. At 30 degrees it is 155 lbf, and when the two ropes are parallel, they share exactly half the load, or 150 lbf. In effect, when the ropes are parallel, 0-degree angle, the anchors do not influence each other at all.

The practical application for this force vector discussion is that every rig master should have a reasonable understanding of the forces he is creating in his rope rescue systems. Fortunately R/Q equipment has built-in strength and design characteristics that compensate for systems that are not technically well engineered. Nylon and polyester stretch, allowing a certain margin of error that is necessary with any system requiring human interaction. The bottom line is do not let force vectors frighten you away from building your rope systems. Shoot for 120 degrees or wider in COD anchor attachments if you really need to change direction and if any of the following apply:

1. you are worried about a questionable, or flimsy, anchor
2. you are anticipating a very heavy load of 900 lbf or greater
3. you think a shock load could befall your system
4. your equipment is less than R/Q for general use and you have no other options

On all shared-anchor force vectors, shoot for 120 degrees or narrower for exactly the same reasons:

1. you are worried about a questionable or flimsy, anchor
2. you are anticipating a very heavy load of 900 lbf or greater
3. you think a shock load could befall your system
4. your equipment is less than R/Q for general use and you have no other options

NOTE: After some practice you should be able to look at an angle and estimate fairly accurately its degree. It helps to remember that a straight line is 180 degrees. Half of that is 90 degrees or a right angle, and 90 degrees is almost any building's angle relative to the ground. Half of 90 is 45 degrees, which can be visualized roughly by making a peace sign (or victory sign, if you prefer) with your index and middle fingers. You can also use a clear plastic protractor (Figure 7–8). They fit nicely into your helmet liner and work well for determining accurate angle readings.

Lower-System Trouble-Shooting Remedies

Lower-system trouble-shooting remedies often are devised to help release jammed prusiks after they receive dynamic loading, or when they are released accidentally by the handler. When prusiks jam or lock you have three options for getting them unjammed or unlocked. Each of the methods are effective, and you will have to

FIGURE 7–8

Orthopedic protractors fit nicely inside a helmet liner and can be used for determining critical angles in all parts of a rope rescue system. (*Courtesy Martin Grube*)

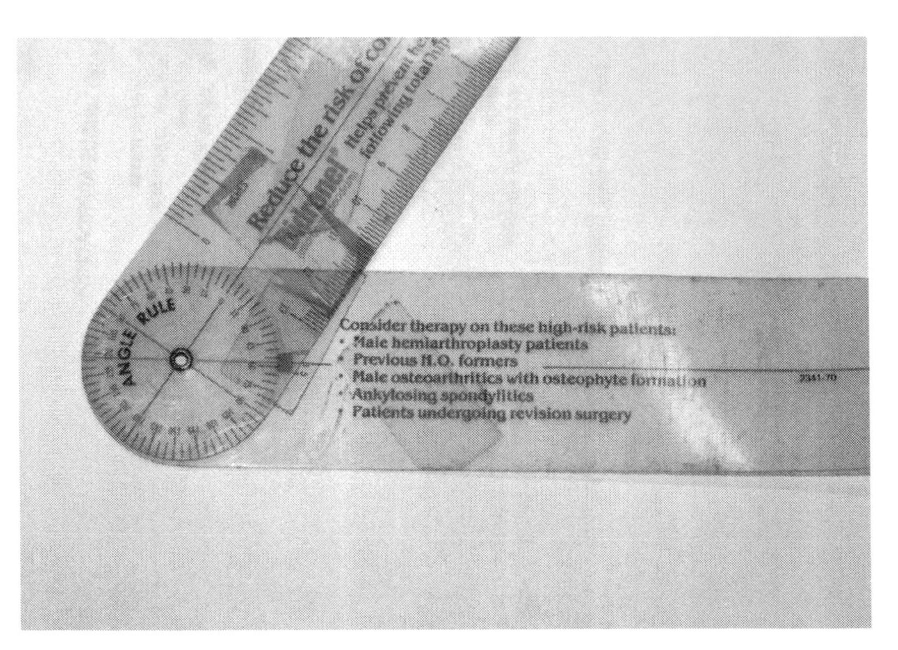

FIGURE 7–9

Hokie hitches built into the lower system wherever there are rope grabs act as shock absorbers and provide a method of releasing rope grabs under tension. (*Courtesy Martin Grube*)

rely on your experience to determine which is right for a given scenario. They are listed in order from good to fair.

► **hokie hitch**

A load-releasing hitch devised to transfer a load safely from one part of a system to another.

Hokie Hitch Load-Releasing Technique. The **hokie hitch (HH)** load-releasing technique (Figure 7–9) is another tool that should be in your rig master's toolbox. It is one of the many ingenious load-releasing hitches (LRHs) that has been devised to transfer a load safely from one part of a system to another. Its primary purpose is to free jammed prusiks, but it also works well as a load-transfer component when passing knots in very long systems. To a lesser degree, it absorbs some shock loading in a system. There are other LRHs that absorb more shock but can allow the load to be released completely. The HH is made from a 25-foot section of 8-mm R/Q cordage that has been tied into a large loop with a double fisherman's bend. It requires three steel locking carabiners. The HH is, in effect, a built-in DCD standing ready to release and transfer the forces in your system. The heart of the HH is the doubled münter hitch that provides the necessary friction to control the descent of a heavy load, up to 900 pounds in the slow lower mode. In their primary function, prusiks on the main line or belay line have been used to absorb system shock. They may have even slipped if very high forces, greater than 2,500 pounds have been encountered. To loosen them for replacement to continue the lower, they must be unloaded and maybe even peeled off the rope.

Building the Hokie Hitch. Building an HH seems a little complicated at first, especially learning it from a book, but with some patience, two people tie it correctly in less than 60 seconds (Figure 7–10 and Figure 7–11).

1. Gather a 25-foot section of 8-mm cordage that has been looped by using the three-wrap double fisherman's bend and the three steel carabiners.
2. Decide which side of the HH will be connected to the anchor and which side will be connected to the load (prusiks, etc.). Remember that the doubled münter hitch is the DCD and always goes to the anchor side.
3. Begin by doubling the cordage so a bight is formed near the double fisherman's bend. This places the mass of the bend out of the way so it will not have to travel through the münter or the carabiners. This bight will be the attachment point for the load side of the HH.
4. Create a doubled münter hitch (Figure 7–10B) in the cordage and attach the anchor-side (Figure 7–10C) carabiner through the working part of the hitch. Roll the hitch back and forth a couple times to make sure that it flip flops through the carabiner properly, and that (Figure 7–10D) you have in fact made a doubled münter and not some other knot.
5. Pull the doubled münter down so it compresses the two carabiners, the load side and the anchor side, until they almost touch. The double fisherman's bend will try to roll awkwardly into the doubled münter when it is fully compressed, which is about 3 inches apart. Now draw the two carabiners apart, capsizing the doubled münter through the anchor side carabiner so it sets in the lowering position. The doubled münter is now ready to accept a load, since it is in the lowering position, not the raising/belaying position, when you unwrap the HH during a lowering operation. The two carabiners should be only about 7 or 8 inches apart if the doubled münter and the double fisherman's are in the proper position. An HH that is 10 or 12 inches long in 8-mm cordage is too slack and should be retied.
6. Find the exact opposite end of the cordage loop and make a bight in the end. Connect this bight to the load-side carabiner (Figure 7–10E). This bight helps to prevent the HH from

FIGURE 7–10 Building the Hokie Hitch

A Determine which side of the hitch will be toward the load, and which side will be toward the anchor. In the photo sequence, the left side is the anchor side and the right side is the load side. The 25-foot section of cordage is doubled and the double fisherman's bend is connected to the carabiner at the load side of the hitch.

B Make a doubled münter hitch in the cordage.

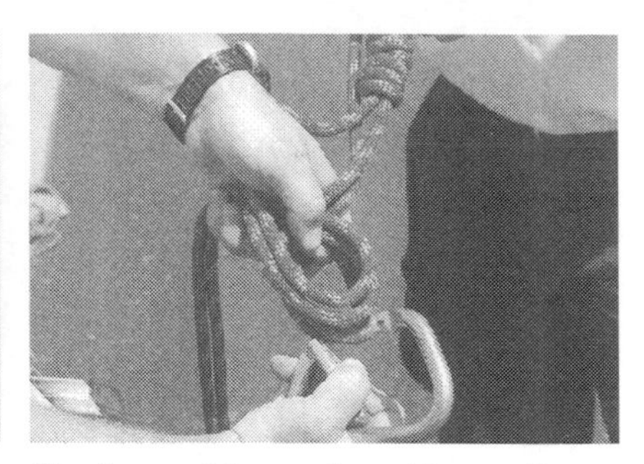

C Connect the anchor-side carabiner into the doubled münter.

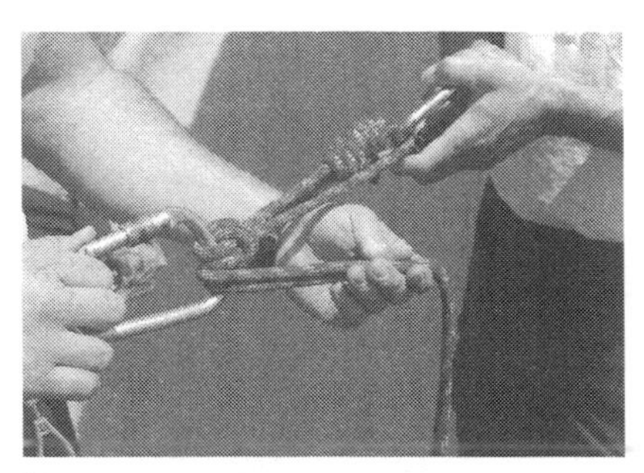

D Roll the doubled münter back and forth setting the rope parts and roll it over into the position that allows it to lower.

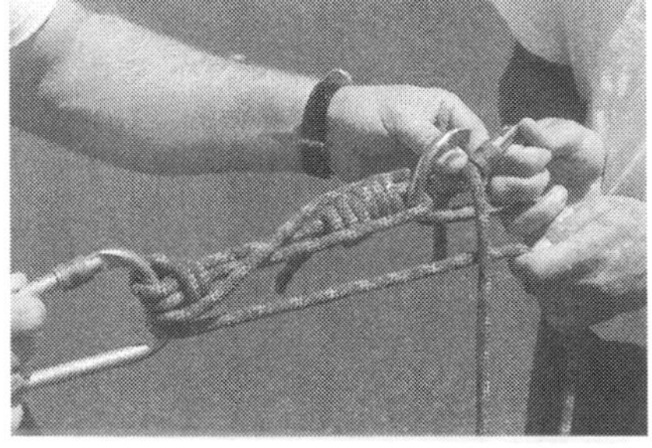

E Lay one part of the long leg of the cordage into the load-side carabiner.

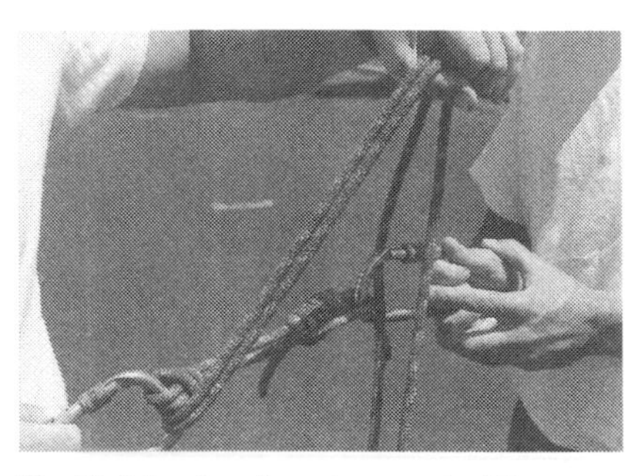

F Pull the two long legs out of the load-side carabiner creating four parts.

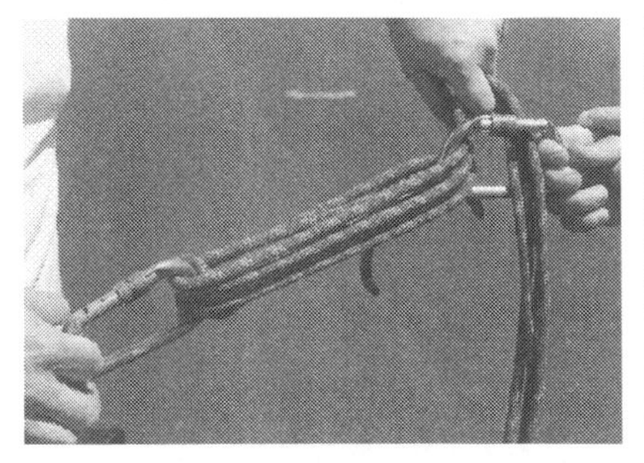

G Pivot the parts through the load-side carabiner so the doubled münter stays closed and in the position for maximum friction for a lowering operation.

H Bring all four parts through the anchor-side carabiner.

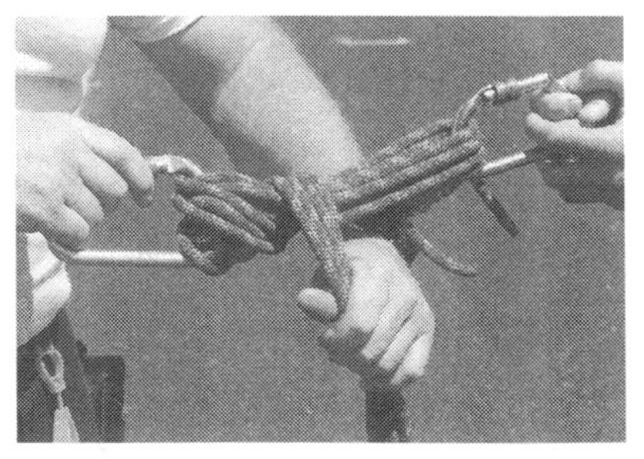

I Bundle the hitch together neatly with two or three turns of the four strands.

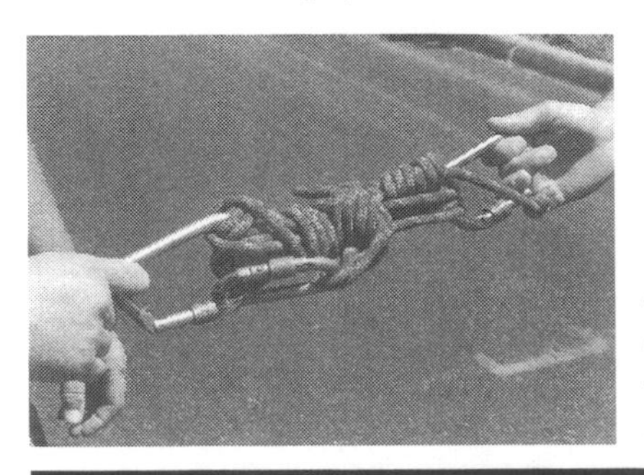

J Finish the hokie hitch by connecting a third carabiner to the remaining two bights in the end of the cordage. Connect the carabiner to the anchor-side carabiner. *(All photos Courtesy Martin Grube)*

FIGURE 7–11 The Hokie Hitch

Illustrations are sometimes easier to understand than photographs.

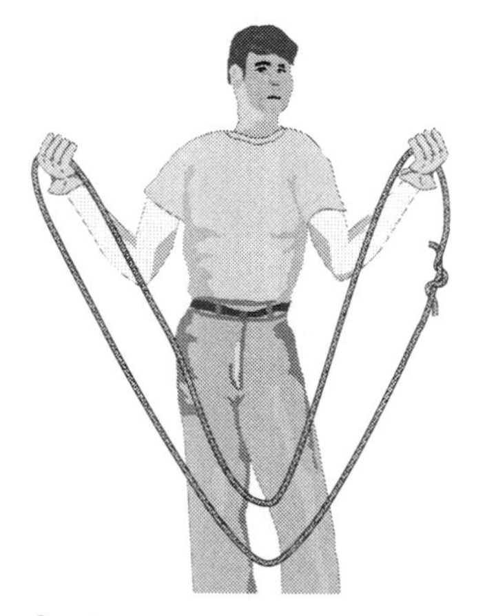

A In the illustration, the right side of the hitch is the anchor side. Rescuer is handling the HH loop.

B The doubled münter hitch for the anchor-side carabiner.

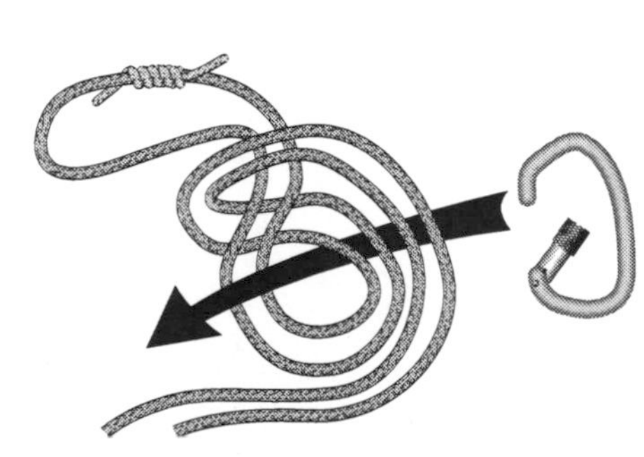

C Anchor-side carabiner attached to the doubled münter.

D Place the long end of the parts into the load-side carabiner. Pivot the end around to create four parts that pass sequentially through both carabiners.

E Bundle the four strands neatly around the body of the hitch.

F Finish with an overhand knot that connects back to the anchor-side carabiner.

G Hokie hitch completed.

overextending and possibly allowing the load to escape. It also prevents the HH from extending much more than 5 or 6 feet, which is all that should be needed for most load-transfer applications. If a longer version is needed for special applications, such as passing several obstacles in one maneuver, 30- and 40-foot sections of cordage can be used to make the HH.

7. Rotate the bight through the load-side carabiner so four strands of cordage come out going toward the anchor-side carabiner (Figure 7–10F). Make sure this rotation closes the münter for lowering, rather than opening it. Also, even out the load-side bight (Figure 7–10G) in the cordage, if necessary, so the four strands are of equal length to keep the hitch neat.
8. Bring all four strands back through the anchor-side carabiner (Figure 7–10H). Feather them neatly around the body of the HH twice to use up the excess cordage (Figure 7–10I). Leave the strands long enough to make a four-strand overhand knot to secure the HH in position.
9. Anchor the twin bights of the four strands from the security knot back to the anchor side with an additional carabiner (Figure 7–10J).

Attaching a Haul System to the Main Line and Releasing Jammed Prusiks You can attach a haul system to the line that has the loaded prusiks and raise the whole thing enough to loosen the prusiks. This may even be the option of choice if you already have a haul system in place to perform a raise. If you have not planned for a haul, you might have planned for only the low forces associated with most simple lowers. A raise increases the forces on your anchors and equipment drastically and may have to be ruled out.

Releasing Jammed Prusiks by Cutting Them Free You can always cut the prusiks free with a pair of crach scissors making sure the remaining load is slowly transferred back onto another part of the system. Once again, cutting out loaded system components is extremely dangerous and potentially fatal. Avoid cutting loaded rescue gear until all other means of accomplishing the task have been considered or tried.

Once the prusiks have locked onto a rope, your first task is to assess any negative affects the loading has had on your system. If you were monitoring the lowering operation closely, and there were no major falls, your system should still be in great operating shape. Remember that a certain amount of bounce and some shock-loading capacity is built into your R/Q rope systems. Next, identify where you want your load to be transferred to. The belay line is often a good choice, but sometimes it is easier to transfer the load forward, closer to the edge, on the main line using tandem prusiks and another HH. Make sure all the slack has been removed from the component that is to receive the load since the HH will expand only about 6 feet until the safety feature prevents any further travel or expansion. Then do the following:

1. Assign someone to operate the HH.
2. Untie the four-strand overhand knot that secures the HH.

3. Always grasp one side of the HH with your free hand and hold it tight to prevent it from loosening before you are ready.
4. Pass the loose strands back through the anchor-side carabiner.
5. Grasp the anchor-side strands to prevent movement and pass the load-side strands through the carabiner. As the surface interface of the strands decreases, the HH may start to loosen and expand.
6. Work the final strands loose so the doubled münter can be worked to release tension and expand the HH. Remember to close the doubled münter to increase friction, and open it to decrease friction. Gripping the strands as the HH expands adds even more friction, if necessary.
7. Allow the HH to expand until the prusiks are loose enough to slide or to remove, as necessary. Watch the maximum length on the HH so the cordage does not max out. If the HH expands all the way out before the load is transferred to another system component, the prusiks will again jam.
8. Rebuild the HH and reset it in the system to use again as needed. With practice, it should take two people only about sixty seconds to build an HH with all the parts at hand. An alternative is to build several HHs and have one ready to put into the system as another one is used up. These can be left in a specific use rigging bag for HHs only with a couple wraps of white medical tape around the body of the HH. The tape will be distorted if the premade HH has been disturbed while in storage or transit, or by another rigging person.

PASSING KNOTS

Passing knots and bends is sometimes required when you have to connect two ropes to get the length needed to finish a lower. The knot/bend you choose will not roll through a rack. However, with some coaxing it will roll through some big-ring rescue eights and, more easily, through tube style DCDs. There will be other obstacles to pass with the bend, too, such as edge rollers, elevated anchor attachments, prusiks, winches, and whatever else is in the system. While you could simply transfer your load to the belay line and rebuild your main line with the knot/bend moved to the front of the obstacle, that technique reduces your load to only a single line. The easiest way to pass a knot through a DCD and other system components is to use the **standby DCD method:**

► **standby DCD method**
Easiest way to pass a knot through a DCD and other system components.

1. Attach tandem prusiks to the main line just below the main DCD.
2. Connect them to another section of rope and reeve the rope to an anchored standby DCD.
3. Release the main line load until the standby system is loaded.
4. Pass the knot through the main line DCD.
5. Remove the standby prusiks from the main line.

The hokie hitch can also be used for passing knots through other system components:

1. Bring the bend or knot as close to the object as possible. Make sure the belay team and people who may be the load are aware that the knot pass maneuver is about to take place. Have all but the slightest amount of slack removed from the belay line. Do not allow the belay prusiks to load unless there is failure from the main line.
2. Have the person working the prusiks on the main line release them and make sure they bite the main line. The load will transfer from the rack, or off the main line, onto the prusik and the HH.
3. Place the knot/bend in front of the obstacle. If it is a rack, simply place the knot just in front of the rack. If the prusiks are close to the rack, the knot/bend can be moved in front of the DCD and the prusiks. If not, the transfer will have to be done twice or especially long HH configurations will have to be used.
4. Expand the HH as described earlier, like releasing a prusik. Make sure to control the load by releasing the tension from the HH slowly. The load will slowly transfer back onto the rack. Make sure the belay prusiks do not catch the load as the rope stretches under load.
5. Reset the HH to perform another transfer of the knot/bend through the prusiks in front of the HH, if necessary. Recertify your system, notify your belay team and the people who may be the load, and continue your lower.

A lowering system (Figure 7–12) simply uses two moving ropes to safely transport a load from an elevation to a lower point. One of the ropes is designated the main line, and the other is the belay line.

FIGURE 7–12

The focus of the lowering system from right to left: (A) Dual rescue lines from bombproof anchor. (B) Double pulley used to spread out the forces of the double bight eights from the anchor. (C) A 10,000-lbf (45-kN) rigging plate. (D) Light-colored rope is belay tied off onto closed end rack; remainder of the rope goes to backup anchor. (E) Dark rope is mainline reeved into closed end rack. (F) Both ropes have tandem three-wrap prusiks connected to the rigging plate. (*Courtesy Martin Grube*)

FIGURE 7-13

(A) Components of the lowering system. (B) Main line lowering components.

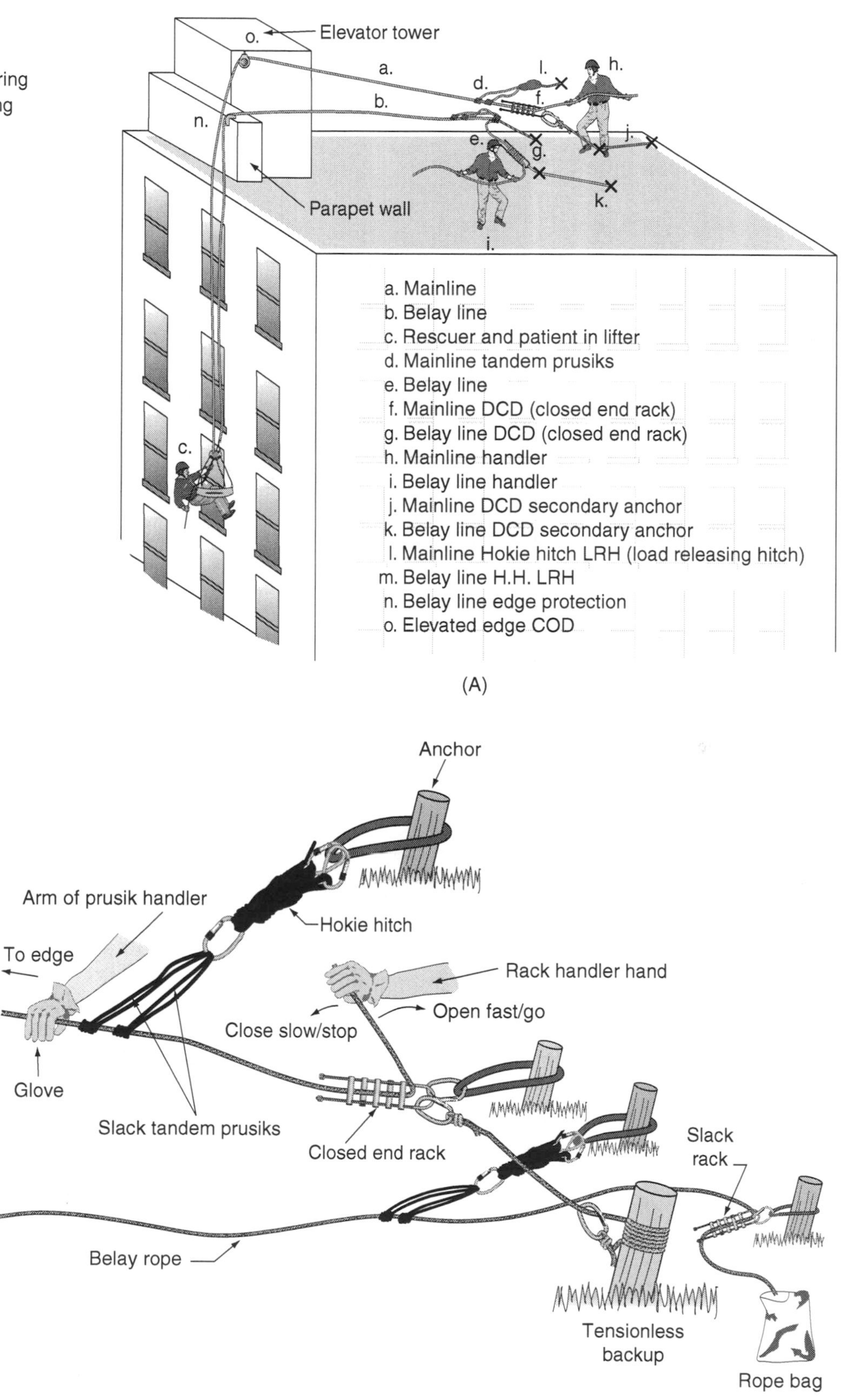

If possible, use different color ropes for quicker identification during the rigging and lowering processes. Half-inch kernmantle is sufficient for each line, unless you have a particularly heavy load. Use 1/2-inch rope for loads anticipated to be in the 600-pound and less range, for example, a patient and one litter attendant. If the load is much heavier than 600 pounds, such as with a patient and *two* litter attendants in a bad trauma case that needs four hands working, consider using 5/8-inch ropes. Make sure the ropes are long enough to reach the ground or some other desired destination. Two ropes can be tied together in a pinch using the figure-eight bend. A discussion of passing knots is included later in this chapter.

Both ropes must be connected to the load, so a figure eight on a bight must be tied in the end of each rope. Make sure the bights are very small and the knots dressed out and pretensioned, and that there is about an 8-inch tail remaining. If the bight is too long, it might be difficult to get the load over the edge due to the limit of travel caused by the location of the knot.

Locate your target and check your anchor sites. If possible, they should be in line with each other. If you are lowering a rescuer to pick-off a person stranded over the edge, your target dictates the location of your leading edge. If your load is a patient who is at the rigging site with you, you will have some options about where to make your leading edge.

Remember that an elevated leading edge makes starting your load, a rescuer or patient and rescuer, much easier and safer. A healthy tree, a bipod, or an elevated rock position near the edge at wilderness sites can work. A tripod, bipod, quadrapod, scaffolding, Larkin Frame, elevator, antenna tower, or almost anything that gets the load up off the edge helps. The elevated leading edge allows you to have the load and the system tensioned and working in a safe area before actually committing people over the edge. The alternative is usually an awkward effort at slowly getting the load over the edge. Loading and patient handling are discussed later in this chapter.

At the elevated leading edge, anchor a pulley for the main line. Calculate the resulting angle created by this COD. The belay line should be placed on an edge roller secured to the edge. In the unlikely event that the elevated COD becomes compromised, the belay line will be in a position to protect the load from falling. This is particularly important if you are doing a manhole entry using a tripod as an elevating device. Tripods tip over occasionally and an almost taut belay line will keep the load from falling as long as it is properly protected from the edge and is not rigged at the apex of the tripod with the main line.

Locate an anchor (Figure 7–14A and Figure 7–14B). You will need to make an attachment to the anchor, usually a sling or presewn rigging strap (Figure 7–14C and Figure 7–14D). The preferred DCD for lowers is the closed end rack, however eight-plates, münter hitches, wraps around a secured standpipe, wonder bars, and other types of DCDs have been used successfully for years. For the purposes of this discussion, the closed end rack will be the DCD of choice.

Anchor the rack and reeve the main line for the lowering operation. If possible, choose another bombproof anchor site that is

FIGURE 7-14

(A) Bombproof anchor attachments. (B) Anchor attachments around padded steel I-beam. (C) Anchor straps extend the focus to a more convenient location. (D) Multi-point anchor system. (*Courtesy Martin Grube*)

closely in line and in front of the main anchor to use as the belay site. If enough people are available, secure tandem three-wrap 8 mm prusiks onto the main line in front of the rack and anchor them to another anchor for use as emergency brakes. Only eliminate these prusiks if you do not have enough people available. They add another source of system redundancy, but usually require someone to work them. You could also place a prusik minding pulley (PMP) in the mainline as a COD and place the prusiks and an HH on the same anchor as the PMP. Adjust the length on the anchor attachment so there is minimal slack, or shock load, between the PMP and the point where the prusiks grab the rope. In this way, they should self-tend and offer an additional safety brake.

Next, secure tandem three-wrap 8-mm prusiks onto the belay line. These are the emergency brakes. Make sure the prusiks are of slightly different length preferably a short and a medium so they will not interfere with each other during the lower. Connect the HH between the prusiks and the anchor. Behind the prusiks and again preferably to another anchor site, attach an additional closed end rack. This pseudo attachment will be used if the prusiks are forced to take the load and the belay line becomes the main lowering line.

Keep the belay line reeved onto the belay rack with three loosely spaced bars. This will allow relatively free movement of the belay rope without hindering the lowering operation. In addition, the rack can quickly be closed and additional bars added if complete mainline failure occurs and the belay prusiks have to be used.

Your team is now ready to certify the system. Again, certification should be done by someone other than the rig master, if possible. Certification should only take a couple minutes, and if there are no real problems, the operations officer gives the go ahead to begin the lower.

Elevate the load into the elevated leading edge, if possible, so the system is loaded in a safe place. Make sure everyone has their assignments, their personal protective gear is secure, and they are in position. Review the rope movement and stop commands. The major commands are *tension,* which means to increase tension or friction in the system, in this case, to slow the lower; *slack,* which means to reduce friction or tension in the system, in this case to begin or continue the lower; and *stop,* which anyone can use when they feel something is not quite right with the system or the lower. In this case, stop means to bring the lower to a complete stop and be prepared to lock-off. There are many other minor commands, such as slower, faster, too much slack in the belay, 50 feet to go, 40 feet to go, and so forth.

Now your team is ready to begin the actual lower. Work the load out over the edge. The main line should have all of the tension in the system. The rack person should be able to control descent with the pressure from the grip of one hand. Make the rack do the work. Remember to close the rack somewhat to increase friction and to open it or remove some bars to lessen the friction and allow the lower to go more rapidly.

The belay line should be kept slightly slack. If a sharp object is encountered, the tensioned line will cut quickly leaving the barely slack belay line to catch the load. Make sure the belay prusik operator is positioned to catch the entire load at any given moment. If the belay line becomes the main line by default, the belay rack person must be ready to reeve the remaining bars on the belay rack and prepare to lower as necessary.

Observe the movement of the lower. Keep the speed steady and only as fast as the load attendant(s) are comfortable with. Always maintain constant contact with the load attendant(s) for that reason. The rack person can watch the pics of the rope travel through the bars of the rack to get a feel for the constant speed of a fluid lower. Remember when starting and stopping the lower that even small amounts shock load the equipment, the load, and the anchors. *Seek fluid movement.*

PATIENT PACKAGING

Patient packaging (Figure 7–15) is the least practiced, but one of the most vitally important life safety skills. "The load" is referred to repeatedly throughout this book. This section is intended to recog-

FIGURE 7-15

Practicing the integrated harness system (IHS). (*Courtesy Chase N. Sargent*)

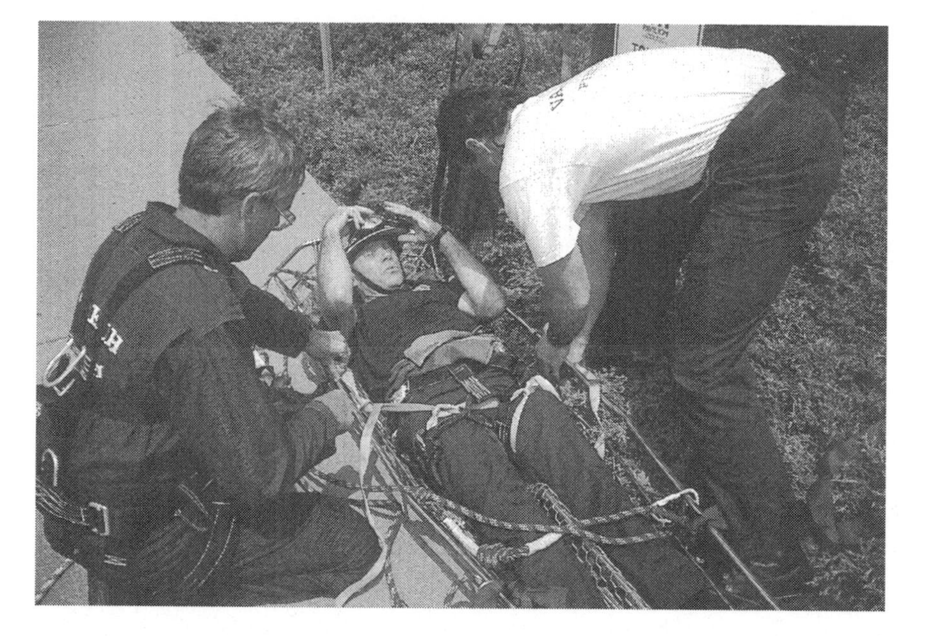

nize that the term "load" refers to a real-live important human being. It also emphasizes the importance of medical stabilization and secure patient packaging and the critical importance of treating patients as if they were the most important people in the world, at least while they are being transported on the rope-based transportation system. One of the advantages of using ropes for rescue is that when used properly, they can provide a much smoother and more medically stable means of transportation than other methods. This fact has eluded many incident managers, partially because as rope rescuers we have not made our cases very well. Fortunately, the improvements in rope, hardware, and technique have helped make the goal of medically stabilized transportation achievable. The difference between truly great and mature rope rescuers and marginal rope rescuers is that the great ones inject an honest feeling of concern and warmth for the patient into the rope rescue. Because they treat patients like they are members of their own family, the rope transportation segment of the overall rescue is likely to be the least traumatic part of the incident.

There are numerous patient packaging devices available. All of them have advantages and disadvantages. Some are excellent for dragging a patient through confined spaces, like SKED; some provide superior cervical-spinal immobilization when lifting such as Life Support Products Half Board (LSP); and still others, like the Kendricks Extrication Device (KED), are superior for use in auto extrication. However, none of them are perfect for every occasion. It is wise to maintain an array of the different patient packaging devices, because for their strengths, there are no substitutes. Still, nothing can replace the utility, price, and dependability of the wire basket type, or simply Stokes litter.

Patented in the late 1890s by Charles A. Stokes, the wire basket litter, like "Coke" is to soft drinks, has assumed the generic name Stokes basket, Stokes litter, or simply Stokes. The modern

NOTE: This discussion of patient packaging presumes the application of reasonable medical care and stabilization for each patient prior to and during Stokes litter application. It is not within the scope of this book to discuss every possible medical condition for which a person might need to be transported in a rescue litter. Certain portions of the IHS, as with any patient packaging technique, must be adapted to patient condition including but not limited to environmental concerns, such as water, wind, and fire, or serious injuries to applied body parts, like pelvic and femur, and/or head, neck, back, chest, and shoulders.

stainless steel mil-spec. version of the Stokes is classified as *litter, rigid, Stokes.* In fact, almost any wire, plastic, or aluminum full-body litter is erroneously called a Stokes. Today's Stokes is almost unchanged from the original that was invented more than 100 years ago. The wire mesh Stokes, supported by a light steel frame with leg dividers has the tool of choice for carrying literally hundreds of thousands of injured persons to safety for more than a century. Because of its simplicity, durability, and cost, the Stokes litter remains the most effective tool of choice for today's rope rescue.

It is vitally important that your patient become a part of the Stokes litter. The litter can be considered exoskeletal to the patient, a protective mechanical force simultaneously providing raising and lowering support and safeguarding the patient from further injury. Other than completely submerging and drowning a conscious and relatively uninjured victim, a patient is almost always safer securely reeved into the litter. There are many horror stories about hurried patient packaging where the patient slipped out of the basket, or shifted in such a manner as to cause further injury. The Integrated Harness System (IHS) as described below, is a highly effective method of securing a patient to a wire basket type litter.

Some discussion of how to get the patient into the Stokes litter is required. It is very important to have a blanket or a presewn tensioned litter bed in the bottom of the Stokes. This makes the litter more comfortable and helps considerably to keep shocky patients from becoming hypothermic. In general, patients are either on good horizontal ground or in some elevated and precarious position. After medical stabilization, if patients are on solid ground, a simple log roll into a tilted basket works fine. Reeve the patient into the IHS as explained below and illustrated in (Figure 7–16). If the patient is hanging on a rope, a narrow ledge, a tree branch or something similar, it is sometimes better to capture the patient into the Stokes before attempting thorough medical stabilization. Regardless, disconnect the leg attachment of the litter bridle closest to the victim. Position the Stokes parallel to the patient—head to head and feet to feet. The Stokes can then scoop the patient. Once the patient is in the Stokes, reattach the litter bridle leg. Be prepared for the additional weight of the patient to increase rope stretch and change your position downward a distance corresponding to the weight. Be sure your riggers at the top are also prepared for the increase in weight. Sometimes it is beneficial for the top side team to do a coordinated mini-raise of the basket during the scooping operation. This can help orient the patient to the litter while counteracting any lowering caused by the additional weight of the patient.

▶ **Integrated Harness System (IHS)**

Combines two modified webbing harnesses integrated with old-fashioned "diamond" lashing techniques.

The **Integrated Harness System (IHS)** combines two modified webbing harnesses integrated with old-fashioned "diamond" lashing techniques. In short, once the patient is lying in the Stokes, one harness is tied over the shoulders and chest to prevent movement toward the head of the Stokes and the other harness is tied over the pelvic girdle to prevent patient movement toward the foot of the Stokes. Then the patient is lashed from foot to shoulders with webbing, and finally the head is secured in a fashion that is suitable for

Integrated Harness System.

Modified Swiss seat webbing, using a modified bowline

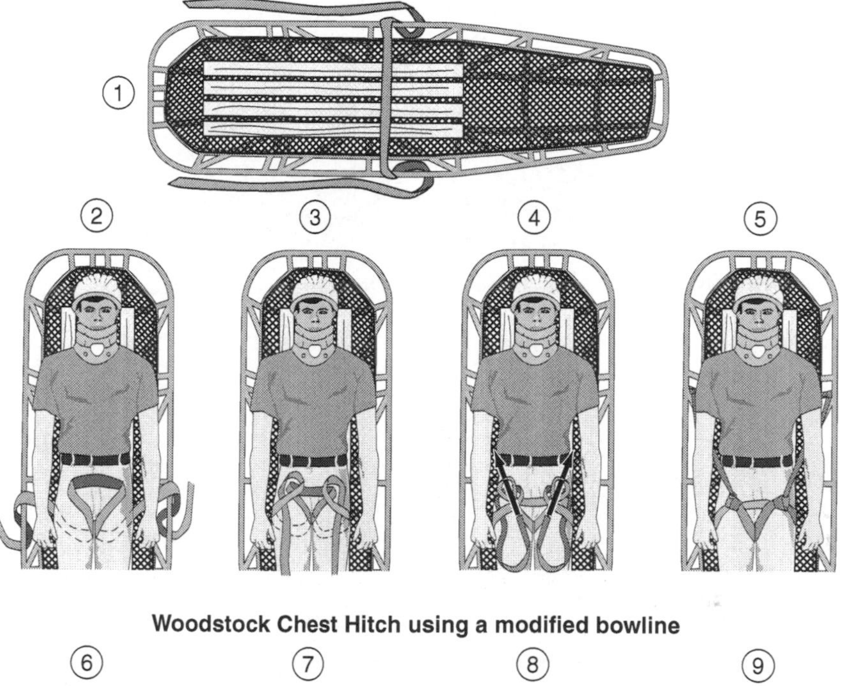

Woodstock Chest Hitch using a modified bowline

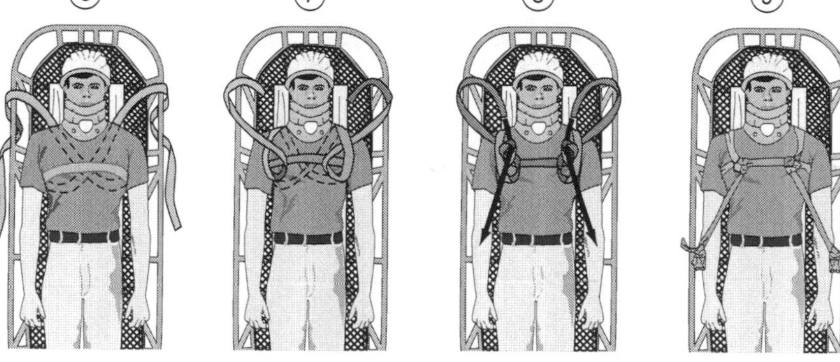

Traditional Criss-Cross lashing using webbing

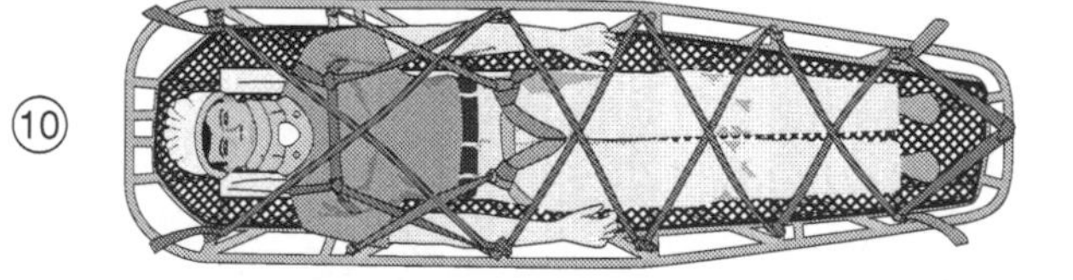

NOTE: Anytime a person is lashed into a litter during an actual emergency or training and then transported through a potentially hazardous medium, such as free air, over water or lava, and so forth, it is wise to have some way to cut them free in the event of an emergency. Crash scissors are the safest tool for cutting webbing without cutting the patient, yourself, or other life-supporting equipment.

the patient's injuries. The IHS, like all good techniques, requires practice but can easily be reeved onto a patient in a Stokes litter by two competent rescuers in less than five minutes.

Securing the Patient in the Stokes Integrated Harness System

1. Secure four 20-foot sections of 1-inch tubular webbing. Two-inch webbing can be used, but it is more time consuming and bulky. Mark the center of both sides of each piece of webbing so it is easy to locate.

NOTE: Step number six can be eliminated if the webbing places force across the soft abdomen. It is important that the harness be secured over the body prominence of the pelvic girdle, like fastening a seat belt in a car. If this option is eliminated, secure the running ends of the webbing to the tie-off point using round turns to rachet the webbing comfortably tight and finish with half hitches. Secure the loose ends of the webbing inside the Stokes.

2. Find the middle of one of the sections of webbing and make a bight in it. Pass the bight under one of the legs, bring it into position between the legs, and cinch it up near the groin of the patient. Pass the running ends of the webbing under each buttock and alternately out and around each upper thigh. Make sure the marked middle bight of the webbing remains accessible.
3. Fashion a no-slip leg loop over each thigh by tying a modified bowline into the webbing. This is most easily accomplished by placing a loose underhand loop on the patient's pelvis so the running end of the webbing lays over the hip-point. Reach through the middle of this underhand loop and pull a bight up from a section of the center webbing. This bight actually emerges from the middle of the underhand loop. Give the bight one-half turn. Now insert the running end of the webbing into this bight, pull out the excess, and cinch down the knot. Always test the knot to make sure it is a modified bowline and will not slip along the webbing in any direction. The leg loop does not have to be tight around the thigh. You should be able to slide three or four fingers between the loops and the upper thigh. Make sure the leg loops are nonslipping and cannot cinch down on the leg and hinder circulation.
4. Complete the leg loop on the other leg the same as above.
5. There are now two lengths of webbing emerging from the modified bowline secured leg loop. Bring the right side length out toward the right rail of the Stokes and secure it to the closest vertical tie-off point that supports and is welded to the rail. This tie-off should be higher than, or superior to the waist of the patient. Do not secure any webbing to the rails of the Stokes litter. This could allow chafing as the combined weight of the litter, patient, and attendant is raised or lowered along a wall. On every tie-off point, secure the webbing with a round or two first. This allows you to gather tension, and rachet the webbing tight onto the Stokes.
6. Bring the remaining ends of the webbing together over the belt line and connect with a square bend. Finish with an overhand knot safety around the taut part of the webbing.

The patient's pelvic area is now securely attached to the Stokes litter. Tied correctly, slippage downward toward the feet is virtually impossible.

Securing the Chest and Shoulders. The upper body litter harness is very similar to the pelvic harness. Once you learn one of them, the other is simple.

1. Place one of the sections of the webbing over the patient's chest. The center mark should be over the sternum, just above, or superior to, the xyphoid process.

NOTE: Always leave the neck area open to prevent choking by the webbing. Much depends on the size of the patient, so use your best judgment when connecting the horizontal chest pieces. While it does stabilize the chest more, it only works if the connection is just about on the nipple line and not above.

2. Work the running ends of the webbing behind the patient's back diagonally, criss-crossing the shoulder blades, and exiting over the shoulders and toward the chest.
3. Make an underhand loop in the running part of the webbing and lay it on top of the horizontal chest piece, similar to the pelvic harness. Reach through the loop and pull enough slack up from the chest webbing to form a bight. Turn the bight one-half turn.
4. Work the running part of the webbing into this bight and secure close but not tight to the chest. Again, test the knot to be sure it is actually a modified bowline by pulling in every direction to see that it does not slip.
5. Take the remaining running ends of the webbing to a tie-off point below, or inferior to, the chest. Use round turns to rachet the webbing comfortably tight and half hitches on the taut part of the webbing to secure it.

The patient's chest and shoulder area is now securely attached to the Stokes litter. Tied correctly, slippage upwards, toward the head, is virtually impossible.

Lashing the Patient to the Stokes. This is simply the criss-cross method used for decades with no angular pressure applied to the feet and ankles.

1. Secure the end of one of the webbing sections to a tie-off post near the bottom corner angle of the litter. Round turns finished by *overhand safeties* work best here. Cross over the patient's lower leg and round turn on the next available tie-off post, making sure not to wrap the webbing around the rail of the litter. Continue criss-crossing and round turning until you have reached the chest.
2. Repeat the maneuver with the other piece of webbing until you have reached the chest. Injuries permitting, connect the two chest pieces with a square bend and overhand safeties horizontally across the chest, or finish the ends with round turns and half-hitch safeties at the chest-high tie-off point.
3. If specialized c-spine immobilization has not already been done, it should take place now. Always secure the head of semi-conscious and unconscious patients. Preferably, the patient will be in a low-profile helmet. Fill the gaps next to the head with towels or pads. Duct tape works well to secure the head to the litter tie-off points.

► litter bridle
The attachment device that connects the stretcher to the system. The bridle spreads and balances the weight of the litter over four or more legs.

The **litter bridle** (Figure 7–17A and Figure 7–17B) provides a means of connecting the Stokes litter to the raise/lower system. Like the rest of the tools in our system, the litter bridle can be very simple or very complex. The two most common litter bridles are the presewn, or manufactured, and the Yosemite. The presewn bridle consists of webbing legs connected at the top by an O-ring and at each of the four points on the litter with carabiners or permanent

FIGURE 7–17

(A) Stokes litter bridle.

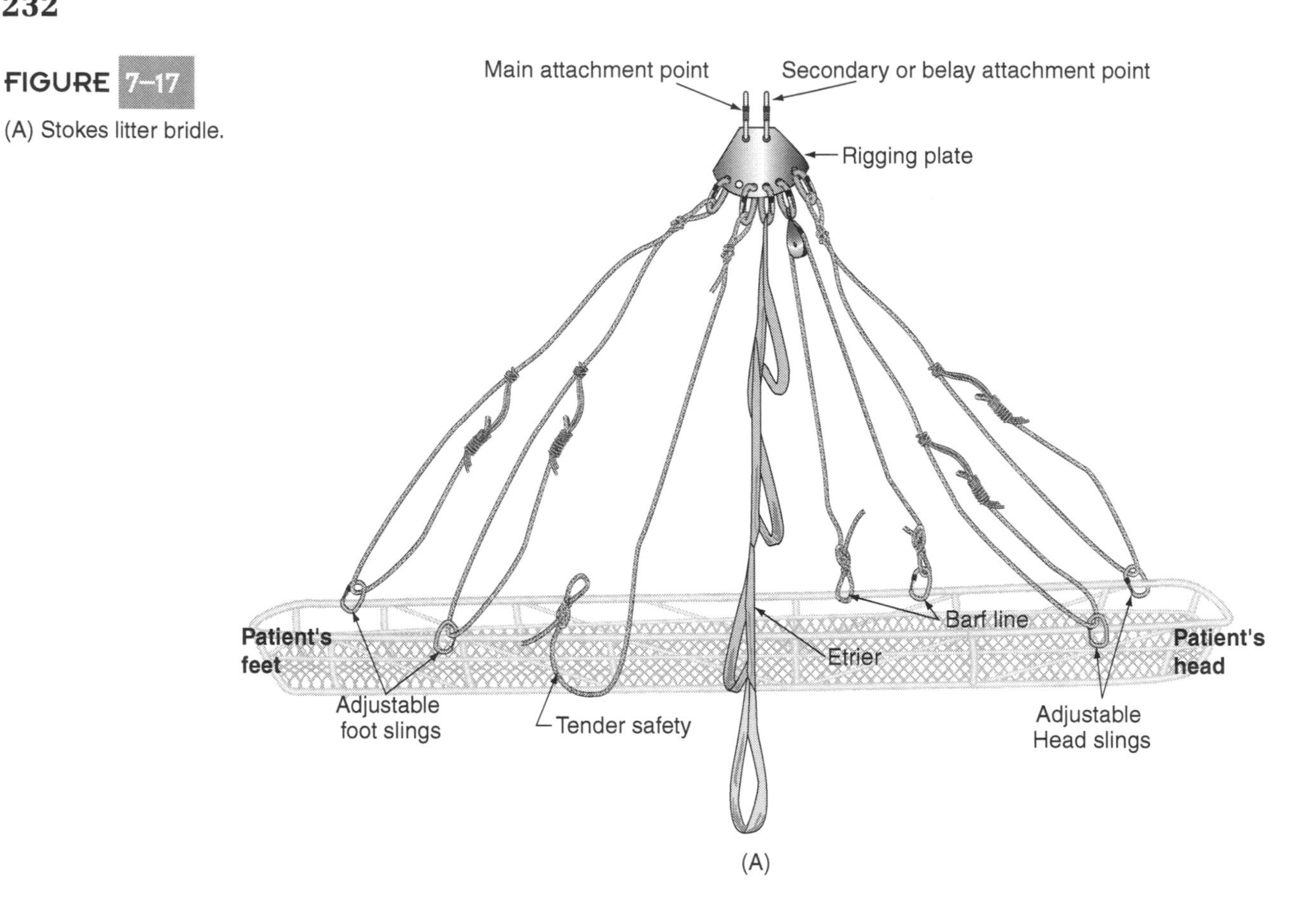

NOTE: If the legs are made from the 7/16-inch rope, which is lighter and smaller, 7-mm cordage provides the best and most versatile bite. If the legs are made with 1/2-inch rope, which is heavier and stronger but possibly too big, 8-mm cordage provides the best and most versatile bite. Also, due to the wide range of characteristics of cordage and rope, homemade Yosemite rigs should be thoroughly tested, loaded, and retested at a safe, low angle before committing them to rescue work. Once broken in, however, they are extremely easy to use and versatile.

snap hooks. Some bridle legs are fixed, that is not extendable, but most presewn bridles have some flexibility in length by incorporating single-pass, drop-forged steel buckles. The presewn bridle is dependable and relatively inexpensive. It is probably the device of choice for intermediate rope rescuers or for teams that do not need the flexibility of the Yosemite rig. The Yosemite rig is the term most commonly used to describe litter bridles that are home made from rope and cordage. They are very adaptable to different situations and are favored by many technical rope rescue teams. The legs of the Yosemite rig retract and extend so the head can be adjusted in relation to the feet in a variety of ways. The legs connected to the head are made from the same piece of 7/16-inch or 1/2-inch rope connected to a rigging plate or O-ring with a small, tight figure eight on a bight. One-inch tubular webbing is fitted over the bight in each of the legs to provide chafe protection. The rope loops through a carabiner attached to the litter and is attached back onto itself using a double-wrapped prusik loop. The prusik loop is connected to the rope via a double fisherman's bend, a doubled 7- or 8-mm cordage interfacing with the 7/16-inch or 1/2-inch rope leg.

The foot legs of the Yosemite bridle are made like the head legs of the bridle and are connected at the top connection.

The top connection for the bridle legs can be as simple as using a drop-forged steel O-ring or forged steel eight-plate. A rigging plate is preferred by many rig masters because it can organize a lot of litter bridle equipment. The legs can be attached to either end, the atten-

FIGURE 7–17 cont'd

(B) Litter bridle and Stokes basket.

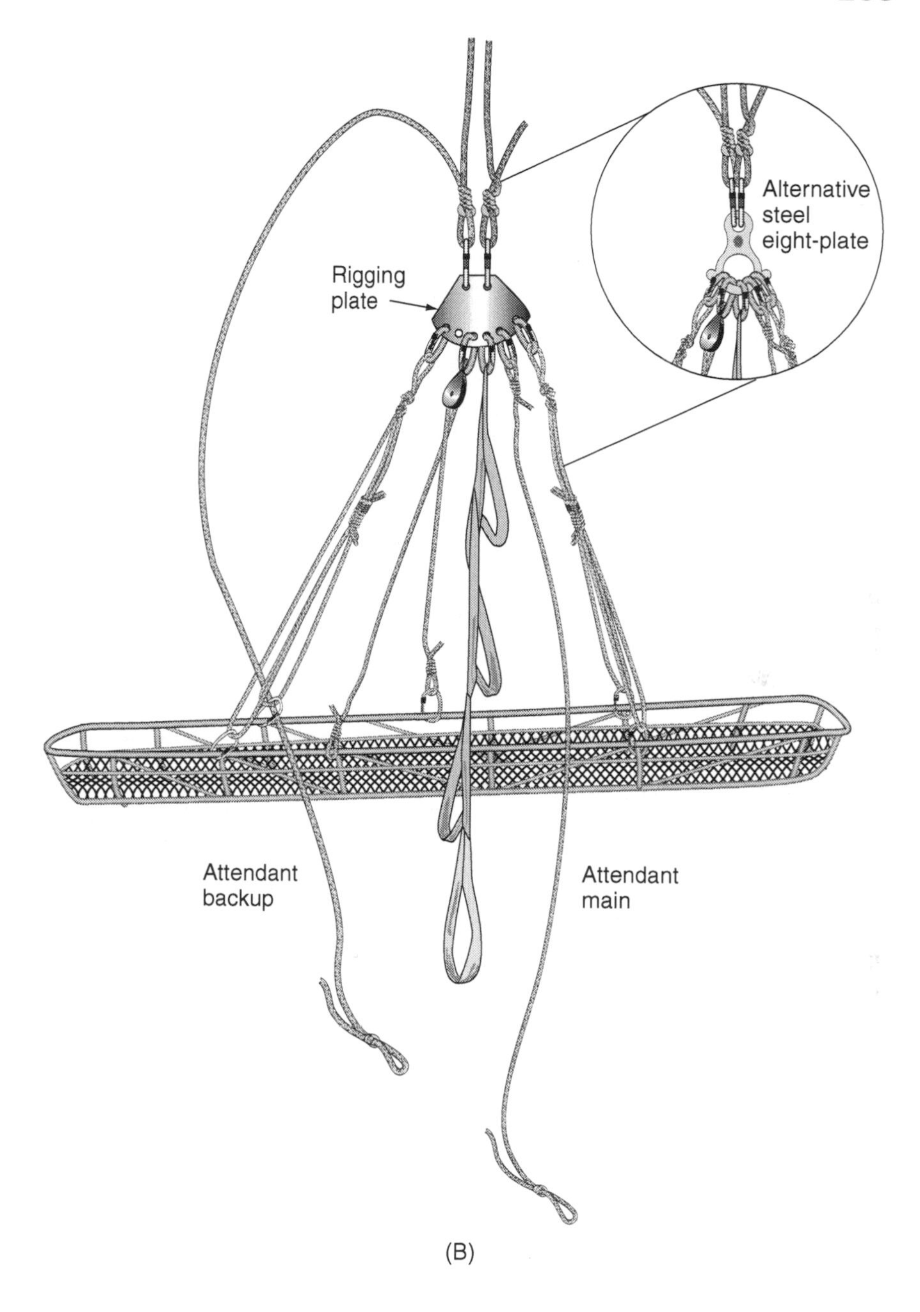

dant to a center hole, and a barf line and IV bag to other holes. The smaller side at the top toward the anchors of the rigging plate has at least two holes, one for the main line and one for the belay line.

When attending a litter and patient during transport, you depend on your team to control your direction and speed. Your job is to smooth out the rough edges of the ride and to tend to patient needs. Position your body so the litter rail is at waist level, which will allow you to walk the litter off of and around obstacles while leaving your hands free to work the patient and the litter bridle as necessary. Do not be afraid to climb around on the litter to accomplish your tasks. Use the rails as hand and foot holds to balance and adjust the attitude of the litter and patient. Be careful not to step on the patient. Learn to make your moves fluid to avoid frightening the patient or aggravating any injuries.

(A)

(B)

FIGURE 7–18

(A) Proper positioning of the litter and patient in the high angle environment. (B) Attending the litter and patient in the high angle environment. (*Courtesy Martin Grube*)

It is sometimes necessary for the litter to travel in the upright position. For example, if the patient is in an area of a ship with a small opening, in a silo or a manhole, or if other obstacles present themselves, the attendant and litter must be made as narrow as possible. Vertical riding of the litter must be kept to a minimum because the litter is not as stable and will seek to be horizontal, the ride is not as comfortable, and more importantly cerebral blood pressure can plummet in shock patients, rendering them unconscious and accelerating brain damage. If medical protocol suggests, the litter can be made to ride Trendelenburg, that is with the head slightly lower than the feet.

The litter attitude can be adjusted in a number of ways:

1. On adjustable leg litter bridles, the legs can be shortened at the head and lengthened at the feet, achieving a nearly vertical position. Stand on the foot rail of the litter to transfer weight from the head bridle legs, so they can be made

FIGURE 7–19

(A) Attending the litter and patient in the low angle environment. (B) Preparing for the wilderness lower. (C) Lowering operation over the rocks. (D) Two attendants working the litter and patient. (*Courtesy Chase N. Sargent*)

(A) (B)

(C) (D)

shorter. Then use your weight to load the head bridle legs and adjust the leg bridle legs out for maximum length.

2. With some practice, the litter can be brought vertical by working the stretcher into position. The attendant should stand on the foot rail of the litter and grab the head bridle legs. By applying body weight to the foot rail and simultaneously pulling hard on the head bridle legs the litter will remain vertical long enough to make the short trip through the obstacle. Bear hug the patient and litter to secure the position and then reverse the move to return to the horizontal position.
3. A less-preferred method uses two attachment points to the litter. It can be rigged so that one line goes to the bridle legs attached to the litter near the shoulders of the patient, and the other line is attached to the bridle legs that attach to the litter near the patient's knees. The ropes are then separate single-line lowering systems complete with their own operators, anchors, DCDs, and safeties. With careful

coordination, the head line can be made static, and the foot line slacked enough to cause the litter to ride vertically. The attendant, who is attached to the head line, floats around to stabilize the ride through the obstacle. There are several problems, however, with this system. While it is a two-rope system, failure of either rope would cause the litter to violently adjust to the remaining rope. Failure of the head rope would drop the attendant and leave the patient dangling upside down by the foot rope. Again, though versatile, this system should be used only after all true two-line systems have been evaluated and eliminated as not as usable for the situation.

KEY TERMS

Lowering systems
Force vectors
Change of direction (COD)
Golden Angle
Hokie hitch
Standby DCD method
Integrated Harness System (IHS)
Litter bridle

REVIEW QUESTIONS

1. Lowers will generally develop _____ system tension than raises.
   a. a little more
   b. less
   c. a lot more
   d. a tremendous amount more
2. Lowers mandate that a majority of your team will be _____ .
   a. occupied working belay system components
   b. establishing multiple anchor sites
   c. building mechanical advantage systems
   d. above the moving functions of the systems
3. In lowers there are _____ and more people with less experience can safely rig lowers than can rig raises or highlines.
   a. fewer components to rig, the systems are simpler
   b. more components, but they are simpler
   c. no required certifications
   d. many more available people
4. In shared-anchor systems, the _____ the angle created by the force vector, the _____ the force is on the anchors.
   a. narrower, greater
   b. wider, less
   c. more obtuse, less
   d. narrower, less

5. In change of direction components, the _____ the angle created by the force vector, the _____ the force added to the system.
   a. wider, more
   b. wider, less
   c. more acute, less
   d. narrower, less
6. In single COD force vectors, the most force you can generate at the point of the angle is _____ the load.
   a. no more than
   b. two times
   c. a force equal to the load plus one half
   d. one half
7. The HH hokie hitch load-releasing configuration's primary purpose is to allow you to _____, but it also works great as a load-transfer component when passing knots in very long systems.
   a. rest your team when necessary
   b. let loose of the load once it has been transferred to safety
   c. free jammed prusiks
   d. release loads that have gotten hung up on obstacles
8. The HH hokie hitch is made using a _____ that has been looped by using the three-wrap double fisherman's bend, and the three steel carabiners.
   a. 25-foot section of 8-mm cordage
   b. 20-foot section of 2-inch tubular or flat webbing
   c. 15- to 20-foot section of 1/2-inch rope
   d. 113-foot piece of 8-mm cordage
9. Once the prusiks have locked onto a rope, your first task is to _____ .
   a. check the accountability of your team
   b. add two new prusiks to the system
   c. connect an additional belay line to the load
   d. assess any negative effects the loading has had on your system
10. An _____ will allow you to have the load and the system tensioned and effectively working in a safe area before actually committing people over the edge.
    a. aerial platform
    b. elevated leading edge
    c. elevated mechanical advantage system
    d. acute change of direction

CHAPTER

8 Rigging and Operating a Raising System

OBJECTIVES

Upon completion of this chapter, you should be able to:

- explain how to engineer a counterbalance system for the purpose of raising a rescue load.
- explain the advantages a counterbalance system has over a mechanical advantage raising system.
- identify three simple machines and discuss their efficiency.
- explain how to discover the approximate efficiency of a pulley.
- discuss the differences between simple and compound mechanical advantage pulley systems and the advantages and disadvantages of each.
- explain how to build four different pulley-based mechanical advantage systems.
- engineer a safe two-rope raising system.
- discuss techniques that help get the raised rescue load up and over the edge.

A DAY ON THE ROPES

The coffer dam spaces of the landing ship dock (LSD) were as deadly as standing in front of a bullet. Gasoline had penetrated the tank through a hole in the steel shell and was sloshing around, creating enough fumes to blow up the ship, the dock, the surrounding businesses, and us. Bobby had courageously advanced through the spaces using his SCBA to find a contract worker who had apparently passed out from breathing-apparatus problems. The problem was that Bobby had to take his SCBA off, pass it through each compartment, and climb through after it. Now Bobby was our problem. We were called after his engine company lost contact with him in the spaces.

It took almost an hour to find Bobby and determine that he was dead. It seems that one of the spaces had been about a foot lower than the previous ones, and his SCBA slipped from his hands down into the space, pulling his mask off as it went. A couple breaths of intensified gasoline vapors sucked the life out of him in just seconds. After he was pronounced by one of our paramedics, we shifted into body recovery mode. That meant we could slow down, think things through, and incorporate lots of system redundancy. Nobody gets hurt doing a body recovery.

The process was slow, but we were moving ahead. Team members, working in pairs and wearing dual regulator-supplied air breathing apparatus, had finally gotten Bobby back through four of the fifteen coffer dam openings. The fifth one was not going to be as easy. Bobby was a muscular 240 pounds, and now the leaking tank had to be ascended on the way back to the point of entry.

The tank was a huge cylinder that was laying on its side supported in the hull of the ship by the coffer dam system. The ascent was only about 15 feet up the side of the tank and into space number five, but three rescuers wearing full protective gear crammed into the confined space just did not have enough strength to get him out. They needed a mechanical advantage (MA) system.

We decided to build a five to one simple MA and send with it provisions for piggybacking an additional pulley onto the 5:1 in case more advantage was needed. Bobby was propped up, and a hasty hitch webbing harness was attached to his torso. Then the inside rescuers had to figure out how to anchor the 5:1 so there would be something to pull against. The ship builder had not been so kind as to weld pad-eyes in all the right places.

In short order, the three decided to expand the 5:1 from Bobby's harness over the top of the tank and anchor it to the steel plate opening of space number five. A carabiner with the gate open was latched over the opening like a hook. It would have to work. The command TENSION was called, and the three rescuers pulled on the haul system. Bobby lurched forward and with some teamwork and serious maneuvering, he was lifted up the side of the tank. Fortunately, the additional pulley was not needed since converting to a 10:1 would have required lighter pulls but twice as much time and twice as many pulls. After the short haul operation, Bobby passed through the remaining openings and on to heaven with considerable ease.

SIMPLE MACHINES AND MECHANICAL ADVANTAGE

Webster's Dictionary will tell you that a machine is an instrument designed to transmit or modify the application of power, force, or motion. A machine's energy source may be a complex chemical fuel like gasoline, diesel, or even rocket fuel, or it may be a simple source like people or gravity. When all the components of a rope system are assembled, a working and relatively sophisticated machine has been designed to accomplish a life-saving task. Specifically, the rope-based machine's purpose is to direct people's energy or gravity toward safely accessing, stabilizing, and transporting people away from peril.

We have learned that it is much easier to pull down rather than to pull up on a rope. A simple machine, a pulley, can change the direction of a rope making our job dramatically easier by allowing us to pull down rather than up (Figure 8–1). **Simple machines** accomplish a task in a single movement. They translate input by people into multiplied force. The exchange is defined in terms of time and distance. Managing a rope system involves altering the effects of gravity on objects with a mass that is less than that of Earth's. Humans have a relatively narrow power window, meaning we can usually exert a given amount of force in a given biomechanical configuration. When pulling an object across the ground, the average person can drag about his body weight a couple of times. This is dependent on many variables such as *surface friction,* which includes mud, ice, snow, dry concrete, wood, grass, slope angle, shoe sole design and *physical condition,* including hand grip strength, leg and back strength, body weight, and adrenaline.

► **simple machines**
Machines that accomplish a task in a single movement. They transalate input by people into multiplied force. The exchange is defined in terms of time and distance.

Conditions permitting, a 200-pound person can *drag* 200 pounds, *lift* about 100 pounds, and, when on belay, *hold* about 50 pounds. These numbers represent the person's power window in different configurations. The number of times these actions can be repeated is equally variable. Unfortunately, a person's power window is not strong enough or repeatable enough to accomplish many of the tasks needed to successfully perform rope rescue. Therefore, rope systems must be engineered to mechanically increase a person's power window.

FIGURE 8–1
Pulley system components are added to ropes to haul loads from one point to the other. *(Courtesy Chase N. Sargent)*

People might not have a lot of power, but they do save time, room, and distance. The rope rescue system components trade time and distance for power.

NOTE: A controlled descent works by anchoring most of the load.

A DAY ON THE ROPES (EXAMPLE 1)

Joe cannot hold a rope loaded with a 200-pound person. The rope is too thin and slippery, and the load is about four times his power window capability. Without Joe the load will hit the ground very quickly and violently. Imagine 200 pounds accelerating from zero to 140 miles per hour at about the same rate as a Lamborghini. Joe realizes that his power window is limited and devises a way to spread out the descent energies over time. A properly anchored DCD attached to the rope and the load will reduce the feel of the load significantly by *anchoring most of the load.* If Joe can reduce the load by four times, it will weigh 50 pounds, which is well within his ability to control. The load gets to the ground at a much slower and safer pace, because the energy is absorbed by the metal DCD and dissipated into the air. In fact, properly rigged DCDs not only reduce the feel of the load four times, they can reduce it to infinity by increasing the friction until the load stops completely.

NOTE: Inclined planes exchange force required for time and distance.

A DAY ON THE ROPES (EXAMPLE 2)

Jolean has to lift a 200-pound box straight up onto a 4-foot-high loading dock, but she can only lift 100 pounds straight up. Since she is alone and the task must be completed, she must engineer a system that will make the box feel about 100 pounds or less. Jolean notices a 16-foot aluminum ramp at the base of the loading dock and arranges it in an inclined plane configuration. She is able to easily slide the box up the ramp and onto the loading dock. She traded some distance and some time to reduce the feel of the box by a ratio of four to one (16-foot ramp/4-foot height). The box is now well within her power window and seems to weigh only 50 pounds.

NOTE: Pulley systems exchange force required for time and distance.

A DAY ON THE ROPES (EXAMPLE 3)

Jolean and Joe are the only rescuers available at a remote park where their friend Jeff has fallen while bouldering and injured his ankle. Jeff is not seriously hurt, but his ankle is sore, and the only way back to safety is straight up the rock face to the trail. Jeff weighs about 200 pounds, and Jolean and Joe cannot lift him without help. Jolean has two small pulleys and some carabiners in her pack, and Joe has a length of rope. Jolean and Joe rig a four-to-one compound mechanical advantage pulley system, anchor one side to a tree and hand the rope down to Jeff, who attaches it to his harness. Jolean and Joe can lift Jeff easily now because he seems to weigh only slightly more than 50 pounds, which is well within their power window. The trade off was time *and* distance. The raise took longer because the combined rope needed to perform the raise was longer than the actual lift.

► **conservation of mechanical energy**
This law states that the total mechanical energy in an isolated system is always constant.

It is helpful to understand one of the most fundamental laws of physics, the **conservation of mechanical energy.** This law states that the total mechanical energy in an isolated system is always constant. In a mechanical system that is isolated from external influences, energy is neither created nor destroyed but may change from one form to another. This law cannot exactly be applied to rope systems well, because we obviously cannot isolate external influences. However, it does help explain in general terms how rope machines alter energy in somewhat the same way energy is changed in mechanical systems. We have seen how this law works in lowers, when descent forces are transformed into heat via friction when DCDs are rubbed by rope. The total energy is the same whether we fly out of the air and go splat or float down slowly like Tinkerbell. The energy simply changes forms.

With raising systems, there is a given task such as raising a load from bottom to top. The total amount of effort required to accomplish the task is called *work*. Theoretically, the energy required to meet the task is always the same—it just changes form when humans carry out the task. The total amount of energy exerted is the difference between theory and reality. That difference can be considerable, even prohibitive, if we do not design our machines intelligently and with efficiency.

One measure of a machine's ability to assist people is called **mechanical advantage (MA).** It can be described as the advantage that is gained by using a machine to transmit force, and it can be calculated by comparing the amount of input force (expressed as 1) to the amount of output force. **Theoretical mechanical advantage (TMA)** is the ratio of output force compared with input force generated by a machine with a complete lack of friction, which is at this time completely impossible. MA is expressed in ratio form. For example, 4:1 implies four units of force coming out of the machine, and one unit of force going into the machine.

Besides time and distance, another trade-off in using machines is the friction generated by interacting parts that reduces efficiency. The **actual mechanical advantage (AMA)** of a machine is found by calculating TMA and subtracting from it the effects of friction, which can be expressed as the **friction coefficient (FC)** (Figure 8–2). While friction is the greatest ally in DCD-controlled lowering systems, it is an energy robber in machines. Understanding three simple machines and their interactions with surfaces and other machines helps greatly in making rescue tool decisions. They are inclined planes, levers, and pulleys.

► **mechanical advantage (MA)**
The advantage that is gained by using a machine to transmit force.

► **theoretical mechanical advantage (TMA)**
The ratio of output force compared with input force generated by a machine with a complete lack of friction.

► **actual mechanical advantage (AMA)**
This is found by calculating TMA and subtracting from it the effects of friction.

► **friction coefficient (FC)**
A measure of the loss of advantage created by the effects of friction on a machine.

Inclined Plane

Simple machines accomplish tasks in a single move. The simplest machine is the **inclined plane,** which is a ramp. Everyday examples are on-ramps to interstate highways, stairs, ladders, wedges, axe

► **inclined plane**
The simplest machine, which is a ramp.

FIGURE 8–2

A simple 2:1 pulley system. The anchor bears half the load and the haul team bears the other half. All pulleys have some friction and therefore the actual mechanical advantage (AMA) is always less than the theoretical mechanical advantage (TMA).

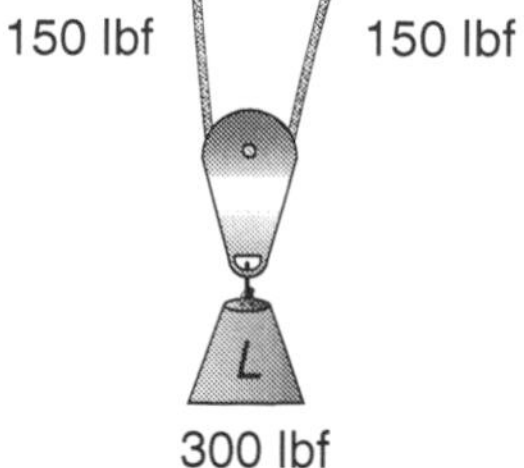

Actual MA = 1.9:1
Theoretical MA = 2:1
Friction coefficient of pulley bearing = 10%, or 0.10

TMA = 2.0:1
AMA = 2.0 – 0.1 = 1.9:1

blades, and door stops. Inclined planes can even be wrapped around a post to make a helix. This creates a machine that applies the mechanical advantage of the inclined plane repeatedly. Wood screws, bolts, machine screws, screw jacks, car jacks, propellers, and even spiral staircases are all helixes. Climbing a mountain trail uses a kind of inclined plane where, again, the work is spread out over time and distance rather than climbing a straight vertical wall.

Inclined planes are the perfect example of how mechanical advantage spreads the workload out over time and distance (Figure 8–3). A 16-foot ramp with a 4-foot rise has a 4:1 rate of rise for a 4:1 TMA. Inclined planes are the least efficient of all simple machines, meaning that the friction developed in gaining the advantage is higher than in other simple machines. This reduces advantage drastically in most cases. A box being pushed up a ramp has a great deal of surface area interacting with the ramp. Friction creates heat as the surfaces rub together. There are many ways to reduce the friction in an inclined plane, usually by using rollers, wheels, and lubricants. Due to their inefficiency, inclined planes are rarely used in rope rescue. Nevertheless, they are an important concept on which to base an understanding of other simple machines.

Lever

▶ **lever**

A combination of four parts: the lever, fulcrum, applied force, and the object being moved.

▶ **Class I lever**

Mechanism where the fulcrum is placed between the object being lifted and the applied force. The force is then applied in the opposite direction of the object being moved.

The **lever** is a combination of four parts: the lever, fulcrum, applied force, and the object being moved, aligned so the applied force is transmitted to another object. Levers can be classified as Class I, Class II, or Class III depending on where the fulcrum, applied force, and lever are placed in relation to the object being moved.

In a **Class I lever,** the fulcrum is placed between the object being lifted and the applied force. The force is then applied in the opposite direction of the object being moved. A teeter-totter is a familiar example (Figure 8–4). The *lever* is the teeter-totter board, the *object being moved* is one of the children on the board, the *fulcrum* is the board support or cross member, and the *applied force* is the

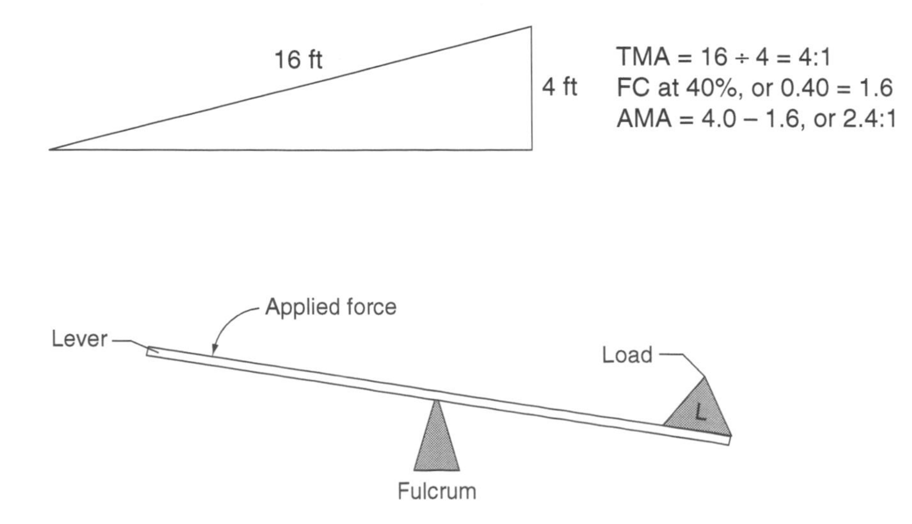

FIGURE 8–3

The inclined plane is notoriously inefficient due to the large amount of surface area generating friction between the object being moved and the surface of the plane.

FIGURE 8–4

The Class I lever; for example, a teeter-totter.

other child. As the fulcrum is moved closer to one side, more advantage is gained. A small child can lift an adult if the fulcrum is moved very close to the adult making the lever, or leverage, very long for the child. Specifically, if you have one foot of the lever on one side of the fulcrum and 10 feet of lever on the other side of the fulcrum, you have a 10:1 mechanical advantage. Theoretically, 100 pounds pushing down on the lever translates to 1,000 pounds being applied to the object being lifted. Unlike the other classes of levers, Class I levers benefit infinitely by extending the lever infinitely. Theoretically, you could lift an aircraft carrier with one finger if you had a lever that was long and strong enough and a fulcrum on which to push.

▶ **stroke**

The distance from the start of lever movement to the finish of lever movement.

The drawback to the Class I lever is its inability to perform its function repeatedly. The Class I lever has a **stroke,** the distance from the start of lever movement to the finish of lever movement, that is determined by the position of the fulcrum, the object, and the lever. Unless there is something to capture the progress made by the lever, any gain will be lost when the lever returns to the starting position.

▶ **Class II lever**

The fulcrum is at the end of the lever, the object being moved is between the applied force and the fulcrum, and the applied force is at the end opposite the fulcrum.

In a **Class II lever,** the fulcrum is at the end of the lever, the object being moved is between the applied force and the fulcrum, and the applied force is at the end opposite the fulcrum. The force is moving in the *same* direction as the object being moved. A wheelbarrow is a classic example (Figure 8–5). The *applied force* is the person carrying the handles, the *object being moved* is the load in the container, the *lever* is the handles and frame, and the *fulcrum* is the axle in the wheel. Again, the mechanical advantage is determined by the relative location of the object being moved to the fulcrum. The closer the object being moved is to the fulcrum, the greater the mechanical advantage, but this has a point of diminishing returns. If the object gets so close to the fulcrum that it is *on* the fulcrum, all advantage ceases.

▶ **Class III lever**

Applies force between the fulcrum and the object being moved.

The **Class III lever** applies force between the fulcrum and the object being moved. The fulcrum is at one end of the lever, the object being lifted is at the opposite end, and the applied force and the object move in the *same* direction. The best example of a Class III lever is a fire department aerial ladder, where the *lever* is the ladder, the *fulcrum* is anchored to the turntable, the *applied force* is from hydraulic cylinders pushing on the bed ladder, and the *object being moved* is the tip of the ladder or platform (Figure 8–6).

FIGURE 8–5

The Class II lever; for example, a wheelbarrow.

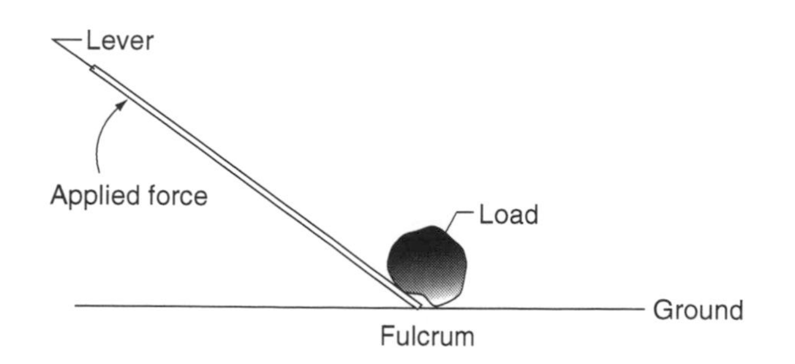

FIGURE 8–6

The Class III lever; for example, an aerial ladder.

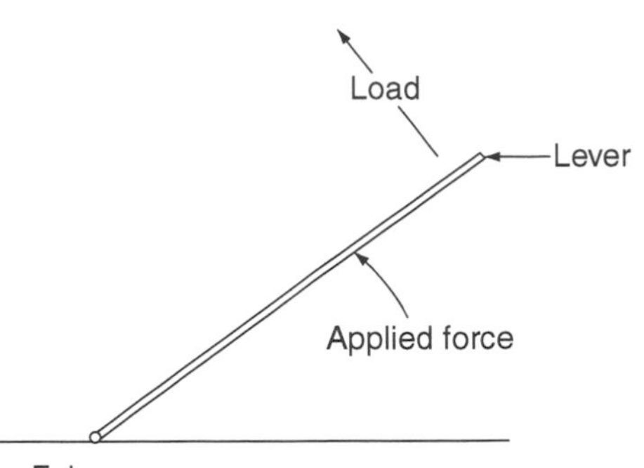

FIGURE 8–7

Comparison of lever components and action.

Levers	Fulcrum	Applied force	Object being moved	Applied Force and direction of object being moved	Example
Class I	Between	End	End	Opposite	Teeter-Totter
Class II	End	End	Between	Same	Wheelbarrow
Class III	End	Between	End	Same	Aerial ladder

No mechanical advantage is gained from the Class III lever. In fact, mechanical ***dis****advantage* is the result as the weight of the lever *adds* weight that the applied force must move. Disadvantage gets progressively worse as the applied force moves closer to the fulcrum. It is only through the near miracle of shear hydraulic force, delicate balance, and raw outrigger stabilization that the Class III aerial lifting lever lifts anything at all. The Class III does have a potentially long stroke, though, and is useful in applications where stroke is needed and sufficient force is available to lift the lever and the object with no additional mechanical advantage. Figure 8–7 shows a comparison of lever components and their actions.

► **pulley**

A simple machine that transfers applied force from one location to another when used in conjunction with a flexible medium, such as rope, chains, belts, and so forth.

The **pulley** is a simple machine that transfers applied force from one location to another when used in conjunction with a flexible medium, such as rope, chains, belts, and so forth. The round disk in a pulley is called a sheave (Figure 8–8), the sides are called side plates, and the device in the center, around which the sheave rotates is called the axle. Pulleys with two or three sheaves are called double or triple sheave pulleys, respectively. Any good double pulley has a center plate and an extension off the center plate called a becket for a carabiner attachment (Figure 8–9A). Some double pulleys do not have beckets. Modern pulleys are descendants of the block and tackle. Early blocks were just that—blocks of wood rounded off to accept and change the direction of a rope. Eventually sheaves, axles, and beckets were added to blocks to make them more useful and efficient. Today's R/Q pulleys (Figure 8–9B) distance themselves from their older cousins by being lighter, stronger, cheaper, and more versatile and reliable than the blocks of yesteryear. R/Q pulleys are made from aluminum, steel, or a combination of both and come in various sizes and shapes. The breaking strength of most pulleys is in excess of 8,000 lbf, and some all-steel versions break in excess of 15,000 lbf. Axles are joined with sheaves using

FIGURE 8–8

A labeled view of a pulley.

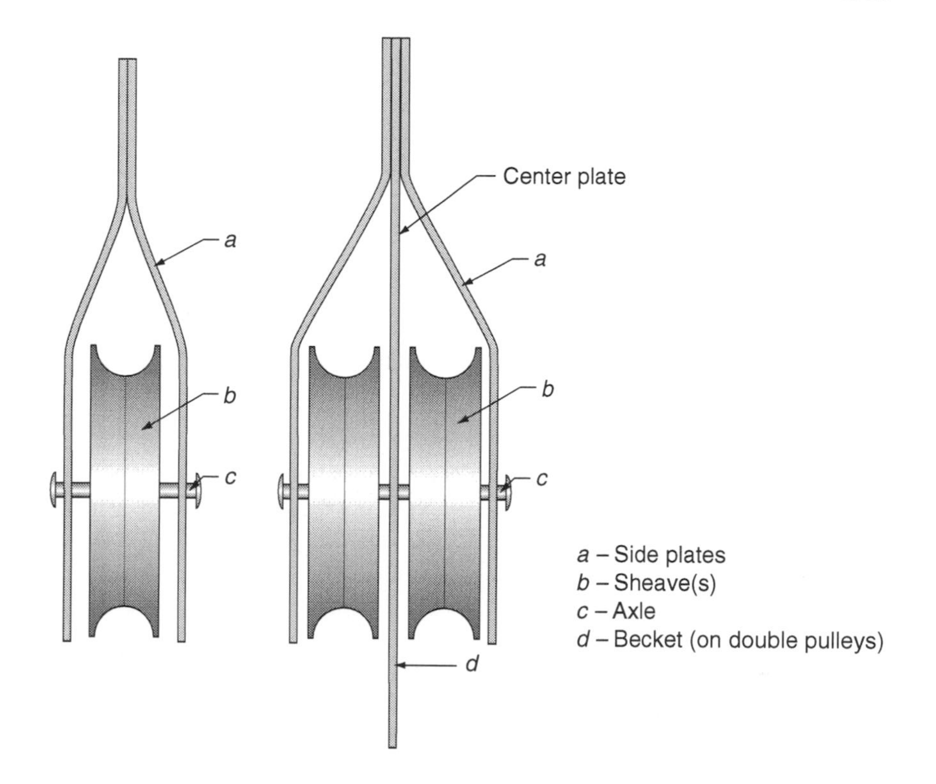

sealed and permanently lubricated bushings or sealed ball bearings. A few pulleys are even made using needle bearings on the axle.

There are some interesting but marginal similarities between pulleys and levers. For example, both have applied force, an object being moved, and a fulcrum or axle. The sheave can almost be seen as a lever similar to a steering wheel with spokes. But pulley and lever comparisons end there because there is a distinct difference between wheels and pulleys. While there is mechanical leverage being applied to the hub of a wheel in increasing and direct proportions to the length of the lever arms, or size of the steering wheel, a pulley transmits force equally on *both* sides of the sheave. Force is not transferred to the hub, as in a steering wheel, or out of the hub, as in a drive wheel. A pulley simply transfers the same amount of force into the sheave (input) as it receives out of the sheave (output) minus any frictional elements, of course. The pulley itself develops no mechanical advantage since the fulcrum (axle) is dead center, like a balanced like a fulcrum center teeter-totter.

All pulleys turn on their axles. When they are used in building a rope system, they are either fixed in position (anchored), or they are moving a load. Identifying whether a pulley is anchored or moving a load is critical in determining the mechanical advantage of a machine and its components, such as a rope system.

► **anchored pulleys**

They are directional only and provide no mechanical advantage.

Anchored pulleys (Figure 8–10A and Figure 8–10B) are directional only and provide no mechanical advantage. Imagine anchoring a pulley to one of the overhead beams in your station. Reeve a rope through the pulley so both ends of the rope dangle to the ground. A COD pulley arrangement like this has many uses but has no mechanical advantage. In fact, it can be considered a *disadvantage*. When a load is being pulled up one side by a rope being pulled down on the other side, twice as much weight is on the pulley as

(A)

(B)

FIGURE 8–9

(A) Double pulley with a becket.
(B) Modern R/Q pulleys come in all different sizes and shapes.
(Courtesy Martin Grube)

the load itself—the force of gravity on the load and the force put on the rope by pulling down on the other side are added together (see the discussion about CODs in Chapter 7).

In addition, there is a little more load on the pulley when it is turning because of the effects of friction in the pulley axle. The FC of a pulley or any machine can be determined by measuring the input (a known amount of weight/work) and subtracting it from the output (how much work is required to actually raise the weight). The difference is the FC can be expressed as a percentage of the known weight. You can estimate the FCs of pulleys, and other devices by rigging up a test jig as illustrated in Figure 8–11 and as explained later. Exact calculations of FCs are almost impossible outside of strict laboratory environments due to the influence of myriad variables. They can include rope and cordage stretch, humidity and temperature, quality and type of measuring device, and who is doing the test and what their expected outcomes are.

FIGURE 8–10

(A,B) Anchored pulleys make no mechanical advantage, but are extremely useful in directing the rope to the desired location. *(Courtesy Martin Grube)*

(A)

(B)

NOTE: The friction coefficient of a pulley can be determined by using the following formula (Figure 8–11): I = input, O = output, FC = friction coefficient, I − O = X, FC = X ÷ O.

Input is considered to be the amount of force delivered to the pulley, enough to cause the load to raise. Output is the load side of the pulley, on the outside of the pulley travel.

For example, if input equals 660 lbf and output equals 600 lb:

I = 660   I − O = X
O = 600   660 I
−600 O
60 X

FC = X ÷ 0
FC = 60 ÷ 600
FC = 0.10

You will need a measuring device. An electronic quartz-based force link or load cell is best or, if that is not available, a dynamometer to insert on the input side of the pulley.

To rig a test jig to calculate pulley efficiency (Figure 8–11A):

1. Hang a pulley on a strong structural overhead beam.
2. Reeve the pulley with 1/2-inch rope and tie a figure eight on a bight in the end of the rope.
3. Strap together 300 lb of your station's workout free weights and connect the end of the rope to the strap with a carabiner. You will raise this known quantity of 300 lb on the input side of the pulley.
4. Rig your measuring device onto the rope on the input side of the pulley. This is the side you will pull down on to raise the free weights and to measure the pounds force required to raise them *plus* any friction added by the pulley. The measuring device can be easily rigged inline with the rope by

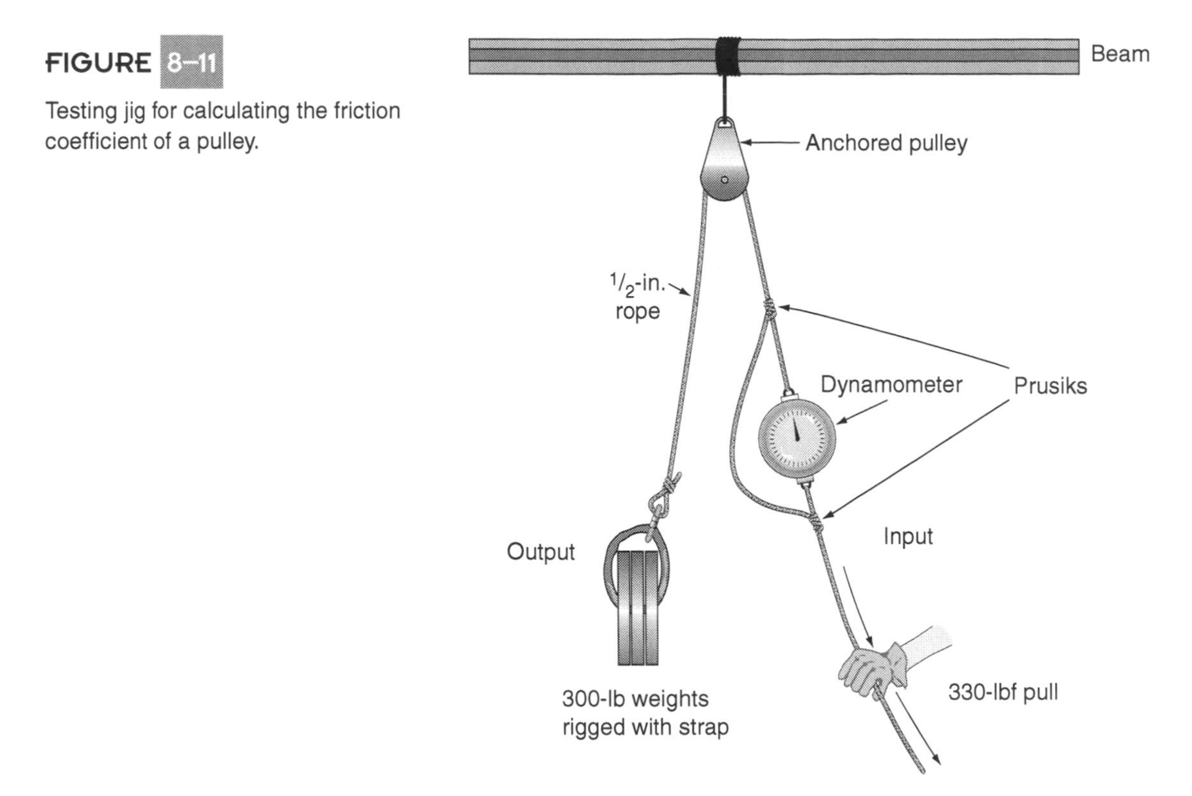

FIGURE 8–11

Testing jig for calculating the friction coefficient of a pulley.

inserting it with prusiks or rope grabs. The load will be transmitted into and out of the dyno or load cell via the prusiks, and the section of the rope between the prusiks will be slack.

5. As you pull down on the rope and the free weights start to raise, have someone read the measuring device while the load is in motion. Subtract the weight of the free weights, or output side (300 lb), from the input force of maybe 330 lbf. The result is 30 lbf. It takes an additional 30 lbf or 10% to raise the load. Therefore, the FC of the pulley is about 10%.

In general, larger diameter bearings spread the load out over a greater bearing distance. This makes them more efficient than small diameter bearings that must bear the entire load on a very small surface area. As the load on a pulley bearing is increased, its friction coefficient also increases. A pulley with an FC of 10% when loaded with 600 lbf may have an FC of 17% when loaded with 1,600 lbf and 35% FC at 5,500 lbf. Eventually it will fail at 100% FC. A graph can be drawn to illustrate the efficiency profile of a pulley by connecting dots representing measurements at different loads. Studying different efficiency profiles allows manufacturers to determine the best combination of bearings or bushings, material and costs for application and customer-specific manufacture.

Friction coefficients add up in all machines, but especially in rope system machines. We always use *more* total energy on the input side than we get on the output side because of the effects of friction and other variables. The actual MA of a pulley system with many pulleys of various sizes, types, and mechanical conditions is impossible to determine on paper, but it can be estimated. On a *non-moving* multi-line, multi-pulley *simple* MA system, such as a 4:1 that is not being pulled (Figure 8–12A), all the ropes and pulleys share the load equally. When the system is in motion, the friction coefficient of each bearing must be calculated in series and added to the ultimate force required to move the load (Figure 8–12B). Even more interesting is an observation of compound systems where each leg and pulley away from the load carries the mechanically derived advantage (less load), so the system must be calculated differently than with simple systems. On compound pulley systems, there is a greater load and a higher FC on pulley bearings nearer the load than on the applied force (input) side. As bearing load increases, so does FC. In pulleys with identical bearings and sheaves, the primary pulley bearing (closest to the load) has a slightly higher FC, perhaps 11%, and the secondary pulley bearing has a slightly lower FC, maybe 10%, due to the lighter sheave and bearing requirement.

Accurate field measurements of AMA can only be measured with instruments by dividing the weight of the load being moved (output) by the force required to move it (input). The easiest way to understand pulley systems is to focus on the TMA and observe and practice input to output ratios.

▶ **nonanchored, or moving, pulleys**

Move in the same direction as the load toward the anchor provided no CODs have been added to the main line. They provide a mechanical advantage by sharing loads among multiple ropes in the pulley system and by the rolling action of the sheave that facilitates the transfer of applied force to the load.

Nonanchored, or moving, pulleys move in the same direction as the load toward the anchor provided no CODs have been added to the main line. They provide a mechanical advantage by sharing loads among multiple ropes in the pulley system and by the rolling action of the sheave that facilitates the transfer of applied force to the load. Moving pulleys provide a mechanical advantage of two to one (2:1) minus any friction that may be in the bearings. The rope going to a moving pulley halves the weight of the load on the pulley. By anchoring one of the rope segments and pulling on the other rope segment, this effectively forces the anchor to carry half the load and you to carry the other half (Figure 8–13). Larger MA pulley systems are engineered by progressively adding more and more parts to the anchor. In a 10:1 MA, the anchor carries nine parts, the haulers still carry only one.

So how does this affect the decisions rig masters make? Rescue situations contain an infinite number of variables, such as the number of rescuers available, their skill levels, the available equipment, the location of the victim, anchors, weather, and so on. Understanding the cause and effect of rescue solutions *beforehand* makes a rescue much more efficient. For example, there are simple and compound 4:1 pulley systems. Both systems are 4:1s, but they have major differences that make each superior under different circumstances. The decisions you make are the difference between a good rig master and a great one.

▶ **rescue quality (R/Q) pulley system components**

Combinations of R/Q pulleys, ropes, and carabiners that develop mechanical advantage in a neat and orderly fashion.

Rescue quality (R/Q) pulley system components are combinations of R/Q pulleys, ropes, and carabiners that develop mechanical advantage in a neat and orderly fashion.

FIGURE 8–12

(A) In this non-moving 4:1 pulley system each part of the system carries one-fourth of the load, or 75 pounds.

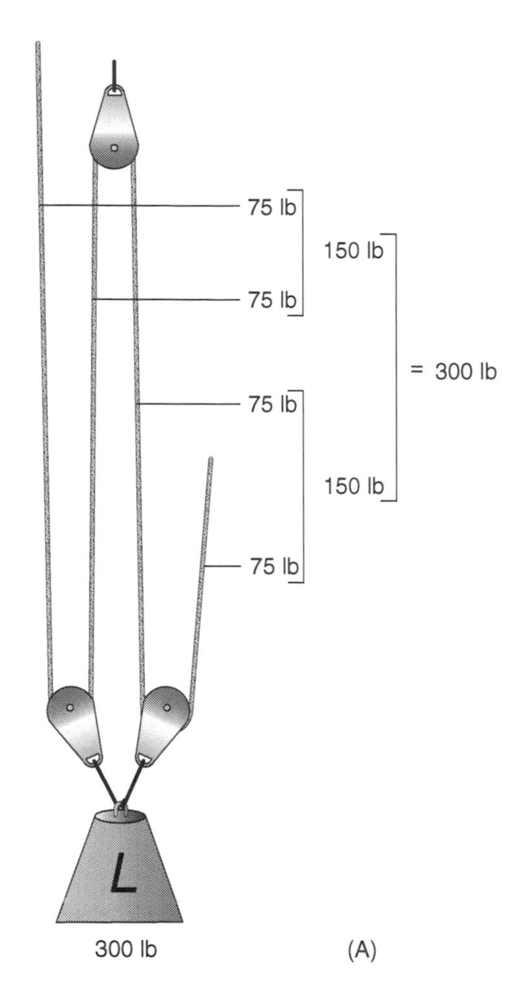

► **compression**

Pulling the pulley system down to its smallest stroke.

► **de-set**

When the system is compressed so it will expand again when using the MA system as the DCD to lower a load rather than raise it.

► **extension**

To pull out a pulley system to its longest strike.

► **re-set**

When the system is extended again during raising operations and another haul segment is made on the main line.

Twenty-One Rules of Thumb for Pulley Systems

The following are twenty-one rules of thumb, as applied to the TMA gained by pulley systems:

1. Effective pulley systems must always have one side anchored and the other side attached to the moving load, known as the anchor side and the load side. There *must* be something to pull against.
2. The longest distance a pulley system can be stretched, the distance from anchored pulleys to moving pulleys, is called its stroke. Usually the longer the stroke, the more useful the MA system.
3. Pulling the system down to its smallest stroke is called **compression.** It is called **de-set** when the system is compressed so it will expand again when using the MA system as the DCD to lower a load rather than raise it.
4. **Extension** means to pull out a pulley system to its longest stroke. **Re-set** is when the system is extended again during raising operations and another haul segment is made on the main line.
5. All anchored pulleys are COD only.
6. Pulley system MA can be expressed as a ratio of the amount of force that must be applied to a haul line in order to move a load

(*continued on page 254*)

FIGURE 8-12 cont'd

(B) In the moving (dynamic) 4:1 pulley system, each part of the system must be calculated in series starting from the anchor and working toward the hauling rope. The amount of load on each succeeding pulley is increased, since all previous pulleys' friction must be overcome.

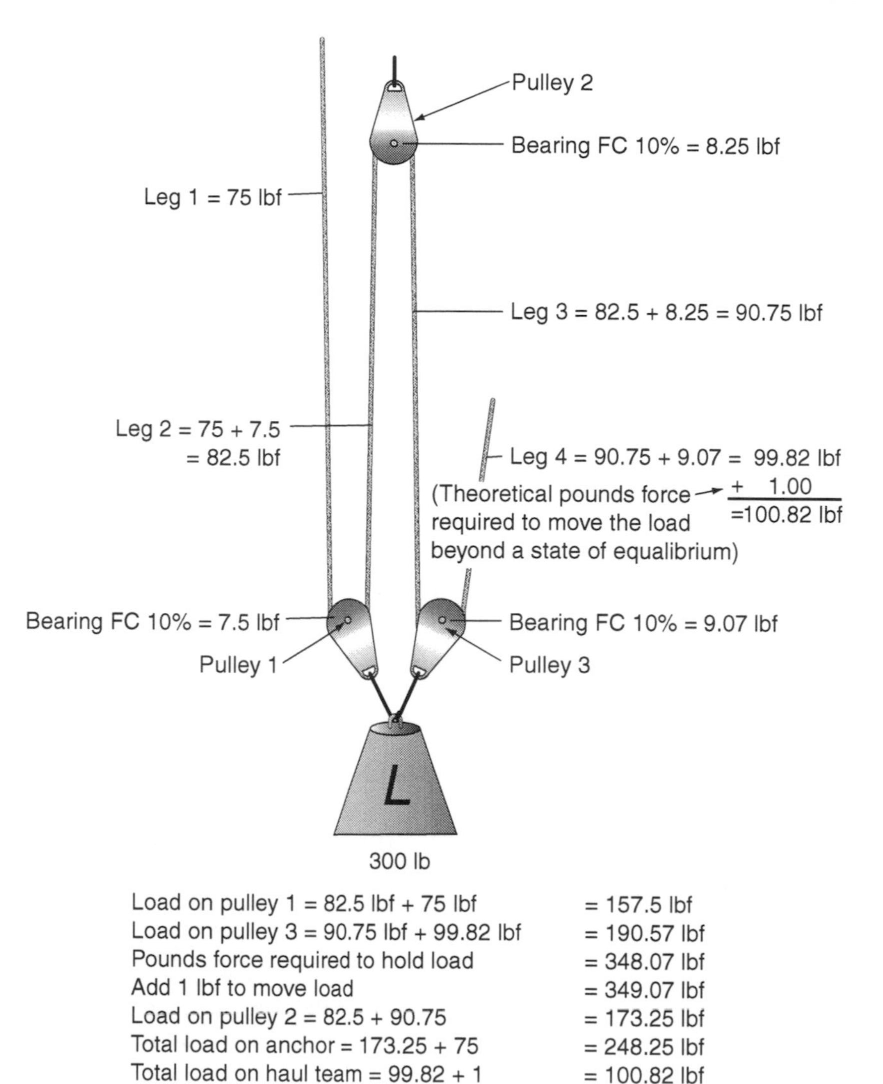

(B)

FIGURE 8-13

Pulleys allow us to change the direction of the rope so that we can share some of the work with the anchor. The more parts of the pulley system that are attached to the anchor, the greater the mechanical advantage, and, proportionally, the less we have to carry.

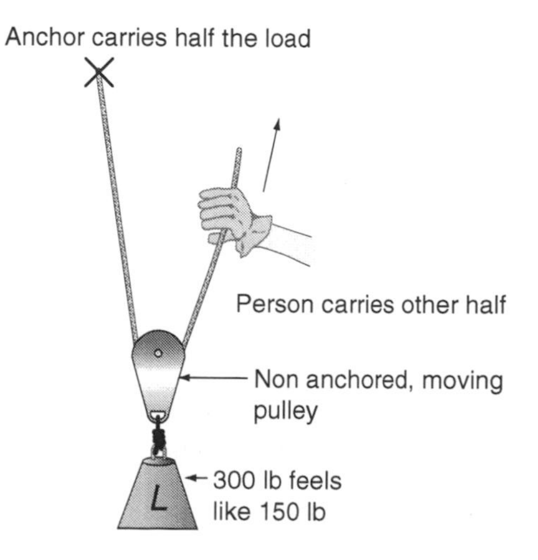

Twenty-One Rules of Thumb for Pulley Systems (Cont'd)

divided into the weight of the object being moved. A 300 lbf load that is being raised by 50 pounds of applied force through a pulley system is being moved by a 6:1 MA since 300 ÷ 50 = 6. Using a rope and some pulleys has allowed the anchor to hold five parts of load or five ropes and the hauler only one part of the load, the haul rope. In other words, 5 + 1 = 6 for a 6:1 MA.

7. Pulleys that move with a load, nonanchored, are simple machines that gain advantage. Side-by-side moving pulleys such as double sheave, triple sheave, or multiple single sheave pulleys at the same point of attachment add mechanical advantage by a factor of two for each working sheave.
8. Pulley systems are either simple, compound, or complex.
9. In simple pulley systems, you can add the number of pulleys moving with the load by a factor of two, to determine mechanical advantage (see number 5). For example, 1 moving pulley = 2:1 MA, 2 moving pulleys = 4:1 MA, 3 moving pulleys = 6:1 MA, and so forth.
10. In simple pulley systems, count the ropes supporting the load to determine the MA. *Be careful not to count any additional ropes that may be reeved to anchored pulleys just to make the direction of the pull more convenient. Remember rule number 4, all anchored pulleys are COD only.*
11. In simple pulley systems, count the MA and COD pulleys throughout the system and add one to get the MA. For example, with 2 pulleys in the system, 2+1 = 3:1 MA; with 4 pulleys, 4 + 1 = 5:1 MA. Conversely, a 4:1 simple pulley system uses 3 pulleys, a 5:1 uses 4, etc., not including any additional CODs.
12. In the normal application of odd-numbered simple pulley systems, the *terminal* end of the haul line rope is attached to the load, 1:1, 3:1, 5:1, 115:1, etc.

Note: When building pulley systems, always start by establishing the **terminal end** of the rope you are using. Tie a compact figure eight on a bight in the end of the rope and identify this important component to your team members. Determining where this terminal end is placed is critical for ciphering and building pulley systems.

13. In the normal application of even-numbered simple pulley systems, the terminal end of the haul line is attached to the anchor, 2:1, 4:1, 6:1, 248:1, etc.
14. Simple pulley systems have a greater stroke than compound pulley systems of the same advantage rating.
15. Compound pulley systems are made up of at least two simple pulleys that are compounded. A primary pulley creates the initial advantage, and the secondary pulley pulls on the advantage created by the primary pulley, effectively multiplying the advantage gained by the first. The nonanchored MA pulleys are multiplied to determine the MA as follows: 2 × 2 = 4 for a 4:1 MA, 2 × 2 × 2 × 2 = 16 for a 16:1 MA, etc.
16. Compound pulley systems, all equipment components being equal, are more efficient than equal numbered simple pulley systems, because they have fewer pulleys and, therefore, less pulley bearing friction to overcome. Total system friction is less than in simple systems.

► **terminal end**

A designation for the very last part of the rope. The terminal end is used to make connections to equipment and anchors.

17. In compound pulley systems, the pulley, or pulley set, closest to the load is called the primary pulley; the second pulley from the load is called the secondary pulley; and the third pulley back from the load, if used, is called the tertiary pulley.
18. When in doubt as to the MA developed by any pulley system, you can measure it to determine MA (Figure 8–14). Carefully measure the amount of rope length that must be pulled through the pulley system to move the load a given distance. The rope length on the haul side of the system is the advantage when compared to the distance the load has moved. For example, if the rope pulled through the system is 30 feet, and the load moves 5 feet, the system is a 6:1 or 30 ÷ 5 = 6. You must also account for rope stretch, which will exaggerate your calculations if a large load and haul force are used.
19. The trade-off in time and distance for pulley systems is the length of the haul line rope that is directly proportionate to the MA gained. For example, in a 5:1 MA, the haul rope will move 5 feet for every 1 foot the load moves (see Figure 8–14).
20. Compounding, or stacking, 2:1 pulleys is the most efficient (less friction coefficient) method of increasing MA. Compounding means that 2:1 is pulling a 2:1 is pulling a 2:1 is pulling a 2:1, and so forth. It is relatively impractical, however, due to the limited stroke of compounded systems.
21. Complex pulley systems are combinations of simple and compound systems. They are useful in some specialized applications where pulleys move toward the load and/or toward the anchor simultaneously in order to move pieces of equipment at different paces and in different directions (Figure 8–15).

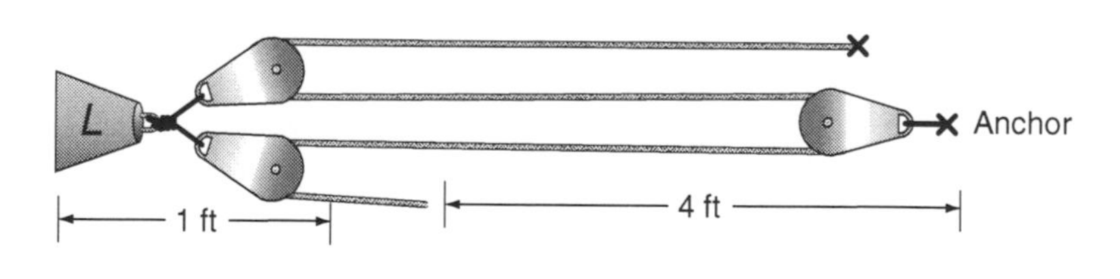

FIGURE 8–14

Pulley systems can be calculated by measuring the amount of rope that must be pulled through the system compared to the distance the load has moved.

Reeving pulleys to add to rope systems for raises is relatively simple. The general rules are as follows:

1. One side of the pulley system is always anchored, and the other side attaches to the object being moved, usually the main line. While this may seem obvious, it works well as an orientation point in unfamiliar circumstances. Pulley systems that have both parts connected to anchors do not work well.
2. When assembling parts, try to use like pulleys with similar FCs, sheave diameters, and mechanical histories. Make sure the sheaves spin freely, and the side plates are not sloppy. Check for axle and bearing damage by listening to the rotation of the sheave and visually check for metal shavings and leaking lubricants that may indicate bearing or bushing

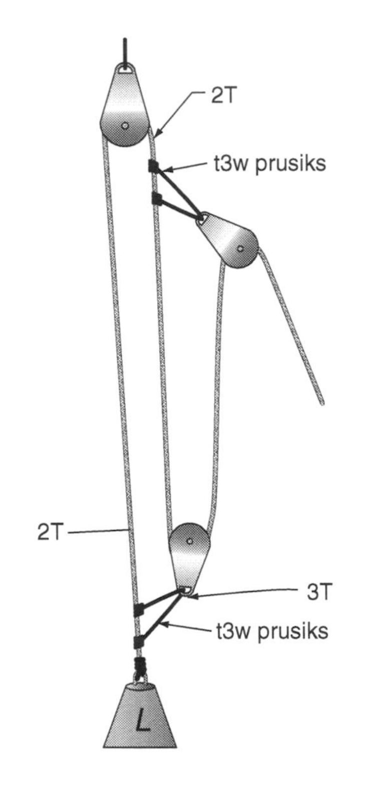

FIGURE 8–15

Complex pulley systems are made up of both simple and compound systems combined. The middle leg of this 5:1 complex pulley system is counteracting one part of the 3:1 simple system (3:1 being pulled by a 2:1 [3 × 2 = 6:1] minus 1 for the self-defecting middle leg equals a 5:1 complex).

failure. Do not skimp on rope selection. Use 1/2-inch R/Q rope with the least amount of stretch possible. Polyester ropes are preferable, if available. Make sure the rope length is suitable for the stroke required. Usually the longest possible stroke is preferred, depending on how much room your team has to maneuver the system and the MA you choose to build.

3. Two or three people can assemble almost any pulley system rapidly with some practice. Most teams can assemble 4:1 simple pulley systems in less than 15 seconds, complete with locked carabiners, tied knots and artificial tension placed on the system. Assemble the pulleys in close quarters, about 4 feet apart so the ropes can be oriented and aligned to prevent friction between the running rope sections. When properly assembled, expand the stroke of the pulley system to whatever stroke length is desired.
4. The main line can be used as the pulley system rope in a pinch, but the safest and most orderly method of adding a pulley system to your rope is called piggybacking. Simply build a separate pulley system and add it, or piggyback it, to the main line with tandem three-wrap 8-mm prusiks. The prusiks have has the advantage of slipping somewhat if the load gets hung up during the raise, and the haul team is overly aggressive. It is better to have excess force slipped through a couple of prusiks at around 3,000 lbf than to destroy part of the system with a mechanical rope grab that damages the rope.

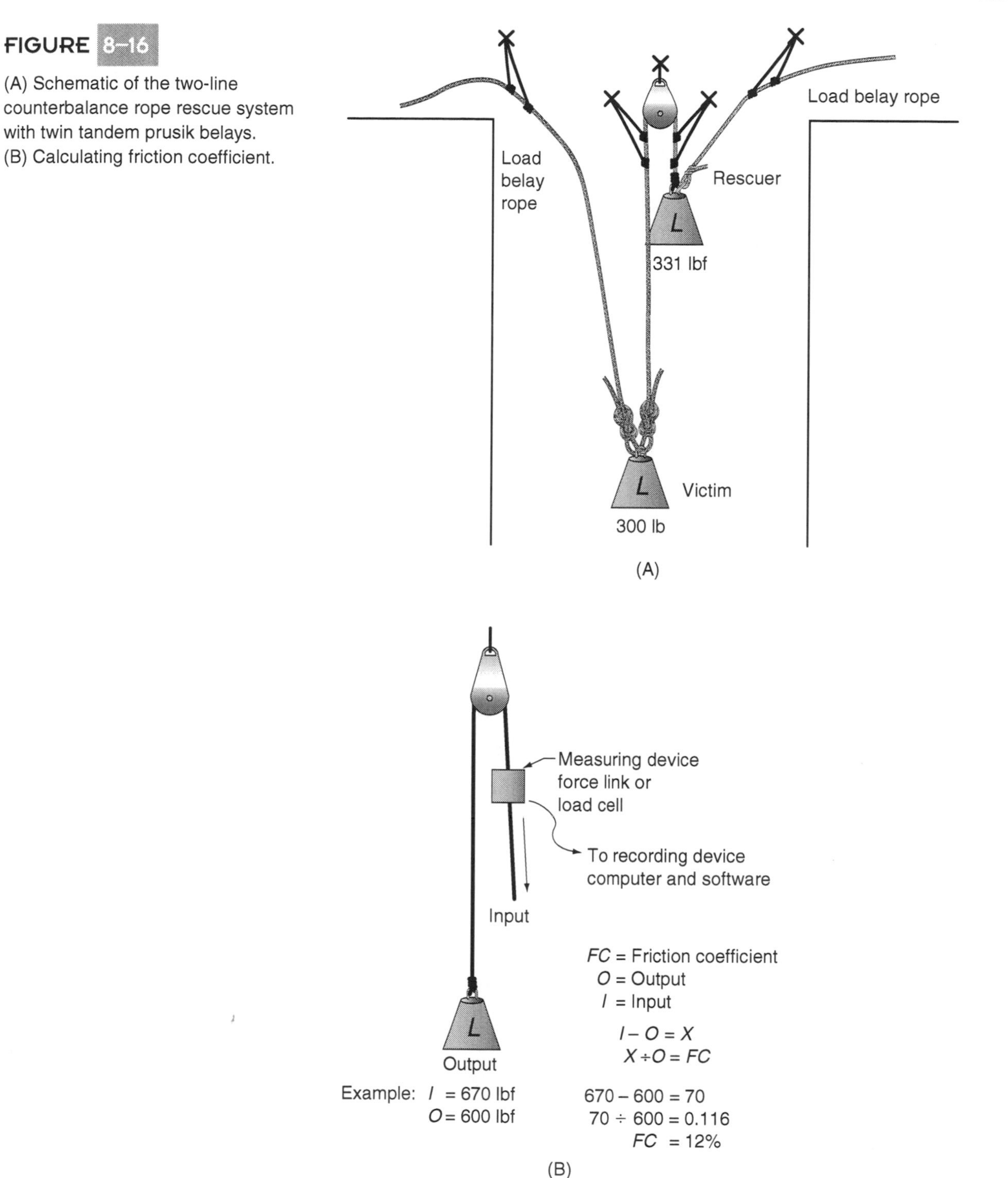

FIGURE 8–16

(A) Schematic of the two-line counterbalance rope rescue system with twin tandem prusik belays.
(B) Calculating friction coefficient.

▶ **counterbalance**

A rope system that uses gravity to perform a raise. A heavy object pulls down on one end of a rope that has been reeved over an anchored pulley. The object connected to the other end of the rope is raised by the countereffect of this object.

Building Specific Pulley Systems

The **counterbalance** (Figure 8–16A) is very useful in confined spaces and in other areas where haul system team movement is severely limited. This might be on board a large ship or at an industrial facility, where a patient has fallen and sustained non-life-threatening injuries but simply cannot climb out of the space. A rescuer can rappel or be lowered to the victim for medical considerations and patient packaging and to attach the counterbalance rope(s). Another rescuer of sufficient weight can be used to raise the patient and then simply climb the ladder or stairs out when the rescue is complete.

No MA is developed using the counterbalance. It is only a COD device constructed as follows:

1. From an elevated anchor attachment, connect an R/Q pulley.
2. Reeve the pulley with R/Q rope and make a figure eight on a bight in the end of the rope to connect the counterbalance load. The counterbalance load can be a rescuer who is slightly heavier than the load to be raised or even bags of equipment. Mylar bags filled with water have been used to precisely control the amount of counterbalance weight. Caution must be taken not to break the bag.
3. Attach the other end of the rope to the load being raised and make sure all the slack is removed. Some calculations should be considered. Unless it is known otherwise, assume the pulley has an FC of 10%. If the load to be raised is 300 pounds, the lbf required to raise the load must exceed 330, since 300 + 30 = 330.
4. All raise and lower systems require system brakes. Attach a set of tandem prusiks on both sides of the overhead COD pulley. The prusiks on the rope going up will jam against the side plates of some pulleys automatically. Prusik minding pulleys (PMPs) with squared-off and relatively narrow side-plates work best for the automatic releasing of the prusiks. If the load on the opposite side of the counterbalance disappears, for some reason, the prusiks will prevent the rope from falling back. The prusiks on the rope going down must be held in a slightly slack position to allow movement of the counterbalance. Again, if the load on the other side comes off the rope or the rope breaks, the prusiks will catch the rope.
5. Although prusiks *will* under almost every circumstance catch the rope, it is best to include a belay line with the counterbalance. Engineer the counterbalance system to have system redundancy whenever possible. The most redundant backup system you could build for the counterbalance would have *twin prusik* (t3w) belay lines—one for each side. A tandem prusik belay would go down, and a tandem prusik belay would come up, each of them separate from the overhead anchor attachment and COD pulley. Each belay line would be completely independent of the overhead anchor and COD and would remain intact in the event of any main line counterbalance system failures.
6. A slightly less redundant technique incorporates a triple-sheave pulley as the overhead COD. The counterbalance system and prusiks are built into one of the side sheaves, and the twin belays are run through the center and the other side sheave. It is absolutely critical to ensure the credibility of the overhead anchor and triple-sheave pulley for this technique. Any failure in that part of the system could be catastrophic.

Building a 2:1 simple pulley system (Figure 8–17A and Figure 8–17B)

1. Attach the terminal end of the rope to an anchor (indicator that it must be an even-numbered system) using a figure eight on a bight and a carabiner.
2. Reeve the pulley and attach the pulley to the load being moved.
3. Apply force to the running end of the rope.

Building a 3:1 simple pulley system (Figure 8–18)

1. Attach the terminal end of the rope to the load being moved (indicator that it must be an odd-numbered system).
2. Reeve and anchor one pulley.
3. Reeve another pulley and attach it to the object being moved.
4. Apply force to the running end of the rope.

Building the Z drag (Figure 8–19A and Figure 8–19B)

The popular Z drag is a 3:1 MA built from the same rope as the main line in a raising system, *instead* of building it from a different piece of rope and piggybacking it to the mainline with prusiks or a good rope grab.

1. Anchor a pulley where you want the far end of the pulley system stroke to be.

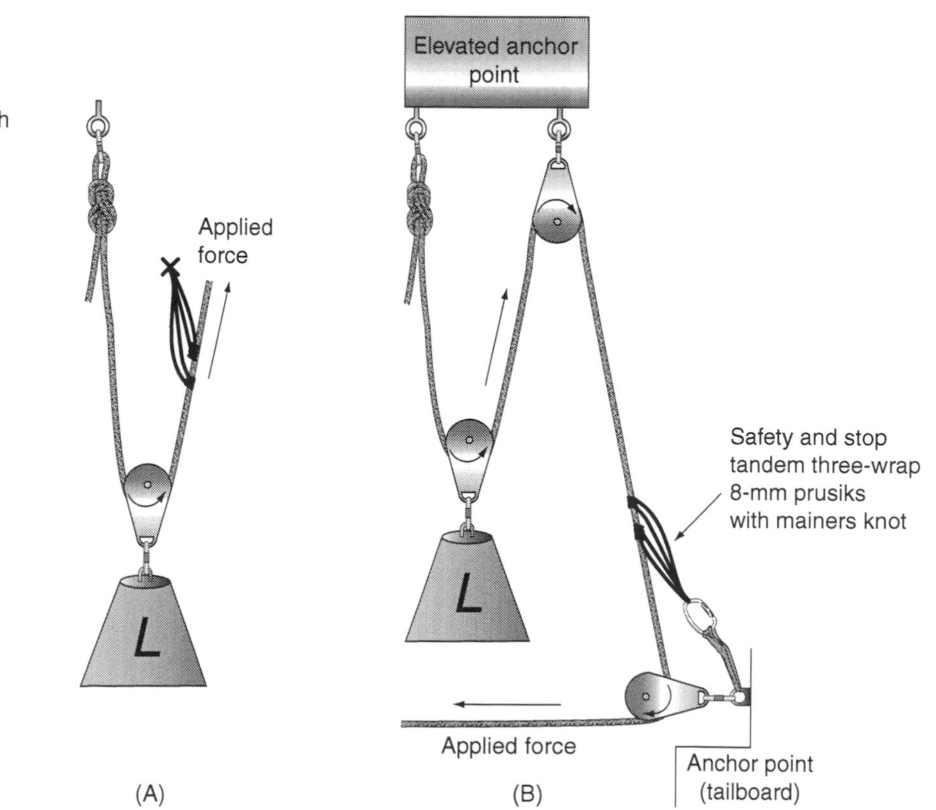

FIGURE 8–17

(A) A 2:1 simple pulley system.
(B) A 2:1 simple pulley system with two CODs.

FIGURE 8–18
A 3:1 simple pulley system.

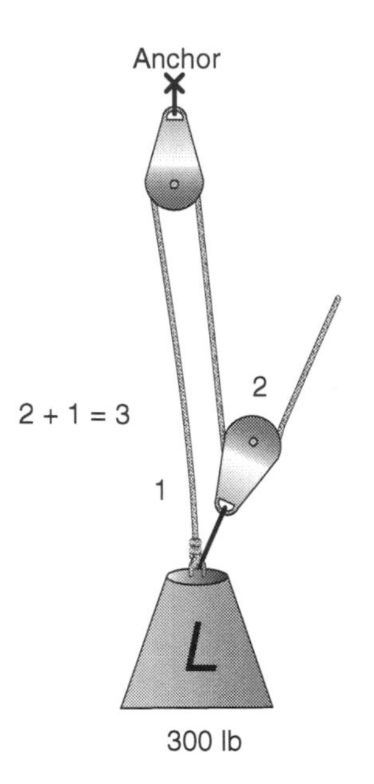

2. Bring the main line rope to the anchored pulley and reeve it, bringing the running end of the rope back toward the edge and the load.
3. Attach t3w prusiks or a good rope grab to the main line on which to connect an additional pulley.
4. Reeve the pulley so the haul rope runs parallel to the other two segments of the main line and back toward the anchor(s).

Building a 4:1 simple pulley system (Figure 8–20)

Double pulleys will be assumed for these instructions; however, side-by-side single pulleys can be substituted when necessary. Side-by-side pulleys pull in tandem effectively (2 + 2 = 4).

1. Attach the terminal end of the rope to an anchor.
2. Reeve one sheave of the load-side double pulley.
3. Reeve the load-side single-sheave pulley.
4. Reeve the other sheave of the load-side double pulley.
5. Apply force to the running part of the rope.

Building a 5:1 simple pulley system (Figure 8–21)

1. Attach the terminal end of the rope to the load. The becket on a double pulley works best and also balances the pulley.
2. Reeve one sheave of the anchor-side double pulley.
3. Reeve one sheave of the load-side double pulley.
4. Reeve the remaining sheave of the anchor-side double pulley.
5. Reeve the remaining sheave of the load-side double pulley.
6. Apply force to the running end of the rope.

► **compounding pulley systems**

Applying the output force from one pulley system to the input force of another.

Compounding pulley systems means applying the output force from one pulley system to the input force of another. It is a pulley system pulling a pulley system. In compound pulley systems, the

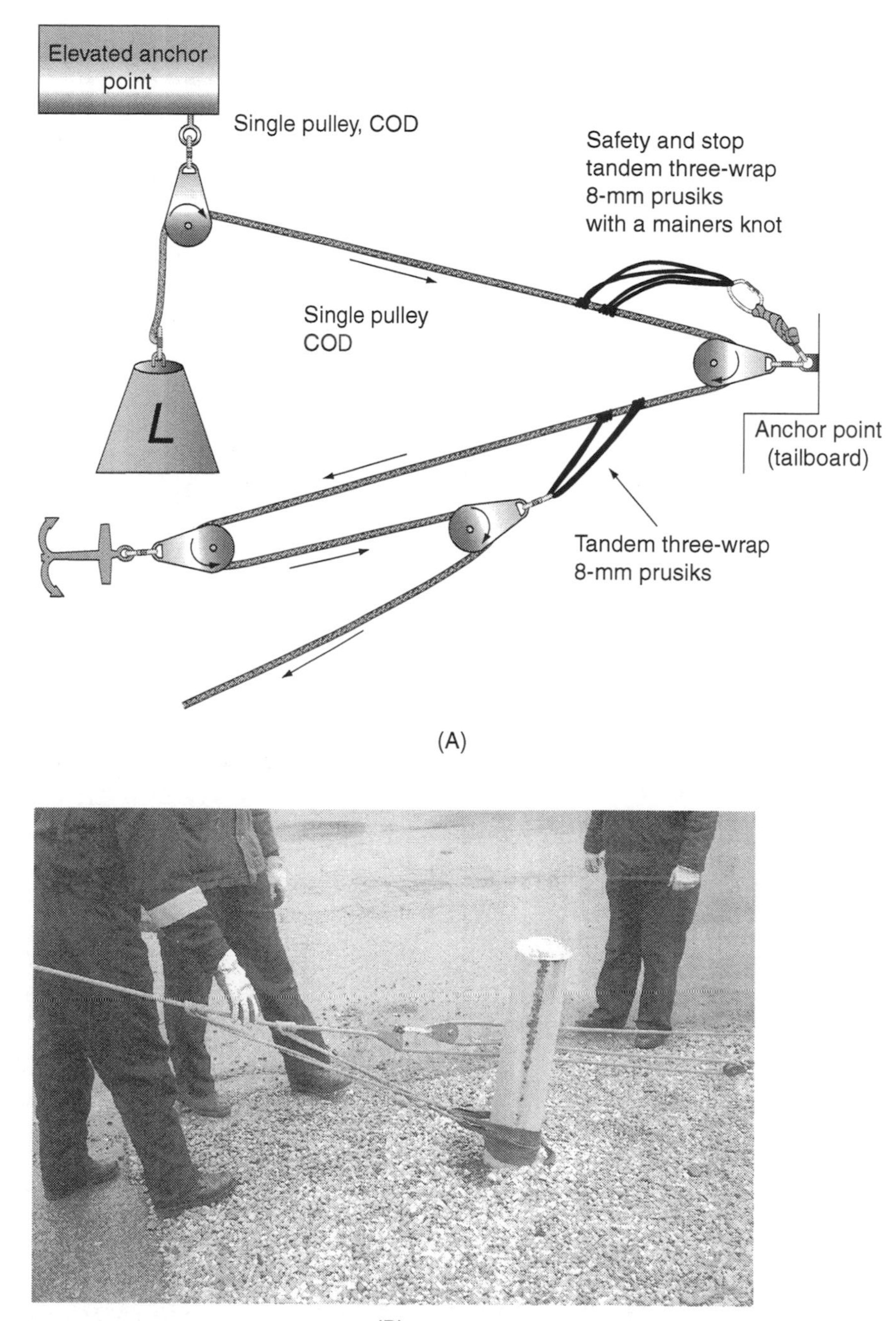

FIGURE 8–19

(A) The Z drag is a 3:1 simple pulley system that is made from the same rope as the main line. (B) The Z drag using a Gibbs rope grab with prusik safeties out front. (*B, Courtesy Craig Aberbach*)

advantage gained by each load-side pulley is *multiplied* to the other pulley, where in simple systems, the load-side pulleys are *added* to one another. In compounded pulley systems, the primary pulley is closest to the load. The primary pulley is essentially a simple pulley developing a 2:1 MA. The secondary pulley's 2:1 is pulling on the input side of the primary pulley, and a 2:1 pulling a 2:1 effectively multiplies the two pulleys. Stated simply, 2 × 2 = 4.

There are notable differences between the 4:1 simple pulley system and the 4:1 compound pulley system (Figure 8–22A). The compound system uses one less pulley, one less carabiner, and one-third less rope and is therefore lighter to carry around. One less pulley equates to less FC and a slightly more efficient system. On the

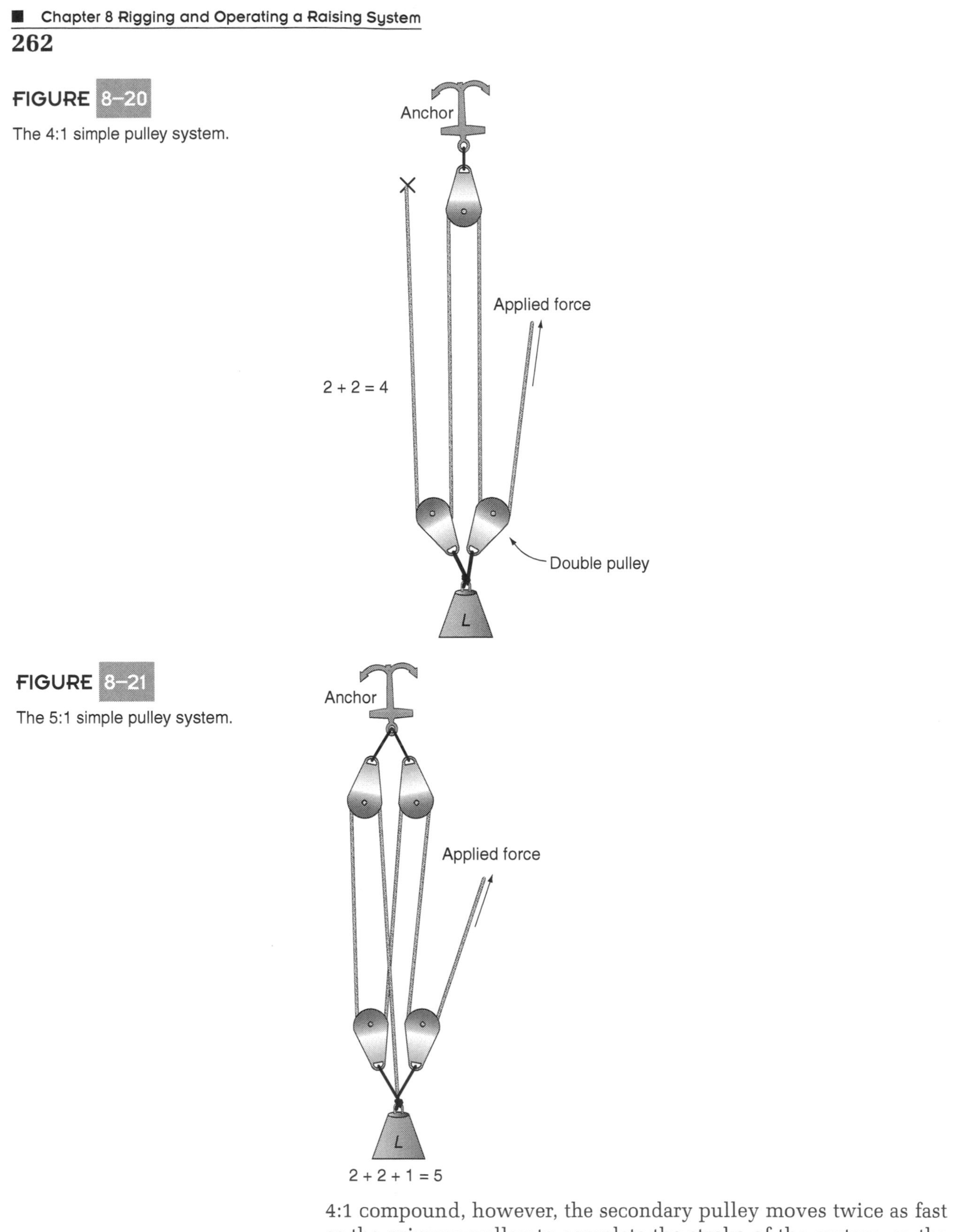

FIGURE 8–20

The 4:1 simple pulley system.

FIGURE 8–21

The 5:1 simple pulley system.

4:1 compound, however, the secondary pulley moves twice as fast as the primary pulley to complete the stroke of the system, so the secondary pulley touches the anchor in only half the distance. The load effectively moves only half the distance so the compound system requires more time-consuming resets. Because the pulleys in the simple system move together, the entire system can compress completely. The stroke is twice the length of the compound 4:1, and fewer resets are required.

FIGURE 8-22

(A) This illustration compares the 4:1 simple pulley system with the 4:1 compound pulley system. In the compound system, the secondary pulley moves twice as fast as the primary pulley halting system compression. The simple system can compress completely, offering a much greater effective stroke length. (B) The 4:1 compound pulley system.

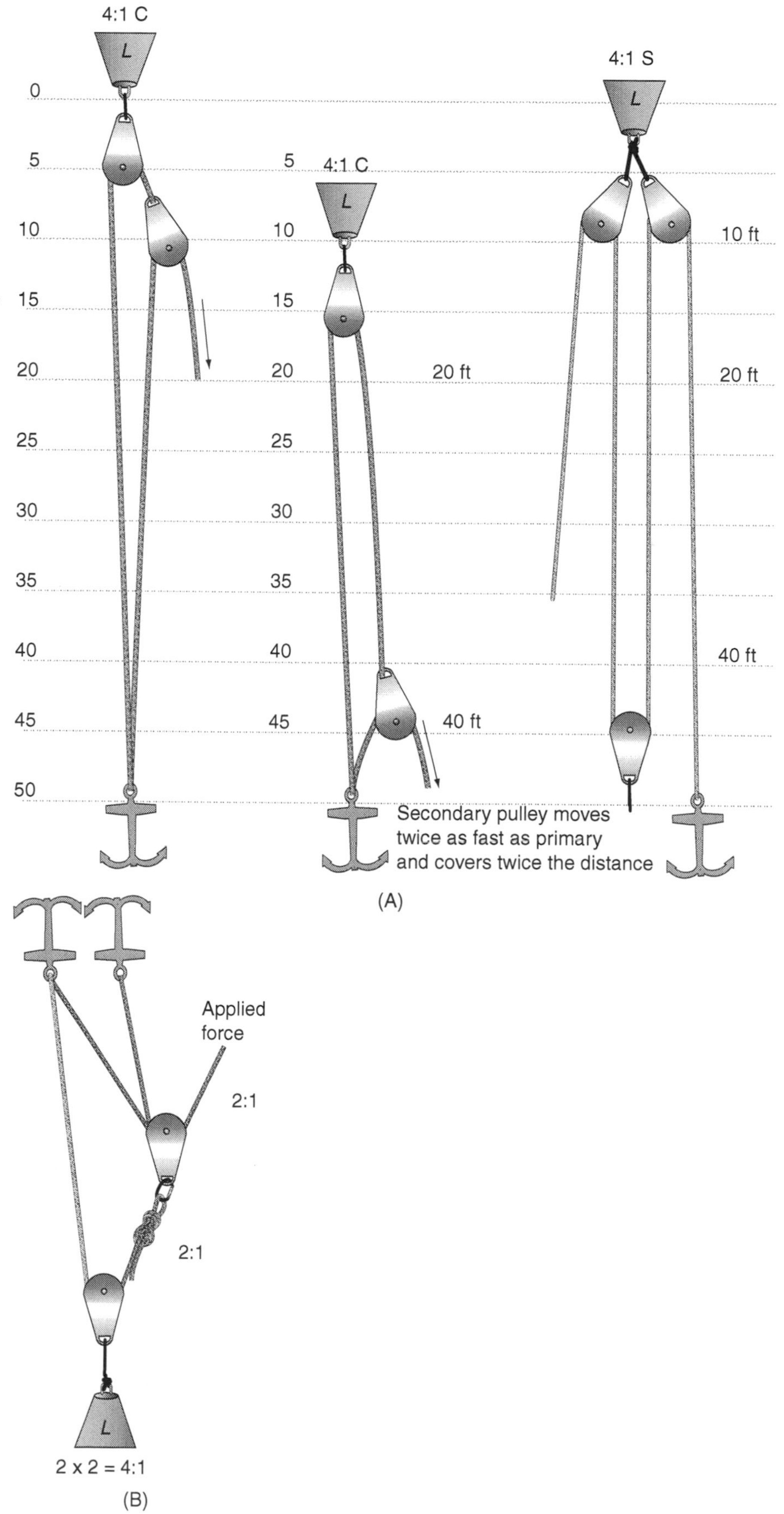

Building a 4:1 compound pulley system (Figure 8–22B)

This technique is different from simple pulley systems and is best remembered by calling it the **middle of the rope technique.**

► **middle of the rope technique**

Most pulley systems start with the terminal end of the rope; this technique starts in the middle of the rope.

1. Decide how much room you have for your haul system stroke. You will want to maximize your stroke, so if you have 50 feet of space to work your pulley system, make your stroke the full 50 feet.
2. Walk off 50 feet of rope from the terminal end, or the primary leg, and tie a figure eight on a bight *in the middle of the rope.* This will be the only anchor attachment. Anchor it.
3. Walk the 50 feet toward the terminal end of the rope toward the load. Tie a figure eight on a bight in the end of the rope that will connect to the secondary pulley. Lay a bight in the rope bringing the terminal end of the rope back toward the anchor a couple feet. Reeve the primary pulley into this bight and attach it to the load.
4. Walk back to the anchor and play the secondary leg of the system back toward the load if it has not already been done. Lay a bight into the rope and reeve the secondary pulley. Connect the secondary pulley into the figure eight on a bight from the primary leg.
5. Apply force to the running end of the rope.

Once you have mastered building pulley systems, you can compound them in any combination to achieve the desired mechanical advantage (Figure 8–23). For example, you can build a 10:1 MA pulley system by first building a 5:1 simple pulley system. Then compound a 2:1 simple pulley system on the output side of the 5:1. A 2:1 pulling a 5:1 creates a 10:1 compound pulley system since 5 × 2 = 10.

Incorporating pulley system components into the raising system (Figure 8–24)

As discussed, pulley systems are usually the most feasible machines to integrate into rope systems to raise loads that exceed the human power window. The mechanics of engineering raises is taken directly from the mechanics of engineering lowers as seen in Chapter 7, except that a pulley system is added to the main line to facilitate the raise.

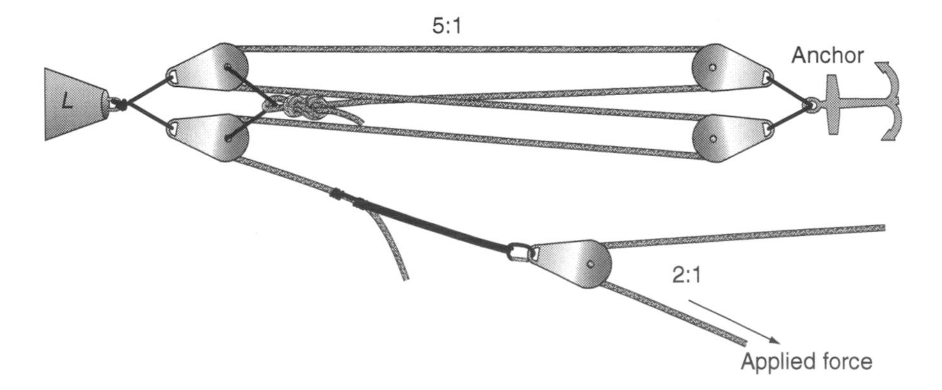

FIGURE 8–23

Compounding pulley systems can achieve tremendous mechanical advantage. In this system a 2:1 is pulling a 5:1 for a 10:1 compound.

FIGURE 8–24

Schematic of the two-line raising system.

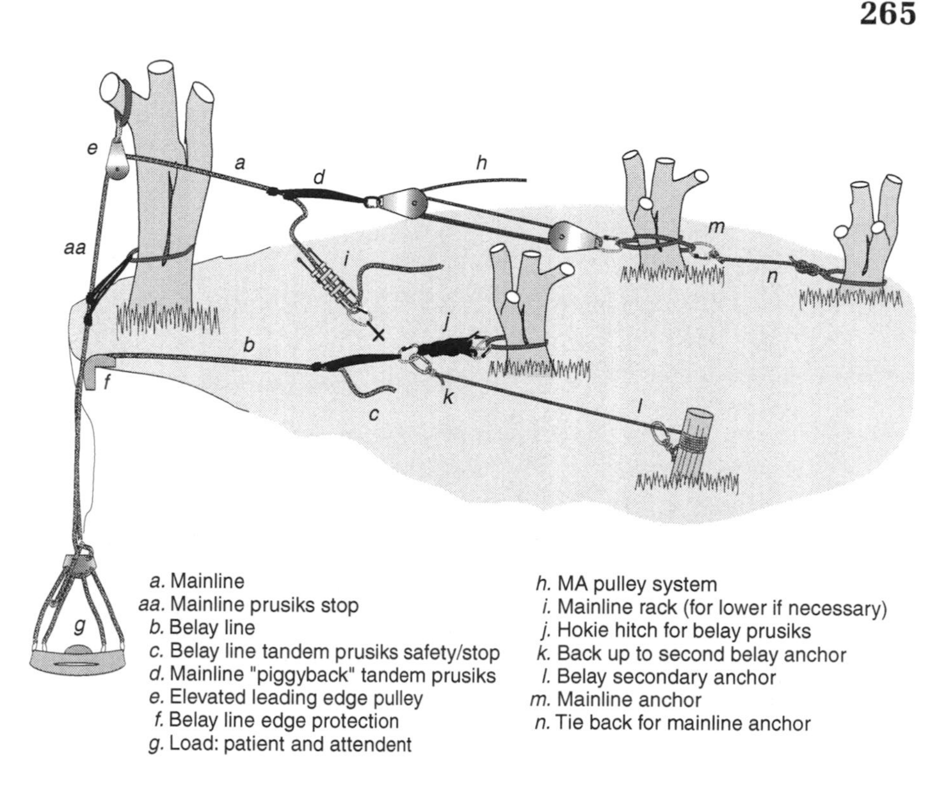

1. The pulley system of choice is added to, or piggybacked onto, the main line. Tandem three-wrap prusiks are placed on the main line to use as rope grabs for the pulley system. The tandem prusiks will slip if the haul team pulls too hard or the load gets jammed. Good R/Q mechanical rope grabs can be used if the rope and rope grab combination is one that slips and regrabs the rope without causing damage.
2. As the pulley system becomes completely compressed, the tandem prusiks or rope grab on the main line and on the load side of the pulley system are locked, as are the tandem prusiks on the belay line. The pulley system is then expanded to its full stroke and the raise is continued.
3. The belay line operates the same as on the lowering system. As the load is raised, slack line is pulled through the tandem prusiks and through the backup closed end rack. Remember that raises create greater stresses on the rope system than lowers do, and the utmost care and observation must be placed on the system components throughout the load. Do not let the load and the system bounce as the haul team works the pulley system.

Calculating Pulley Systems Using the T Method

One method of calculating mechanical advantage is to use the tension or T method. The T method involves adding the parts of the pulley system that carry tension and make mechanical advantage. Starting from the hauling rope add the tensioned ropes that are connected to moving pulleys for theoretical mechanical advantage. Examples are shown in Figures 8–25A to G.

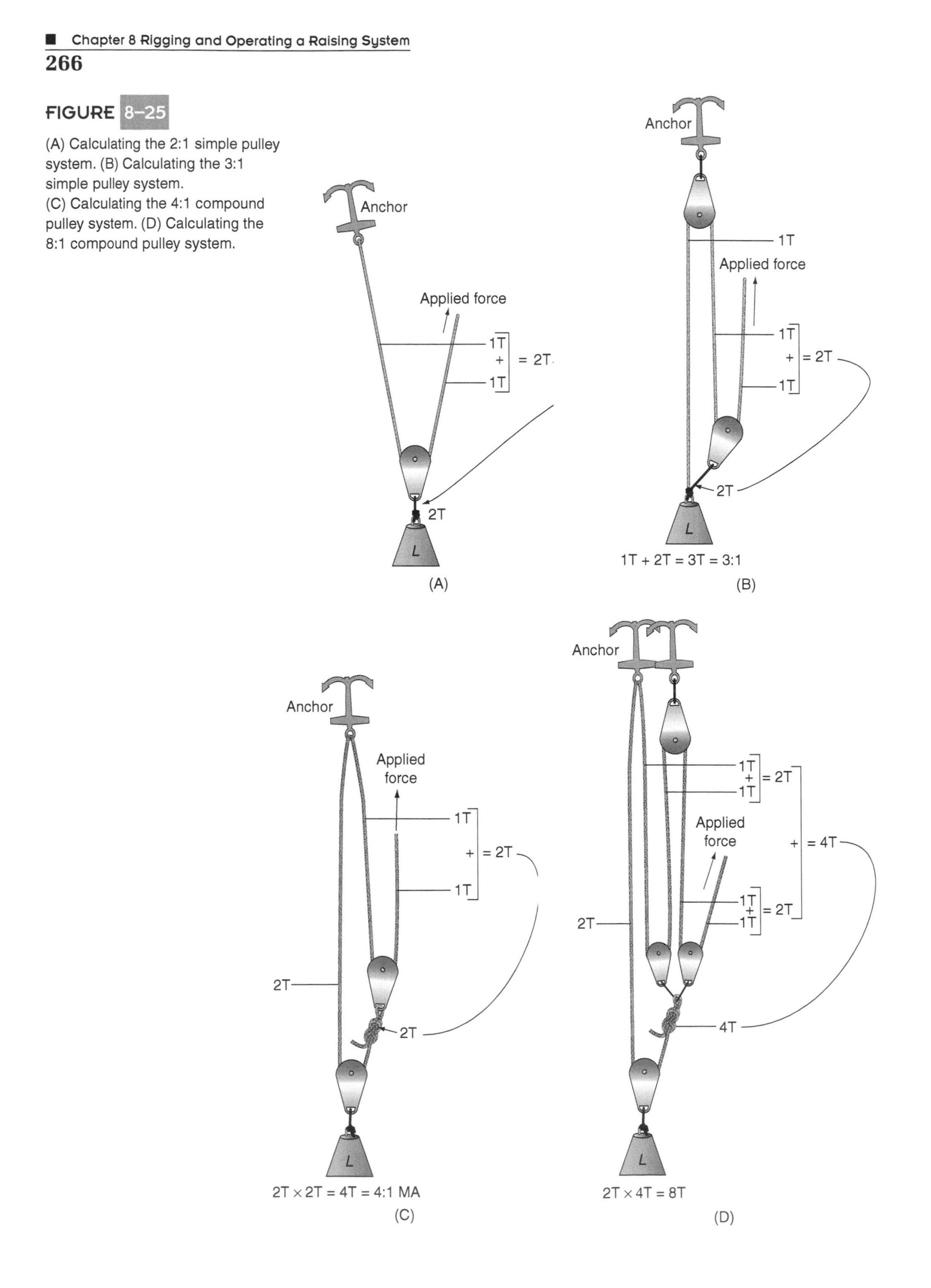

FIGURE 8–25

(A) Calculating the 2:1 simple pulley system. (B) Calculating the 3:1 simple pulley system.
(C) Calculating the 4:1 compound pulley system. (D) Calculating the 8:1 compound pulley system.

FIGURE 8–25 Cont'd

(E) Calculating the 5:1 complex pulley system. (F) Calculating the 7:1 complex pulley system. (G) Calculating the 3:1 complex pulley system.

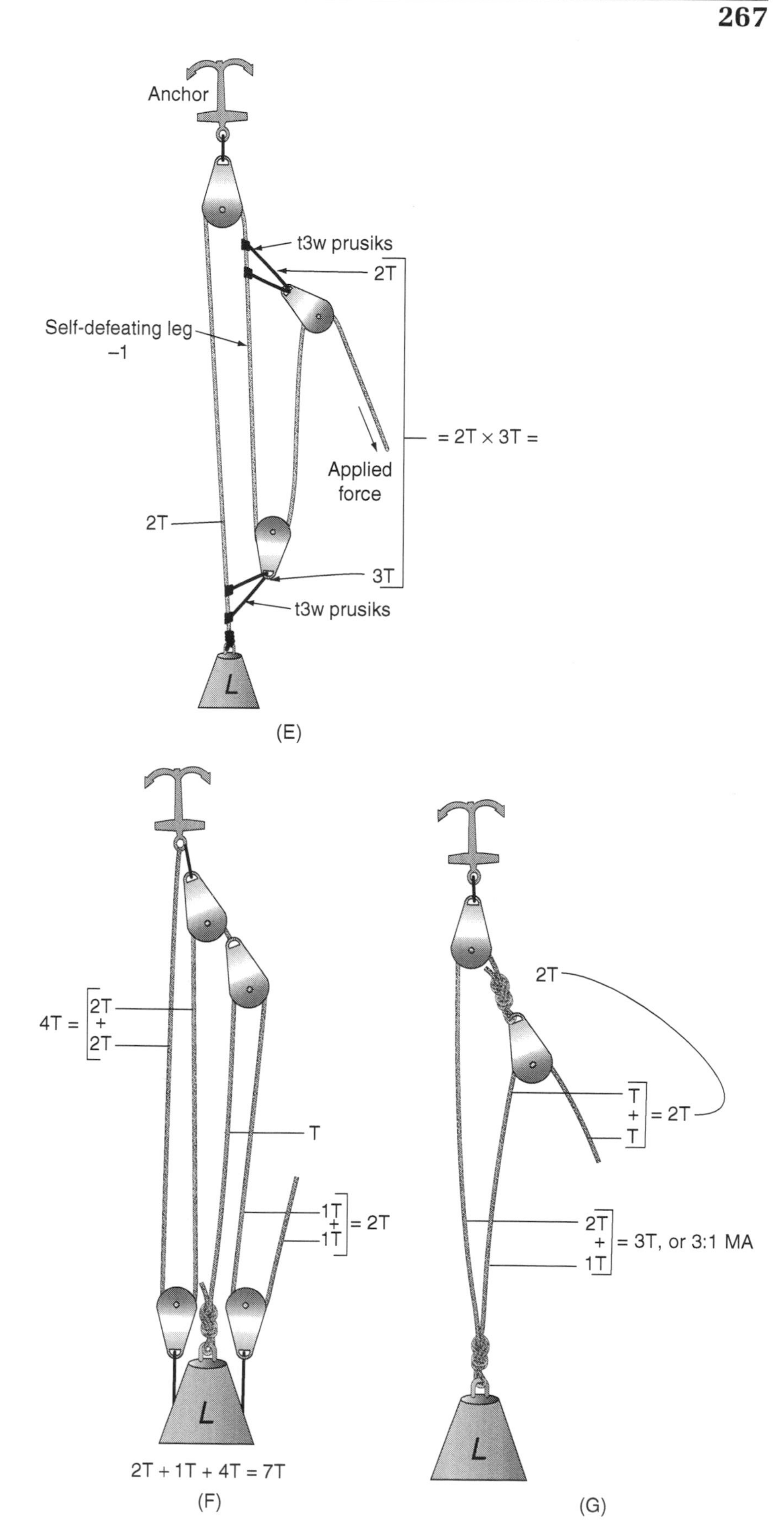

■ SUMMARY

Raising systems are exactly like the lowering systems we discussed in Chapter 7, except a pulley system is added to, or piggybacked onto, the main line and the object (with the exception of counterbalances) is to lift against the forces of gravity. Overcoming the effects of gravity is easily accomplished by using pulley systems that place a majority of the load onto an anchor, so hauling only a fraction of the load is all that is necessary to move the load upward. Raising systems develop higher overall forces than do lowering systems, and therefore extra caution should be used when engineering raising systems.

■ KEY TERMS

Simple machines
Conservation of mechanical energy
Mechanical advantage (MA)
Theoretical mechanical advantage (TMA)
Actual mechanical advantage (AMA)
Friction coefficient (FC)
Inclined plane
Lever
Class I lever
Stroke
Class II lever
Class III lever
Pulley
Anchored pulleys
Nonanchored, or moving, pulleys
Rescue quality (R/Q) pulley system components
Compression
De-set
Extension
Re-set
Terminal end
Counterbalance
Compounding pulley systems
Middle of the rope technique

■ REVIEW QUESTIONS

1. Managing a rope system is about altering the effects of gravity on _____.
   a. rope system components
   b. people in a position of peril
   c. friction producing devices
   d. objects with a mass less than that of Earth's
2. In a system that is isolated from external influences, energy is neither created nor destroyed _____.
   a. but it can be eliminated altogether using the right tools
   b. but certain types of energy can be renewed
   c. but may change from one form to another
   d. and that is why every attempt to balance rope systems should be made
3. The actual mechanical advantage of a machine is calculated by figuring the TMA and _____.
   a. subtracting from it the effects of friction
   b. dividing by 2 for each pulley
   c. adding the friction coefficient
   d. multiplying by 2 for each pulley

4. _____ are the least efficient of all simple machines.
   a. teeter-totters
   b. pulleys
   c. inclined planes
   d. levers
5. A simple machine, the lever is a combination of a bar or the lever, a fulcrum, and _____, aligned in such a manner as to transmit force to another object.
   a. a wheel
   b. the load
   c. an inclined plane
   d. an applied force
6. The Class I lever has the fulcrum placed _____ and the applied force, and that force is being applied in the opposite direction of the object being moved.
   a. somewhere between the object being lifted
   b. near the end of the lever
   c. dead center of the lever
   d. near the axle
7. If you have 2 feet of the lever on one side of the fulcrum and 10 feet of lever on the other side of the fulcrum, you have a _____ mechanical advantage.
   a. 5:1
   b. 10:1
   c. 20:1
   d. 50:1
8. A _____ is a classic example of a Class II lever.
   a. aerial ladder
   b. teeter-totter
   c. shovel
   d. wheelbarrow
9. The Class III lever applies force in the middle of the lever. The fulcrum is at one end of the lever, the object being lifted is at the opposite end, and the applied force and _____ direction(s).
   a. the object move in opposite
   b. the object move in the same
   c. the lever move in opposite
   d. the lever move in the same
10. The round disk in a pulley is called a _____.
    a. wheel
    b. pulley
    c. sheave
    d. block

11. A double pulley has a center plate, and the extension off the center plate for a carabiner attachment is called the _____.
    a. tackle point
    b. swivel
    c. sheave
    d. becket
12. Machines allow us to place work into our power window by spreading the work out _____.
    a. to other members of the team or a fuel source
    b. over time and distance
    c. evenly across the equipment components
    d. so our bodies can handle it more safely and more easily
13. As the load on a pulley bearing is increased, its friction coefficient _____.
    a. also increases
    b. is multiplied by a factor of 2
    c. decreases
    d. is reduced by a factor of 2
14. Simple pulley systems have _____ than compound pulley systems of the same advantage rating.
    a. fewer pulleys
    b. a greater stroke
    c. shorter ropes
    d. a shorter stroke
15. When in doubt as to the MA developed by any pulley system, you can _____ to determine mechanical advantage.
    a. weigh it
    b. load it
    c. compress it
    d. measure it

CHAPTER

9 Tensioned Rope Systems: Filling in the Plane

OBJECTIVES

Upon completion of this chapter, you should be able to:

- explain situations where angled-tensioned rope systems are the tool of choice for accessing and moving patients to safety.
- explain situations where near-horizontal tensioned rope systems are the tool of choice for accessing and moving patients to safety.
- explain edge conditions and safety considerations that are unique to tensioned rope systems.
- describe methods of getting the rope to the other side.
- explain how to tension a rope system.
- describe four ways of determining when the tension is right for the operation.
- describe automatic safety features of tensioned rope rescue systems.
- engineer an angled-tensioned rope rescue system.
- engineer a near-horizontal tensioned rope rescue system.

A DAY ON THE ROPES

Twin Falls in southwest Virginia is locally famous for three things. *First,* it is the watershed for more than 200 square miles. The mirror-image falls drains tons of water into a violent convergence of person-eating whitewater. *Second,* at just less than 500 feet, the steeply angled canyon walls that give the falls their name were dangerous and impossible to navigate without using both hands and both feet simultaneously. *Third,* it was home to some of the smoothest sippin' moonshine ever created. And one more thing—my dog Darth took a 500-foot header down the east waterfall several years earlier. There's no telling how he survived.

There is an old road just a couple hundred yards from the east side—close enough to hear the rumble of the falls if your truck engine is shut off. There are two ways to reach the top of the west side. The shortest and most dangerous down climbing the east canyon wall, rock hopping across the river, and climbing up the west canyon wall. The longest and safest is to hike around the north end, hop the feeder streams, camp out overnight, get up, and hike the rest of the way to the west side. It seems there was some liquid motivation, crystal blue persuasion, as it were, that made some people want to get to that west side very bad.

A man named Folker (not his real name) broke his femur climbing out of a little grotto about two miles further back from the west side and to the south. Folker needed to get to the hospital almost thirty miles away. It happened that our state Heavy and Tactical Rescue Team was doing some maneuvers off the same hospital that Folker needed to get to. We were asked if we could help with the traditional ten- or twelve-hour rescue trek out of the west-side area. We agreed.

By the time we arrived, a hasty team of locals with radios, a Stokes basket, and some medical supplies had started the more dangerous but quicker route down the canyon walls. We were ordered to wait for assignment. We suggested to the incident manager that a highline spanning the falls might be the tool of choice for getting Folker back to the safety of the east side of Twin Falls. It would be a much better ride for a man with a broken femur than the tortuous slip and slide down and then up the canyon walls, and the rescuers would suffer a lot fewer bumps and bruises. And the alternative trip around the falls could take almost a full day by Stokes and human litter haulers. Finally, since we were training anyway, why not set up the highline? It beat sitting around in staging! He wisely agreed. We were in luck.

The span from east side to west side was a little over 600 feet. Since we had a 600-foot section of 5/8-inch rope, we assumed the divine god of technical rescue was on our side. The altitude of the load at midspan over the raging waters would be about 275 to 300 feet figuring a catenary angle of about 130 degrees. Now all we needed was a rope stretcher, some luck, and a little time.

Six team members down climbed the east canyon scree carrying a minimum of equipment and playing out a spoil of 550 cord, parachute cord known for its tensile strength of about 550 lbf. The plan was to get them to a suitable site on the west side and use the 550 cord as a messenger line to haul the big 5/8-inch line across.

The traveling team made the top of the west side in good time, but our biggest obstacle had not presented itself yet. By radio, we learned the hasty team had stabilized Folker's injuries and was moving toward the west side. They would be there in about ninety minutes. The IM had informed them of our plans, and if we could get the highline in the air and tested in time, everyone would be better off, especially Folker. If we could not get it rigged and tested in time, we would be bums because the hasty team was diverting somewhat to get to our team's position on the west side.

Stretching 650 feet of 550 messenger cord through hundreds of tree branches and innumerable brushlings was proving to be more than difficult.

The rest of the team had all of the system components—haul system, carriage, horizontal control lines, anchors, and attachments—ready to tension the highline and transport Folker to the east side. All that was needed was a trackline to move Folker east.

After considerable wrangling, pulling, and tree climbing, all but one piece of the 550 cord was airborne. Dan and Charlie, disgusted, tore off down the east canyon wall toward the offending maple. Dan carried a machete, and Charlie was sporting a Stihl 044 with a 30-inch blade. Smoke rose, the trees shook, and suddenly the 550 cord leapt free into the sky. Within seconds, the big 5/8-inch rope and the control lines emerged from the tree line and arched gracefully toward the west side.

The rest was a cakewalk. The tensioned rope system was set, test loaded, and ready to ride about ten minutes after the hasty team arrived with Folker. After a quick medical re-evaluation, the Stokes spider rig was attached to Folker's Stokes. Harold, one of our team paramedics, was chosen to ride across the roaring, watery chasm with Folker. For a rope rescuer, it was a thing of pure beauty. On the receiving east side, out of the rumbling mist, emerged rescuer and patient. One was smiling ear to ear. The other, Stokes-bound, was looking like it all might be a dream. The ride from west side to east side consumed about eight minutes instead of eight hours. Folker's injuries healed so well that, rumor had it, within five months he was back hiking the high road in that neverending search for the perfect crystal blue persuasion.

TENSIONED ROPE SYSTEMS

Tensioned rope systems, or **highlines,** are extremely useful in filling in the areas that cannot be reached using purely vertical raise, lower, and rappel systems (Figures 9–1 and Figure 9–2). As rope engineers our transportation systems spring from the restraints of the linear into the expanses of planar geometry. Virtually any point between and below two good anchors is accessible using tensioned rope systems and the variations.

Floating a patient across a dangerous span is quicker, easier on rescuers, and, handled properly, much gentler on patients' injuries. After all other means of egress have been considered and eliminated,

▶ **tensioned rope systems, or highlines**

Rope rescue transportation systems where a trackline rope is anchored, crosses a span, and is pulled tight enough to safely support the weight of a rescuer, any patients, all ancillary system components, and any associated span noise.

FIGURE 9–1

Tensioned rope systems allow access from point to point outside of the purely vertical environment.

FIGURE 9–2
Tensioned rope systems are transportation devices that can safely and efficiently move searchers, rescuers, patients, and equipment from one place to another. (*Courtesy Martin Grube*)

evacuation of people trapped in highrise fires and structural collapse can be accomplished quickly and easily using tensioned rope systems. Spanning rivers with highlines effectively bridges river banks, sometimes eliminating miles of parallel river land traverse that causes wear and tear on rescuers and patients.

Tensioned rope systems are different from the other systems described in this book, because the rope is anchored on both ends. A considerable amount of tension is placed in the rope to allow it to function as a track. This track then is "ridden" on by a trolley consisting of pulleys, and pulled side to side with control lines.

As with all systems, tensioned rope systems must start, with bombproof anchor sites that are aligned with the intended target. It helps to think of the target as being a spot imbedded somewhere in the plane that is created by the anchors, the ground, and the estimated catenary angle of the highline rope itself. Sometimes a small amount of sideways deviation can be placed in a highline using a tagline. For general purposes, though, highlines have two dimensions, which still allows us to service a vastly greater area than with a purely linear rope system. The target is like a dot anywhere on a piece of paper. We simply have to engineer a system that allows us to safely travel to the target and then to a place of relative safety.

span
The distance from anchor point to anchor point.

After locating your aligned anchors, getting the line to the other side is an important second step. There are three major considerations in selecting a method of getting the rope to the other side. The first consideration is span. **Span** is the distance from anchor point to anchor point. With a span of only 10 feet, you could simply throw the rope to the other side. But if the span is several hundred feet or more you will need considerably more time and equipment to connect the two sides.

The second consideration is **span noise,** or all the factors that create variables in the ropes from anchor point to anchor point, including rope weight, trees, rain, waterfalls, wind, convergent volunteers, etc. It is usually easier to project, by throwing or shooting, a line from the higher elevation to the lower elevation on nonhorizontal highlines. Trees and brush suck up slack rope like a magnet, and whitewater tugging on a rope can be considerably heavier than an anticipated gravity and wind load.

▶ **span noise**
All the factors that create variables in the ropes from anchor point to anchor point, including rope weight, trees, rain, waterfalls, wind, convergent volunteers, etc.

The third is the **situation,** and whether it is an emergency or training. As with all rescues, a risk benefit analysis of methods must be performed. Shooting a 45-caliber line gun dart into the side of a building that is on fire with lots of trapped occupants is understandable. Shooting a 45-caliber line gun dart into the side of a building during training, where time and lives are not in jeopardy, is harder to justify, especially if you are using someone else's building.

▶ **situation**
A position with respect to the condition and surroundings of an occurrence: a real-time "live" rescue has different operational factors from a training scenario.

Several options, all with various advantages and disadvantages exist to get the rope to the other side:

1. With a team safely placed on each side via hiking, climbing, helicopter insertion, or using an elevator, have each side lower a rope to the ground. Someone on the ground ties them together and the roof crews pull them up. Be sure to have a way of capturing progress since the shear weight of the rope and any span noise will make it hard to hold up. A set of tandem prusiks anchored on each side make excellent progress capturers until the final anchor connections are established.
2. Project a pilot or messenger cord from one side to the other. A **pilot cord** is a strong but lightweight piece of cordage (1 or 2 mm diameter) that can be projected some distance to the other side. On spans greater than about 200 feet, a heavier cord will be needed because the tensile strength of the pilot cord can be exceeded when the main line is added. A **messenger cord** is stronger and somewhat heavier, 3 to 7 mm depending on span requirements, than the pilot cord. Once the pilot cord is safely across, the messenger line is connected and pulled across. The messenger line can carry the main line and control lines across the span without breaking.

   Projecting the pilot line across the span can be done in several ways:

   a. *Throw it.* Attach a pilot cord to a heavy object and heave it across. Obviously, this has limitations and depends greatly on athletic ability, strength, agility, and luck. With practice, you can get about 100 feet using a 20-ounce lead fishing sinker and 2-mm cord, and by faking the cord so it will play out in the easiest manner.

▶ **pilot cord**
A strong but lightweight piece of cordage (1 or 2 mm diameter) that can be projected some distance to the other side.

▶ **messenger cord**
Stronger and somewhat heavier than the pilot cord (3 to 7 mm).

► **faking**

A nautical term meaning to lay rope out for the most efficient deployment and to prevent kinks and entanglements.

► **playing**

An ancient nautical term meaning to direct and work the line to a desired location.

NOTE: All projectile firing devices are potentially hazardous to the user, bystanders, or people on the receiving end of the projectile. Always follow the manufacturers' recommendations for use of the equipment and seek qualified training before using any projectile firing device.

Faking is a nautical term meaning to lay rope out for the most efficient deployment and to prevent kinks and entanglements. Faking boxes were spiked with dozens of parallel wooden pegs, and cord was carefully laid onto the pegs. The projectile was then uninhibited by kinks and tangles in the rope as it came out of the box. **Playing** a rope or cord is another ancient nautical term meaning to direct and work the line to a desired location. Both faking and playing are terms adopted by the fire service, as in "play the hose stream on the side of the building" and "fake the hose into the hose bed so it will come off easily and not become entangled."

b. *The crossbow or compound bow method.* You can attach fishing reel with radiator clamps to a crossbow or compound bow and extend your reach another hundred feet or more. A lead-weighted, blunt-end aluminum rod projectile carrying monofilament fishing line as the pilot cord makes a relatively safe and effective spanner.

c. *Compressed-air line gun.* In recent years there have been some nice advances in compressed-air activated line guns. The gun itself is simply a steel or aluminum tube with two chambers arranged inline. The one in front is the barrel for the projectile, while the smaller chamber in back acts as a staging area for an adjustable amount of compressed air. Attached to the outside of the barrel is a 3,000-psi compressed-air cylinder and a small regulator and valve assembly. On top of the barrel is a welded track that accepts the cartridge of pilot cord, which is in turn attached to the hard rubber projectile. To fire, the staging chamber is filled to anywhere from 50 to 250 psi, or whatever the manufacturer suggests, depending on the desired range. When the trigger is depressed the air from the staging chamber expands against a wooden or rubber piston that forcibly ejects the projectile from the barrel. Some compressed-air gun manufacturers claim distances of more than 400 feet with the pilot cord attached when conditions are perfect. It is reasonable to expect 200 to 300 feet with some accuracy in field conditions.

d. *Ballistic line guns* use gunpowder to project a rod or rubber ball carrying the pilot cord. Some of the larger caliber (45/70) guns can fire the projectile and cord more than 800 feet. There are some hand-carried ground-fired special warfare devices capable of throwing a pilot line more than a mile.

3. There has been some success using large *model rockets* to haul pilot cord. Using two- and three-stage solid rocket propellent engines, the devices trail a metal swiveled leader cable to prevent burning the synthetic pilot cord. Distances of more than a mile have been reported, but accuracy is still a problem.

NOTE: Once a messenger cord or rope has been connected successfully to both sides, consider sending all the rope elements across at the same time. Horizontal and vertical movable control lines should go across the span with the trackline, if possible, in one coordinated maneuver. This may not always be possible if the span is very long. The added weight of the control lines could make it impossible to draw all the rope up onto the receiving side at once. Span noise can interfere with the added ropes, also, complicating the raise. However, when possible, hauling over all the lines at once is a big time saver.

4. *Helicopters* have been used to haul very long highlines from anchor to anchor. Some military versions even eliminate the need for pilot and messenger cord by simply hauling all of the lines in one trip. Quick-release safety features allow the pilot to lose the rope load should there be unexpected crosswinds or an entanglement from the ground. Helicopters can haul track and control lines for distances of one mile.
5. *Boats and swimmers* can be used to ferry the rope across under relatively calm water conditions.

When the rope has been positioned at both sides of the planned rope system, the team must tension the trackline(s) an appropriate amount. Tensioning is accomplished by building an MA (probably in the 4:1 range), anchoring one end, and attaching the moving end to the trackline with t3w prusiks. After notifying the other side about the tensioning action, the haul team gradually pulls the trackline into position. The *appropriate* amount of tension is always an interesting point of discussion. For the purposes of this text, it will be assumed that the team has reasonable confidence in their equipment and their ability to build the system. It will also be assumed that they are using R/Q equipment; trained, confident, and competent team members; bombproof anchors; and there will be no air strikes from military aircraft, militant rocketeers, or earth shattering meteors.

It might help to refer to the foundation established by the lowering and raising systems described in Chapter 7 and Chapter 8, which was basically an anchor, a main line, a belay line, and some sort of motivation. Tensioned rope systems are really quite similar, except that both ends of the rope are anchored. It is almost like taking a raising system diagram from Chapter 8 and tilting it 90 degrees, except that the main line is tensioned and called a trackline, and the load is belayed from both ends as a backup in the unlikely event of trackline failure.

Take a 1/2-inch R/Q rope and anchor one end of it, preferably using a tensionless, or full-strength, tie-off. Get the rope, or trackline, to the other side using one of the methods described earlier and temporarily tie it to something substantial. Build and anchor the MA system of your choice. A 4:1 or 5:1 simple pulley system works well. Piggyback or otherwise attach the pulley system to the trackline using t3w prusiks. Anchor and arrange another set of t3w prusiks to accept progress in the trackline as the pulley system is compressed.

As in all engineering challenges, reduction and control of the variables stabilizes anticipated outcomes. Rope, being inherently flexible, relatively stretchy, and subject to numerous environmental influences, presents an endless array of variables that prohibit anticipating outcomes exactly. Educated approximations are practical and realistic given certain parameters. Following are some useful parameters for tensioning rope systems:

1. When in doubt, let it out. The less tension in a rope, the greater the remaining safety margin before rope failure. With

1,000 lbf in a 9,000 lbf rope, there is an 8,000 lbf buffer zone, or a 9:1 safety margin. With 500 lbf in a 9,000 lbf rope, there is an 8,500 lbf buffer zone, or an 18:1 safety margin.

2. Tension the rope only as tight as needed to carry out the transportation mission. If you have plenty of people and equipment, you can easily haul the rescue load up a slack highline when tensioning requirements are questionable.
3. Tension with a measuring device when possible. Electronic-quartz load cells, force links (Figure 9–3), or even calibrated flex arm beam dynamometers (Figure 9–4A, Figure 9–4B, and Figure 9–4C) can measure the approximate tension in a rope system. They make great teaching and learning devices and help to more accurately determine safety margins and many polymer variations.
4. Always tension using a rope grabbing device that will slip long before your trackline fails. T3w prusiks calibrated to trackline type and diameter make wonderfully forgiving tools for tensioning rope systems. R/Q mechanical rope grabs work well under some circumstances. Be sure to know the difference. After tensioning, leave both ends of the trackline tensioned into anchored t3w prusiks as shock absorbers with 18 to 24 inches of slack. Bomb them off by connecting them to the bombproof anchor using the strongest available method.
5. A good starting point is 500 lbf of pretension for a single rope system. A good medium tension for most loaded highlines with less than a 500-foot span and heavier loads of 300 to 600 lbs is about 750 lbf. The high-end tension for loaded highlines of increasing span and load requirements is 1,000 lbf.

ANGLED HIGHLINE

► **angled highline**

A tensioned rope system with a substantially higher anchor on one side than the other.

The term **angled highline,** sometimes called a telpher line, is borrowed from suspended cable car systems. It is simply a tensioned rope system that is anchored at a high point and a low point, creating a measurable angle relative to the ground and to the object making the higher point. To help understand some of the basic features of angled highlines, it helps to examine the geometry involved (Figure 9–5). Imagine a building with a straight side and a flat, well-maintained

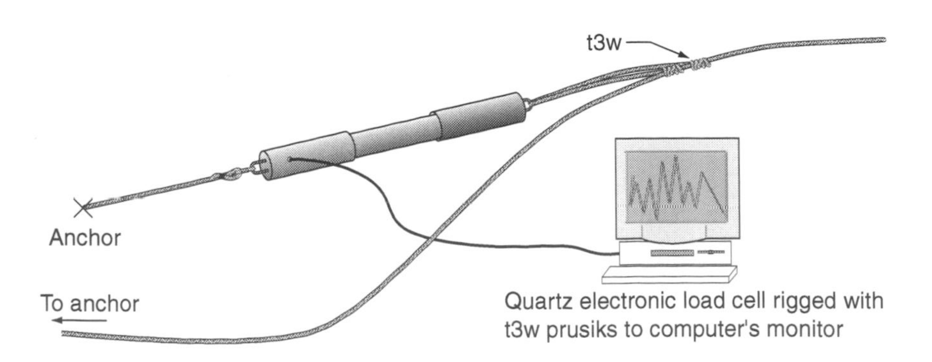

FIGURE 9–3

Electronic force links and load cells can accurately measure tension in rope systems and transmit data to recording devices for analysis.

FIGURE 9-4

(A) Flex-beam dynamometers can mechanically measure approximate rope system component forces. (B, C) Dynamometers are excellent teaching tools to help student rescuers "see" system forces as they learn tensioning concepts. (*Courtesy Martin Grube*)

Flex-beam dynamometer rigged into a rope system to measure rope tension

using tandem three-wrap prusiks

(A)

(B)

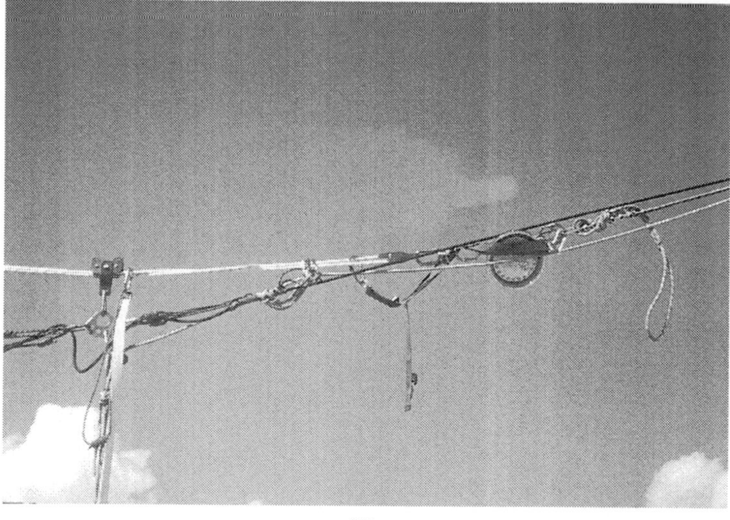

(C)

yard stretching out from the building. The angle at the intersection of the building and the yard is 90 degrees, which is a right angle. A rope stretched from the top of the building to a point in the yard some distance from the base of the building creates two more angles. We have created a right triangle with the three sides being the building, the ground, and the rope. The angles of a triangle always add up to 180

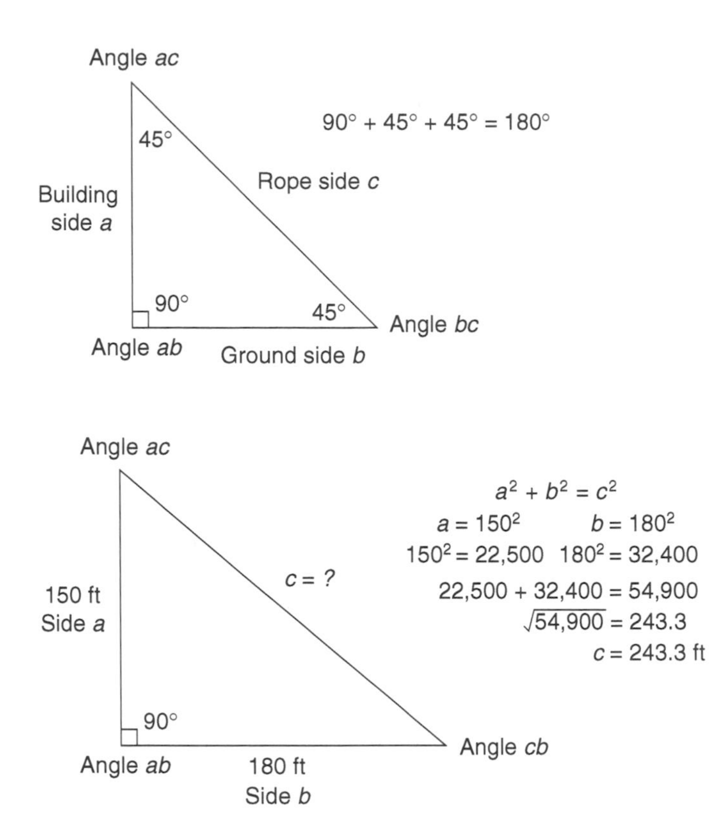

FIGURE 9–5

Angled rope system geometry. An angle rope system can be seen as a triangle with the sides being *a* (the building), *b* (the ground), and *c* (the rope). The three angles equal 180 degrees.

FIGURE 9–6

Rope length *c* can be calculated by taking known values: the building height *a* (150 feet) and the ground *b* (180 feet). $a^2 + b^2 = c^2$.

degrees, and the angle (a/b) created by the intersection of the building and the yard is 90 degrees. The remaining angles (a/c and b/c) must add up to some combination of the remaining 90 degrees.

The unknown length of the span from the top to the ground on an angled highline can be determined by using the geometric formula for calculating the hypotenuse of a right triangle. The hypotenuse is the side of a right triangle opposite the right angle, which, in angled highlines, is always the rope. For the purpose of approximating rope length, it will be assumed that the rope is free of the effects of gravity, making it straight. The answer is derived by using the formula $a^2 + b^2 = c^2$. For example (Figure 9–6), the building height (a) equals 150 feet and the distance along the ground from the building (b) is 180 feet, so the formula reads:

$$a^2 = 150^2 = 22{,}500 \quad b^2 = 180^2 = 32{,}400 \quad 22{,}500 + 32{,}400 = 54{,}900$$

Now you just have to find the square root of 54,900 (for heavens sake use a calculator) which is 234.3. Therefore, the span rope must be at least 234 feet long. Another example (Figure 9–7) is the top of a 630-foot cliff. The only good ground anchor is 700 feet from the base of the cliff so

$$630^2 + 700^2 = 396{,}900 + 490{,}000 = 886{,}900.$$

The square root of 886,900 = 941, therefore the rope to the ground will need to be at least 941 feet long.

The exact trackline length when loaded will be somewhat longer than a simple hypotenuse. With nylon rope, stretch accounts for most of the additional needed length. But with less stretchy polyester ropes, a more exact estimation of trackline length is important. You

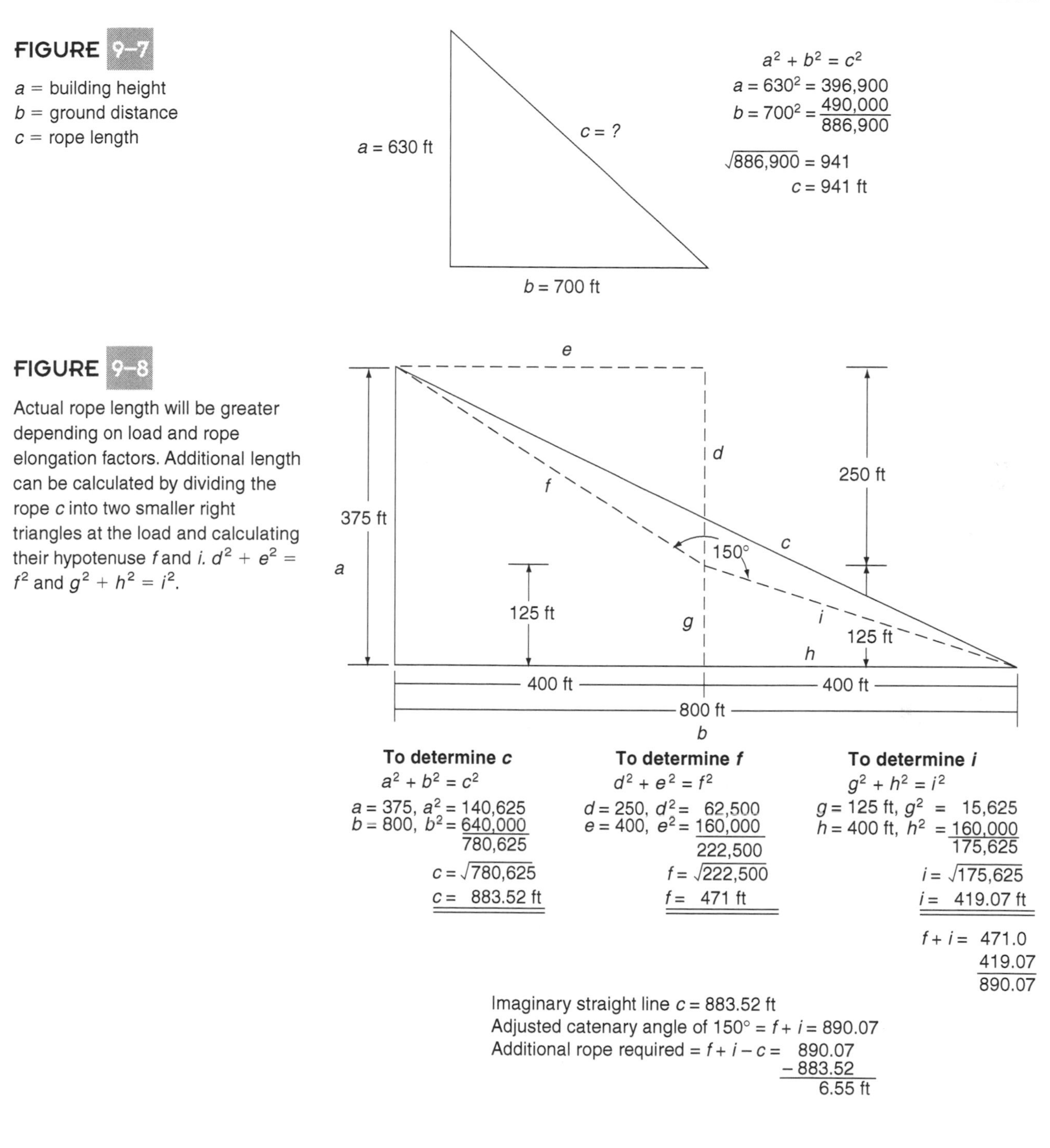

FIGURE 9–7

a = building height
b = ground distance
c = rope length

FIGURE 9–8

Actual rope length will be greater depending on load and rope elongation factors. Additional length can be calculated by dividing the rope c into two smaller right triangles at the load and calculating their hypotenuse f and i. $d^2 + e^2 = f^2$ and $g^2 + h^2 = i^2$.

will need to calculate the additional catenary angle caused by gravity and by the downward force created by the load riding the trackline. The **catenary angle,** is the angle created in a flexible linear medium by the effects of span noise, including gravity, the rescue load, wind, etc. If you want to be more exact, this additional length can be calculated by bisecting the trackline into two right triangles (Figure 9–8) and adding the two hypotenuse figures together. In the example, the vertical (a) is 375 feet, and the horizontal (b) is 800 feet, making the straight hypotenuse (c) 883.52 feet. First, estimate the angle created in the rope after tensioning and with a load on the rope. In the example, 150 degrees is used. Then bisect the main hypotenuse

► **catenary angle**

The angle created in a flexible linear medium by the effects of span noise.

and calculate imaginary right angles, where the trackline in both instances is the hypotenuse. Where c = 883.52 feet, f = 471 feet, and l = 419.07 feet. Adjusted trackline length for a 150 degree catenary is 890.07 feet, or an additional 6.5 feet. As the catenary angle decreases due to increases in load requirements and/or because of a slacker trackline, rope length will become greater.

One of the biggest advantages in using *angled* highlines (Figure 9–9) is that there is usually a lot of latitude regarding trackline tension. Angled tracklines do not have to carry as much load as horizontal tracklines because the upper control line bears much of the load. Because less tension is normally better in rope rescue systems, establishing a safe landing zone is an important part of calculating angled line placement. The landing zone (LZ) will be close to the ground anchor only if the trackline is unusually tight and the load unusually light, or perhaps in an uncontrolled free ride. It is much better to offload people, including accident victims, well ahead of the ground anchor systems and away from system operators. Heavy loads will have an LZ closer to the structure or cliff face that is your top anchor than will lighter loads. A good rule of thumb is to tension your angled highline so the load lands safely in the lower one-third of the system length closest to the ground anchor.

► **control lines**

Used to move the trolley along the length of the trackline(s), and they act as the belay in the event of tracking failure.

Control lines must be added to manage the load traveling on every trackline, horizontal or angled. Control lines are used to stand ready to accept the load in the unlikely event of trackline failure, and to control the rate and direction of travel along the path of the trackline. The control ropes are normally the same diameter as the trackline and are anchored at both ends into t3w prusiks. They are also anchored at the top in the lowering closed end rack and at the

FIGURE 9–9

Angled tracklines do not have to carry as much load as horizontal tracklines because the upper control line bears much of the weight of the rescue load. (*Courtesy Martin Grube*)

bottom in the manned but loosely rigged rope-receiving closed end rack. In the event of trackline failure, the load drops into the control lines, and the t3w prusiks catch the falling load from the trackline and arrest the fall. The sacrificial prusiks will, under most circumstances, catch the load with only minor abrasion of the rope surfaces. If major slippage occurs, the racks are in position to receive the load and bring it to the ground safely. Line management is of the utmost importance to make sure slack is consistently removed from the control lines as the upper one is let out and the lower one is taken in. Hand-tensioned control lines will minimize the shock load forced upon the system if the trackline fails.

It is wise also to rig (Figure 9–10A and Figure 9–10B) the control lines free of the overhead anchor attachment that creates the elevated leading edge for the trackline. In the event of elevated anchor failure, the control lines will be better positioned and

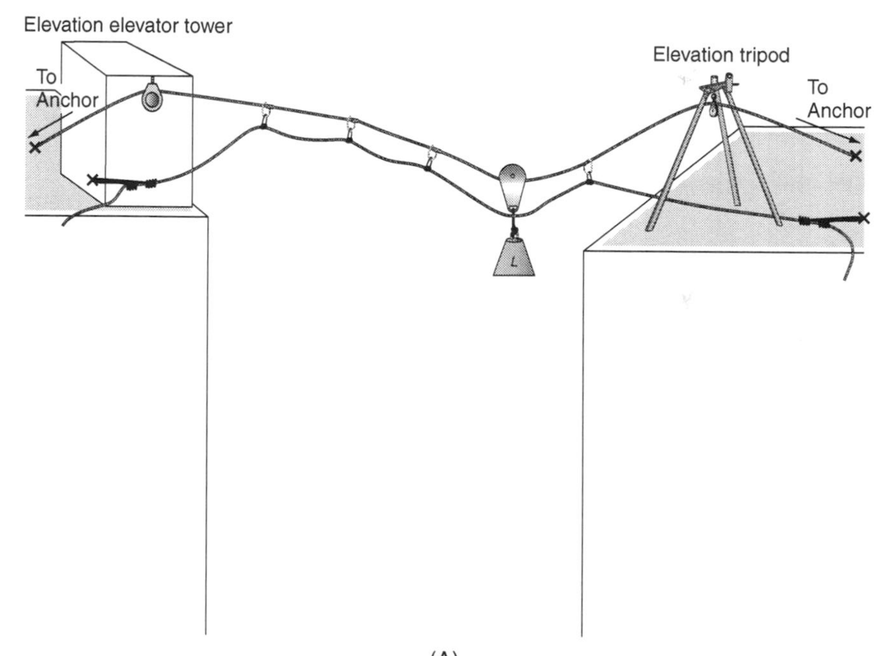

(A)

(B)

FIGURE 9–10

(A) Highline with elevated leading edge—control lines not elevated. (B) Tripods can make excellent elevated leading edges but must be tied back to the anchors for stability. Additionally, the control lines should be rigged outside the influence of the tripod when possible to eliminate excessive dynamic loading in the event of tipover or component failure. (*Courtesy Patrick Anderson*)

NOTE: On angled highlines, it is easy to overlook the necessity of the bottom control lines. In fact, they are a little bothersome, because someone has to manage the lines and put on and take off hangers. Remember, the control line's main purpose is to catch any possible failure in the trackline. Trackline failure with only the top control line in place would result in crashing into the wall or ground.

prepositioned to receive any falling load. It is much like using a tripod as the elevated attachment point for a belay line when lowering a rescuer into a manhole. If the belay line is rigged into the tripod, and the tripod fails or is tipped over, the rescuer is subjected to an enormous fall force. However, if the belay line is rigged free of the tripod and edge protected at the manhole rim, it is better positioned and prepositioned to receive a falling load. The falling rescuer only feels the force of any slack that was allowed to be in the belay line. Make sure this possibility is recognized and the edges are appropriately padded to protect the control lines should they have to arrest a falling load.

At the top of the angled highline, a closed end rack is anchored and operated to make the speed of descent along the trackline compatible with the orders from the system manager. The rack person's responsibility is to have complete control of the load from fast, slow, and stop to catching the entire load in the event of trackline and unlikely t3w prusik failure.

► line hangers, or festoons

Cordage and carabiner rope riders that lift the weight of the control lines up onto the trackline.

Line hangers, sometimes called **festoons,** are cordage and carabiner rope riders that lift the weight of the control lines up onto the trackline (Figure 9–11A, Figure 9–11B, and Figure 9–11C). Hangers are best made with 550 parachute or 3 mm utility cord and lightweight carabiners. Double wrap (double girth hitch) the control lines with the hanger cord and clip the carabiner over the trackline. The carabiner rides smoothly down the trackline and easily bears the weight of the control lines. Positioning hangers every 25 or 30 feet neatly festoons the control lines close to the trackline and helps keep the slack out in the event of a failure.

FIGURE 9–11

Line hangers keep the weight of control lines up and on the trackline(s), reducing span noise and maintaining them in a position of readiness for belay and lowering operations. (*Courtesy Martin Grube*)

(A) (B) (C)

HORIZONTAL HIGHLINE

► **horizontal or near-horizontal highline**
Rope transportation systems anchored at or near the same height.

The **horizontal** or **near-horizontal highline** (Figure 9–12A and Figure 9–12B) is sometimes called a tyrolean traverse, a name derived from the Tirol region alpinists of Austria in the early 1800s. Highlines are the most useful in spanning gaps between mountainsides, buildings, caves, and industrial locations where the trip down, across, and back up is considerably more time-consuming and dangerous than riding a rope across.

Among the first considerations when designing a near-horizontal highline is appropriate anchor sites on each side (Figure 9–13A, Figure 9–13B, Figure 9–13C, and Figure 9–13D). Remember that horizontal-tensioned rope systems place the greatest potential forces on equipment and anchors of any rope rescue system made. Therefore, anchor sites certainly cannot be compromised.

1. Choose bombers that line up with the intended target if at all possible. If they do not line up perfectly, the highline can be pulled sideways somewhat using a directional pulley jig.
2. Always try to choose at least two good anchors that are in line with each other, the target, and the other side anchors.

FIGURE 9–12

(A) Near-horizontal highlines provide access capabilities in the urban environment that can exceed traditional and expensive mechanical devices. (B) Traditional and nontraditional tactics can work in harmony to effect rescues. (*Courtesy Martin Grube*)

(A)

(B)

FIGURE 9–13

(A, B) Bombproof anchors for tensioned rope systems on buildings can sometimes be located through ventilation ducts on structural components.
(C) Tracklines can be directed to the most desirable positions using COD components. (D) Bombproof anchors can be extended to a focus point. Here, a rigging plate organizes multiple components. Note backup anchor ropes bypass the rigging plate. (*Courtesy Martin Grube*)

(A)

(B)

(C)

(D)

While this may sound like highline anchor paradise, it is usually easy to achieve with a little ingenuity and by bringing remote anchors into position using directional pulleys if necessary.

3. Do whatever is necessary to elevate the leading edge of the highline (Figure 9–14A and Figure 9–14B). The most perfect highlines have remote anchors from the edge and an elevated position at the edge. Trees, elevator towers, guyed gins, bipods, tripods, scaffold bucks, industrial structures, and the like all help to get the trackline up and elevated. Elevating the leading edge of the highline allows the load, the patient and attendant, to be safely loaded onto the system before being released into the span.

If span noise factors are reasonable, and you have a suitable site, great equipment, and plenty of people, try for a catenary angle of 120 degrees when loaded. Recall the discussion in Chapter 7 about force vectors and critical angles, especially the section on the

FIGURE 9–14

(A, B) Elevating the leading edge is critical in engineering safe highlines. Elevated leading edges create a safe loading and landing area for rescuers and patients. (*Courtesy Patric Anderson*)

(A)

(B)

Golden Angle of 120 degrees. Tensioned rope systems that achieve 120 degrees of catenary angle at midspan are fairly harmonious throughout.

Following are some observations about the 120 degree highline (Figure 9–15):

1. The downward force on the trackline at midspan equals the forces imposed on the anchor system on each side. If you have a 300-pound load at midspan, each anchor will be supporting 300 lbf. Effectively, each anchor receives 150 pounds from the load and 150 pounds from the other anchor, so 150 + 150 = 300.
2. Triangulation of the forces all around makes it easy to estimate what is happening to the system as the load is moved from side to side. Also, if the load is increased on the trackline, the load on each anchor is increased correspondingly (Figure 9–16).
3. At 120 degrees, you can comfortably float as many as three people if necessary on the trackline. Assuming the load is 900 pounds, and all of our general use equipment including the rope is rated at 9,000 pounds minimum, this creates a 10:1 safety margin.

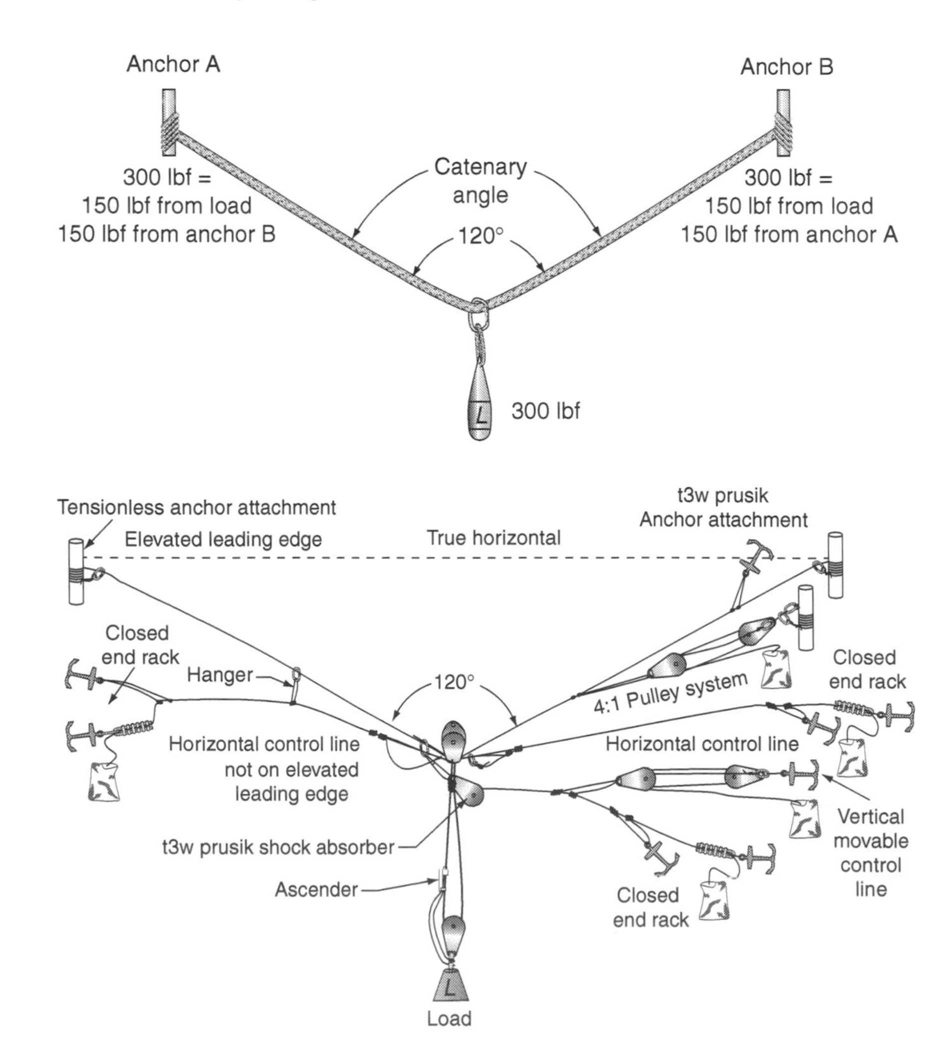

FIGURE 9–15

With a catenary angle of 120 degrees caused by a 300-lbf load, each anchor has 150 lbf from the load and 150 lbf from the opposite anchor.

FIGURE 9–16

Single trackline schematic.

4. The catenary angle of 120 degrees is not too steep to haul the load coming up the angle. Much steeper angles of 90 or 60 degrees make the haul up the angle almost like a straight vertical raise. With only three or four haulers and a 4:1 pulley system, 120 degrees is workable.

Exceeding 120 degrees to 130 through 160 degrees when loaded at midspan should only be considered when the alternatives are even more dangerous, such as dragging the load through a raging river or boiling lava. Perhaps you even have a very light load but only one person to pull the load up the catenary. Remember, when an extremely wide catenary angle is mandated, the solution is always to float more trackline rope (see page p. 295 of this chapter).

The catenary angle, once again, is the angle created in a flexible linear medium by the effects of span noise, such as gravity, the rescue load, wind, etc. Settling on a reasonable catenary angle is one method of justifying the ends in order to accomplish the means. For example, much of the physical performance of a tensioned rope system is dependent on the resulting catenary angle. It is sometimes wise, therefore, to use the necessary angle as a justification for the tension in a rope system. If your team is working a river rescue, and the elevated leading anchors allow for a very high span-to-height ratio or the highline dips into the raging waters (for example, span 100 feet and height 30 feet for 3.3:1 ratio), then a very tight (tensioned) highline and the corresponding lowering of the safety margin may be justified. In that case, doubling, tripling, or even quadrupling the trackline will significantly raise the working load of your system. The catenary angle can be made very wide, so everyone stays dry, without a resulting loss in safety margins because the additional rope in the air shares the load.

On the other hand, if the span-to-height ratio is very low (for example, span 30 feet and height 100 feet for a .33:1 ratio), then such a tight catenary may not be required. The angle and, consequently, forces can be reduced if other variables like equipment availability, personnel time, and so forth are compromised.

Tensioned rope systems are more span dependent than height dependent. The variables concerning height are less important than those concerning span. Compare two 50 foot tensioned rope systems between two highrises. One is 50 feet off the ground and the other 500 feet off the ground. Neither has as much span noise as two tensioned rope systems that are both 500 feet off the ground, one of which has a span of 50 feet and the other a span of 500 feet. Span noise in tensioned rope systems includes but is not limited to terrain, trees and brush, water hazards, wind, rain, structural components, items and people sent out on the rope system, the elevation difference between sides, and gravity.

The span-dependent characteristics of tensioned rope systems also include specific rope properties. The weight and elongation factors of the rope spanning the gap of a highline is of enormous importance when calculating tensioning requirements. At 7.4 pounds per 100 feet of rope, the rope weight of a 100-foot span is only 7.4 pounds and at 4% elongation, the rope is 104 feet long. In contrast, the rope weight *alone* on a 1,000-foot span highline is almost 75 pounds—more if the humidity is high and much more if it is raining. Elongation at 1,000 feet adds a desired or undesired length of more than 40 feet.

Elongation is a combination of two main factors, rope construction and fiber selection. It is important to understand how much the rope will stretch when tensioned and loaded. NFPA 1983-95, Chapter 5-1.3 states, "The minimum elongation of all new life safety rope shall be not less than 15% at 75% of the breaking strength . . ." and Chapter 5-1.4 states, "The maximum elongation of all new life safety rope shall not be more than 45% at 75% of the breaking strength. . . ." In effect, it means an NFPA 1983-compliant life safety rope must stretch between 15% and 45% when pulled to 75% of its breaking strength.

In practice, most manufacturers construct R/Q life safety ropes with a remarkably straight stress/strain curve that easily falls into these parameters. A nylon 1/2-inch rope will almost always stretch about 2% with 300 lbf applied, about 4% with 600 lbf applied, and so on to about 42% at 6,300 lbf or 75% of its breaking strength of 9,000 lbf. Polyester has somewhat less stretch at about 1.5% with 300 lbf, 3% with 600 lbf, and about 31.5% at 6,300 lbf or 75% of its breaking strength of 9,000 lbf. The difference between 31.5% and 42% can be significant when engineering rope systems. For example if you are building a 5:1 pulley system, you might choose polyester at 31.5% rather than nylon at 42%, so your haul team will not have to fight against the stretchiness of the nylon. And if you are building a highline that requires very little slack due to span noise, polyester might be the choice. Consider also that you are building a highline where the catenary angle can be a comfortable 120-degrees with a single rescuer on the line of 300 lbf. On the return trip, however, with a rescuer and a victim, the load exceeds 600 lbf. If you choose polyester to build the trackline, the force on the anchors is going to increase markedly because the lack of stretch almost maintains the original 120-degree catenary angle at about 110 degrees. If you choose nylon, the increased weight will decrease the catenary angle more significantly down to about 90 degrees. In effect, nylon is remarkably forgiving in this circumstance, stretching under the increased load and decreasing the catenary angle respectively. Therefore, force on the anchors is not increased as much as is the force on a less stretchy rope. When the load goes up, the angle comes down. This is not to say that polyester should not be used for this hypothetical highline. In fact, a wonderful teaming arrangement is produced by using polyester in the trackline and closely monitoring the tension and catenary angle, and using nylon for the potentially shock-absorbing control lines. The polyester does not need to be tensioned nearly as much as the stretchy nylon to produce a usable trackline. And the nylon t3w 8-mm prusik interface stands ready to catch the load if the trackline fails. It will also absorb more shock than polyester will. A good rig master considers the alternatives and makes educated decisions based on all the elements present.

NFPA 1983

Standard on Fire Service Life Safety Rope and System Components

NFPA 1983–2001 (Proposed) *Standard on Fire Service Life Safety Rope and System Components*

NFPA 1983–2001 (proposed) suggests less stretchy parameters than the 1995 edition. Apparently Federal Test Standard 191-A has been abandoned, a victim of antiquity and advancing technology. The Cordage Institute (ropecord@aol.com) uses a test method of 3% to 10% elongation at 10% of the breaking strength, which is a nondestructive, recognizable, and repeatable test method. With some modifications, NFPA 1983-2000, Chapter 5-1.3 says, "The minimum elongation of all new life safety rope shall not be less than 1% at 10% of breaking strength when tested . . ." and "the maximum elongation of all new life safety rope shall not be more than 10% at 10% of breaking strength when tested. . . ." If this is approved as proposed, it means that a 9,000 lbf mbs rope when pulled to 900 lbf will stretch at least 1%, i.e., a 100-foot rope at 900 lbf should be at least 101 feet long. A 9,000 lbf mbs rope when pulled to 900 lbf shall stretch no more than 10%, i.e., a 100-foot rope at 900 lbf should be no more than 110 feet long.

Additionally testing a rope to 10% of its breaking strength is much safer on testing personnel than testing a rope to 75% of its breaking strength.

The trackline supports most of the load in horizontal and angled highlines. The control lines move the load back and forth along the length of the trackline, and when engineered properly, serve as a backup to potential trackline failure. Because the horizontal line effectively alters the vertical effects of gravity, there must be some way to move the load carefully down the catenary angle and back up the other side. We also need to ask ourselves The Question—what happens if this part disintegrates? What happens to the load if the trackline is cut and fails catastrophically? The control lines, properly rigged, will catch and support the load until it can be manipulated to a safe location.

The horizontal highline's control lines are rigged (Figure 9–17A and 9–17B) exactly like the angled highline's control lines. They are anchored into manned t3w prusiks and into loosely rigged but manned closed end racks at each end. As the load travels the catenary angle of the trackline on the starting side, the rack is adjusted to allow controlled descent. The t3w prusik operator is positioned in front of the rack to slow down too much descent speed or unlikely trackline failure. The rack operator remains ready to close the rope angle and stop the control line descent if the t3w prusiks slip too far as they eat up descent energy.

► **trolley**

The wheeled device that rolls along the trackline carrying the weight of the load vertically and the force of the control lines horizontally.

The **trolley** is the wheeled device that rolls along the trackline carrying the weight of the load vertically and the force of the control lines horizontally (Figure 9–18A and Figure 9–18B).

There are dozens of different trolley configurations, and they all have pros and cons. The simplest way to build a trolley is to choose a strong, 9,000 lb t/s, sealed ball bearing unit with long side plates. There are some excellent single, double, and triple long side plate sealed ball bearing pulleys available. Some knot-passing style pulleys with very wide single wheels, and long, wide side plates

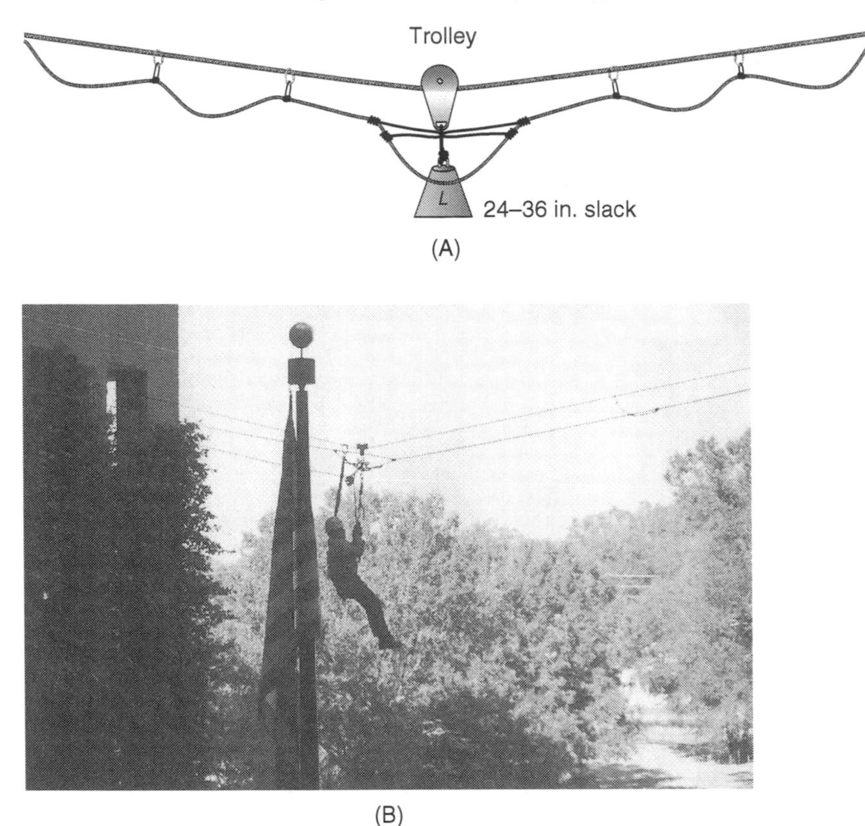

FIGURE 9–17

(A) Single line control line (no knots). (B) Single trackline with both horizontal and vertical control lines. (B, *Courtesy Patric Anderson*)

also make good trolleys. Long side plates make room for auxiliary connection places and increase the angle of approach of the trackline. Short side plate pulleys are great for making MA systems, but because of the limited approach angle of the rope, the side plates tend to rub the trackline when used as a trolley. Side plate rubbing not only abrades the trackline rope, but also can make retrieving the unloaded trolley a nightmare, particularly on angled highlines.

NOTE: The wheel and axle is an entirely different, simple machine with a different function from a sheave and axle. When an inverted pulley is used as a wheel and axle, the round part is called a wheel and not a sheave.

The control lines can be attached to the trolley in one of two ways. The first uses a single control line end to end and is preferable when you have one that is long enough. The trolley is anchored to the middle of the control line(s) using t3w prusiks. There should be 24 to 36 inches of slack built into the rope to allow the prusiks some room to absorb shock energy before the rope in the center is loaded (Figure 9–17A). The rope is continuous with no knots that could fail if massive shock forces were to hit the rope. The second method (Figure 9–18C) has the same prusik shock absorbing feature as the first, but the ends of two ropes are connected to the trolley with overlapping figure eight on a bight follow through. This technique helps back up the knots with the overlapping function and is practical when you do not have a rope that is long enough to use as a single piece. Regardless of which trolley system you choose, it is important to back up the main pulley components with a trailing tether attached to the trackline with a pulley or even a carabiner. This component backs up any failure in the trolley system. It should always be on the trailing side of the trolley so it will not jamb into the pulley wheel and slow or stop traverse progress.

FIGURE 9–18

(A) Dual trackline trolley with both horizontal and vertical control lines.
(B) Dual trackline trolley with load released from static tether for vertical lowering operation.
(C) Trolley with two separate control lines, each one encircling the carriage rig with an eight follow-through and t3w prusiks pulling the load. (*Courtesy Martin Grube*)

(A)

(B)

Trolley
Trackline
Hanger
Figure eight on bights follow through

(C)

FIGURE 9–19

A slackline highline to control the vertical friction of the highline rope.

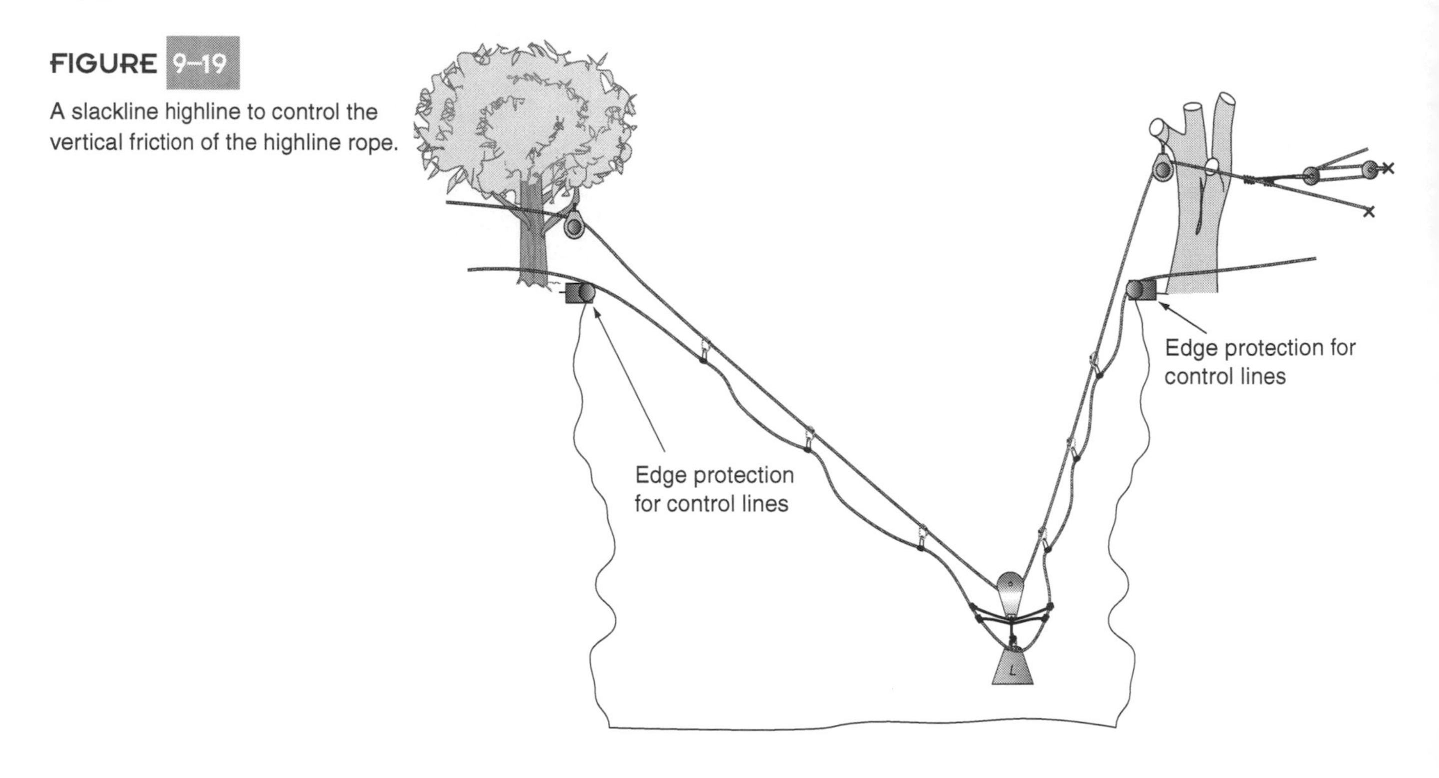

► **vertical function**

A term describing the raising and lowering of the rescue load while on a highline.

► **slackline highline**

Involves releasing or increasing the tension in the trackline of a tensioned rope system to raise and/or lower the load on the line.

The **vertical function** of highlines is accomplished in one of two ways. The first technique is called a **slackline highline** (Figure 9–19) and involves simply slacking off the trackline, decreasing the catenary angle, and controlling the descent into the span with the control lines. This usually has the advantage of lowering tensioned rope forces and increasing equipment component safety margins. Its disadvantage is that it creates a near full-force mechanical advantage pulley system on the receiving end to raise the load up the steep angle. This can be problematic when the loads are heavy, there are too few rescuers for the haul team, haul team access is limited, and there are multiple victims to remove.

The second way to accomplish vertical function with the highline is to rig a vertical movable control line or lines (see the following *Note* and Figure 9–20). This is easily done by attaching another control line, vertical movable control (VMC) line to the trolley by hanging a pulley attached to the trolley pulley. Then do the following:

1. The VMC line is reeved into the pulley and the load is attached to the terminal end of this line.
2. The VMC is controlled on one side by rack lowering or by a pulley system raising the line, as in any other raise or lower.
3. The line should be connected to the hanger system of the horizontal control lines with an additional carabiner and cord to keep the line weight from affecting the load.
4. When the load is being ferried from side to side, the load should be suspended from a tether anchored to the trolley pulley. This eliminates the VMC as a life-loaded consideration when only horizontal movement is desired.

FIGURE 9–20

Vertical movable control line allows for raising and lowering the load anywhere in the span.

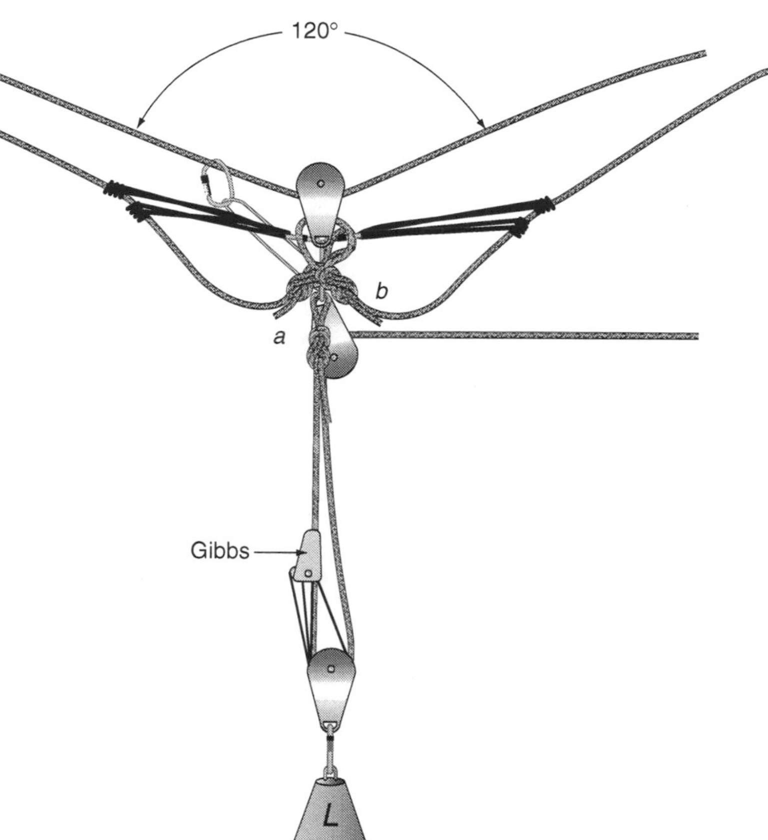

NOTE: In this method of vertical lowering, only a single line has been described as the VMC. Double VMCs create true system redundancy in this component. They require a double pulley at the trolley, double the hanger connections, double the rope and hardware, double the weight, and, on very long spans, double the headaches. If a very long span and equipment limitations prohibit doubling the VMC, the slackline vertical function should be considered the first alternative.

5. When vertical movement is necessary, the end opposite the VMC controlling side anchors the horizontal control line. The VMC is tensioned slightly to raise the load from the tether. When the attendant detaches the tether, the load is lowered into the VMC.
6. The VMC is then operated as a raise/lower system until the load, including rescuer, patient, and equipment, is placed safely back into the trolley tether for the nearly horizontal ride to safety.

Sometimes circumstances dictate a very tight highline due to a particularly heavy load with a correspondingly wide span or an excessive amount of span noise. In this case, the load can be more safely carried by using additional tracklines. While numerous multi-trackline systems have been engineered, the most practical is actually an elevated MA pulley system but with both sides of the system anchored.

The dual trackline system (Figure 9–21A and B) is actually a 2:1 simple pulley system with the advantage of automatically self-equalizing the forces on the ropes and the anchors. To build the 2:1 trackline:

1. Bomb off one end of the trackline rope.
2. Find the middle of the rope and send it across the span using one of the methods described earlier.
3. Build the mechanical advantage system of your choice and piggyback it onto the running end of the trackline for tensioning.

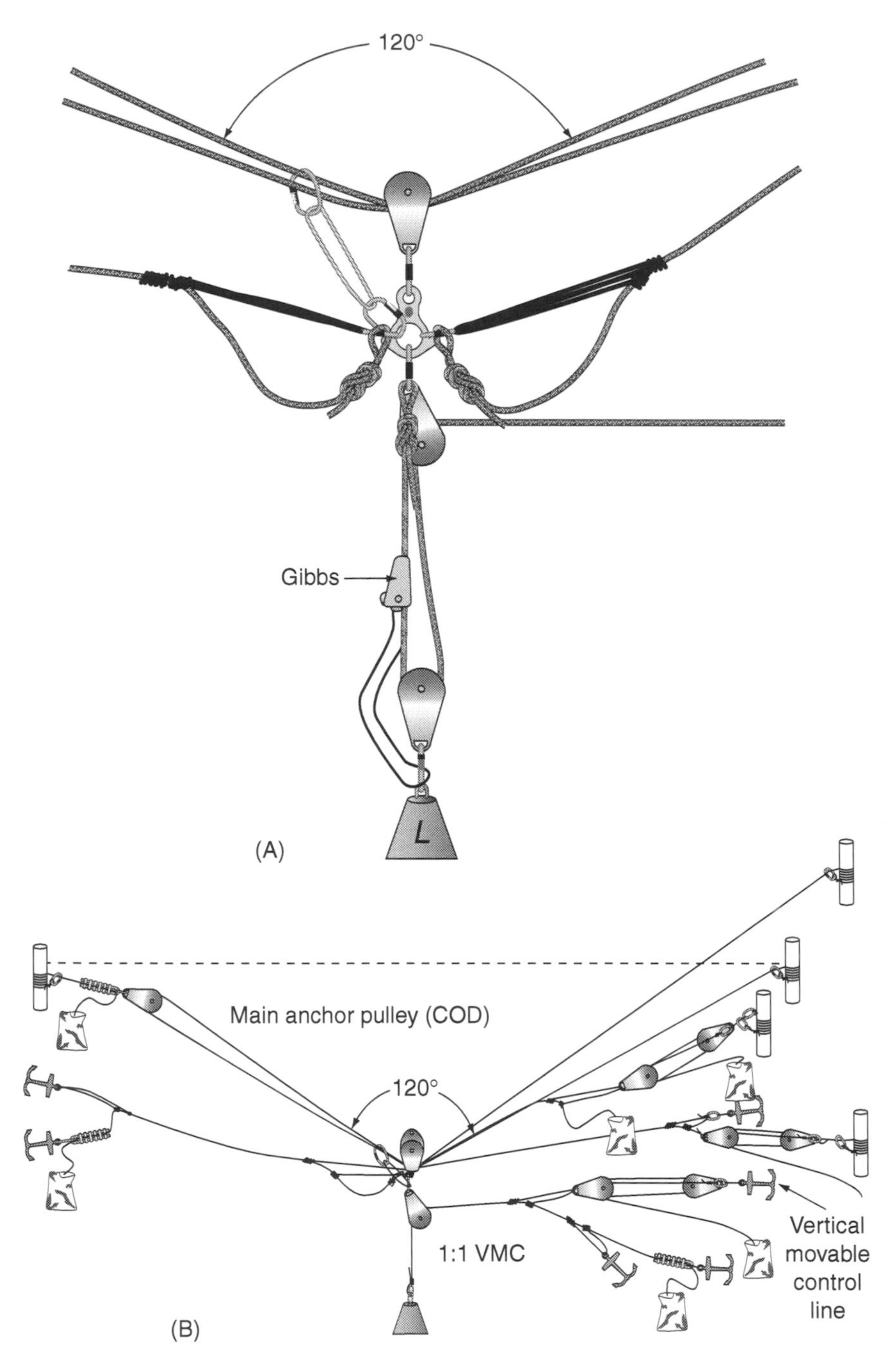

FIGURE 9–21
(A) Dual self-equalizing tracklines with wide pulley and 2:1 VMC.
(B) Dual trackline tensioned rope system components.

4. Bomb off the COD pulley on the opposite side (Figure 9–22). Again, try to elevate both sides of the highline for an easy-load leading edge.
5. Tension the running end of the rope with the MA pulley system to the desired catenary angle.

Consider that the COD pulley in the trackline is, in fact, a mechanical *disadvantage*. If the haul team applies 500 lbf to the running end of the rope, 500 lbf will be transmitted all the way through the COD pulley and back to the anchored side, meaning *1,000 lbf* is applied to the COD pulley during the tensioning process. This tension should not be alarming considering the general strength of the recommended components and anchors. The force on the COD pul-

FIGURE 9–22
Dual trackline self-equalizing pulley and dynamometer. (*Courtesy Martin Grube*)

ley will be equal to the sum of the load of the *two* anchors on the opposing side. This is important to remember because while you have floated twice the rope, and while the same load will not reduce catenary angle as much as a single rope of the same tension, more load is transmitted to a single point of attachment, the COD carabiner attachment point. You are effectively load sharing. Half the load is on one side of this highline, but the load is intensified on the other half. Again, this should not be alarming providing you recognize this fact and engineer your system appropriately. First, in this system always use a 12,000 lbf to 15,000 lbf pulley that will prevent the carabiner attachment point from egging out in all but the most severe shock loads. Next, as always, make sure you have at least *two* bombproof anchors. Then, because of the additional rope, not nearly as much tension is needed to keep the load at a desired level. Where 750 lbf was needed for a single line, only about 500 lbf at the COD will be needed to achieve the same angle with the dual tracklines—250 lbf on each line. Finally, as illustrated in Figure 9–23A, add a t3w and carabiner safety jig to the trackline and bomb it off to a separate anchor site. The safety jig will catch any trackline *or* COD pulley failure and dissipate some fall energy along the way.

► multi-trackline highlines
Highlines with more than one trackline.

Multi-trackline highlines of three, four, five, and six lines can be achieved by simply floating these simple MA systems to make the tracklines. Some considerations include the following:

1. Make sure you have plenty of rope. A 100-foot span, four-line trackline will require more than 400 feet of rope.
2. You will not need a MA system to tension your rope because the pulley system is your trackline and your tracklines are your pulley system.

Making the Dual Trackline Highline Safety Jig

For the dual trackline highline safety jig, you will need two sets of jigs (Figure 9–23B). They are added to the highline, one on each side of the COD pulley. To make one set:

1. Cut one 11-foot and one 13-foot section of 8-mm cordage.
2. Connect the ends of each section into a loop using the three-wrap double fisherman's bend. This makes one XL and one XXL loop. Remember that a 5-foot length makes the small prusik loop, a 7-foot length makes the medium, and 9-foot length makes the large prusik loop. When tied, the XL loop will be 4 feet long and the XXL loop will be 5 feet long.
3. Wrap the XXL prusik onto the trackline with the standard three wraps.
4. Roll a three-wrap configuration onto your finger and thread the half rope part of the jig into the wraps. This takes some practice, so be patient. Dress it out so the wraps bite just like any other prusik.
5. Repeat steps three and four with the XL prusik. Make sure the shorter prusik is positioned closer to the COD pulley than the XXL is and in front of the XXL on the jig rope. All should be inside of the XXL jig.
6. Tie a small figure eight on a bight into the end of the jig rope. This will be the carabiner attachment that is laid onto the trackline as well.
7. Anchor the running end of the jig into a bomber.The jig has four or six points of energy absorbing capacity and eight on a bights should the prusiks slip too far. This feature ensures backup in the highly unlikely event of trackline failure. The carabiners ensure backup in the highly unlikely event of COD pulley failure.

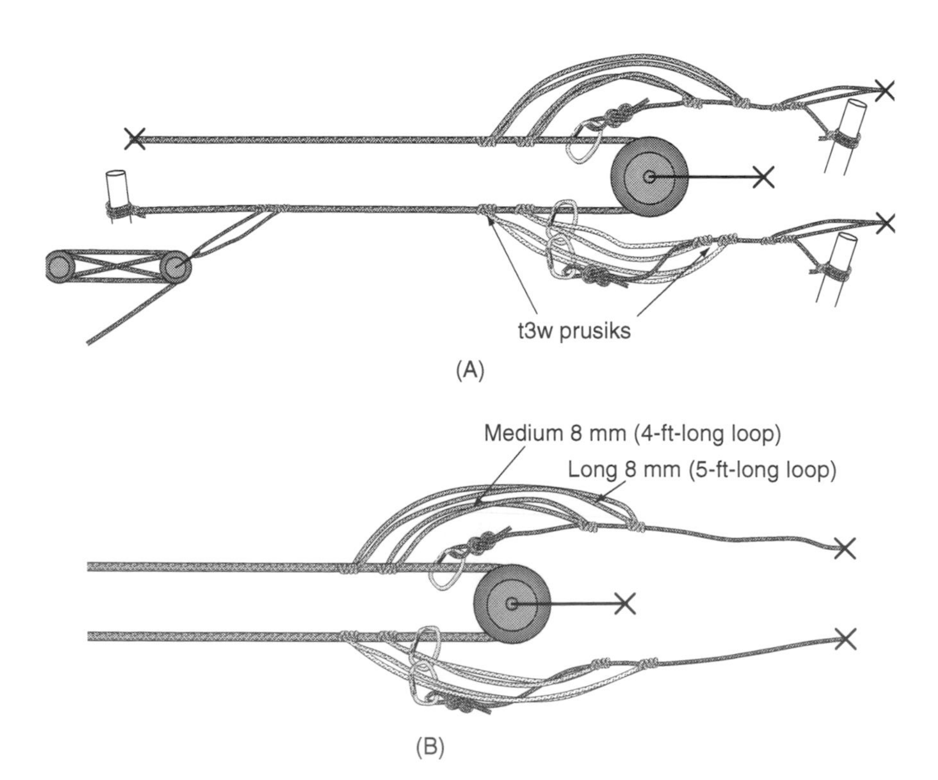

FIGURE 9–23

(A) Schematic of dual trackline system with equalizing pulley backup. (B) Dual trackline safety jig schematic.

3. Your trolley must be configured to accept multiple lines. You can use a double-wheel pulley on dual-trackline systems and a triple-wheel pulley on triple-line systems. At the time of publication, four wheel and up pulleys have not been manufactured in large quantities. Multiple line systems can easily be trollied by using very wide-wheeled, knot-passing pulleys.
4. Very little tension is going to be needed in the haul line to raise the tracklines to the desired levels. Two-hundred lbf on a 4:1 is going to produce a little more than 700 lbf actual MA at the anchor points.
5. Serious span noise is going to be magnified with multiline systems. Wind and rain will increase rope weight/force with each line you float.

■ SUMMARY

While vertical rope systems easily access victims in a straight up-and-down environment, tensioned rope systems allow rescuers to access a victim anywhere in a plane as defined by the two highest anchor points and the ground. Tensioning ropes is a safe and effective method of raising loads outside of the vertical and linear path of simple raises and lowers, but special efforts must be made to understand all seen and unseen forces in the systems. Most importantly, tensioned rope systems can make the transition of moving injured patients from places of danger to places of relative safety dramatically less traumatic.

■ KEY TERMS

Tensioned rope systems, or highlines
Span
Span noise
Situation
Pilot cord
Messenger cord
Faking
Playing
Angled highline
Catenary angle
Control lines
Line hangers, or festoons
Horizontal or near-horizontal highline
Trolley
Vertical function
Slackline highline
Multi-trackline highlines

■ REVIEW QUESTIONS

1. When engineering highlines, there are three major considerations in selecting a method of getting the rope to the other side. They are:
   a. vertical height, rope diameter, and rope weight
   b. victim location, medical condition, estimated length of time
   c. span, span noise, and situation
   d. anchor position, vertical height, rope diameter

2. Once the pilot cord is safely across, _____ is connected and pulled across.
   a. the messenger line
   b. the trackline
   c. the control line
   d. the tagline
3. As in all engineering challenges, _____ stabilize(s) anticipated outcomes.
   a. stronger materials and tools
   b. a good blueprint
   c. better educated engineers can
   d. reduction and control of the variables
4. A good starting point for pretension of a single trackline rope system is _____ lbf.
   a. 250
   b. 500
   c. 750
   d. 1,000
5. The angle created in a flexible linear medium by the effects of span noise, such as gravity, rescue load, wind, etc. is called _____.
   a. the critical angle
   b. the span angle
   c. the calculated angle
   d. the catenary angle
6. A good rule of thumb is to tension your angled highline so the load lands safely in the _____ of the trackline length closest to the ground anchor.
   a. exact middle
   b. upper one-third
   c. lower one-third
   d. last several feet
7. Control lines are used for two specific reasons, to _____, and to control the rate and direction of travel along the path of the trackline.
   a. stand ready to accept the load in the unlikely event of trackline failure
   b. hold excess weight off the trackline
   c. control excess load requirements caused by rescue loads
   d. dampen, or control, the effects of shock on the trackline and anchors
8. On highlines, it is wise to rig the control lines free of the _____ for the trackline.
   a. overhead anchor attachment that creates the elevated leading edge
   b. haul system used
   c. trolley system that is used
   d. other equipment elements used

9. If span noise factors are reasonable and you have a suitable site, great equipment, and plenty of people, try for a catenary angle of _____ when loaded.
   a. 60 degrees
   b. 90 degrees
   c. 120 degrees
   d. 150 degrees
10. Elongation is a combination of two main factors, _____.
   a. weather and rope weight
   b. rope length and diameter
   c. load and catenary angle
   d. rope construction and fiber selection

CHAPTER

10 The Future

OBJECTIVES

Upon completion of this chapter, you should be able to:

- discuss rope rescue systems that reach the third dimension.
- discuss next generation equipment and materials.
- discuss the immediate future of rope rescue.

INTRODUCTION

History can give us some hints about the future. Looking at rescue's past fifty years and the exponential growth and improvement in special operations, including special people, special training, and special equipment, should create some idea about what we have to look forward to in high-angle rescue. Extrapolating our learning and technology curve over the past fifty years would be a highly complicated job requiring much research and data analysis. However, some predications can be made more easily. For example, it can be supposed that ropes used for rescue will eventually find themselves a victim of antiquity. Certainly mankind will find a way to defeat the effects of gravity without using the combinations of ancient and modern rope skills, or the tenuous effects of air being beaten into submission by a rotor wing. Perhaps one day rescuers will mount aerioflotation devices to locate, access, stabilize, and transport targets to safety. Searching for good anchors and understanding practical rigging concepts will no longer be necessary. Fall protection might be replaced with a pager-like device attached to the belt that will instantaneously calculate altitude changes based on input from satellites placed in geosynchronous orbit 16,000 miles above the earth. Triangulating on the falling pager, an antigravity field would be initialized and the fall would be arrested in less time than it would take to snap your fingers.

No doubt the art of rope rescue will fade away, perhaps to be rediscovered by nucleoentologists of a distant century. Perhaps our culture will fall victim to some sort of catastrophe and in the process of recovery, rope skills will again become necessary. Advanced technology will then find itself a not-to-distant cousin of the art of rescue by rope, ultimately, we are all hanging by a thread.

THE FUTURE

In the immediate and the not-to-distant future, rescue operations will be entering their heyday. Regulating agencies are approving parameters that give rescuers authority to pursue training and real-time rope rescue scenarios with confidence and without fear of litigation. Instead of an eccentric and esoteric tool beneath the crescent wrenches, hammers, and screw drivers in the tactical tool box, rope rescue will become as normal as defibrillating a heart or pulling a Class A foam valve on a pump panel. The need for rope rescue services continues to grow as industry expands, the population moves into the sky, and parks and wildernesses become a common place to visit.

A look at rope rescue of the future can be organized into four categories—*hardware, software, systematic,* and *cultural.*

Hardware most certainly will become harder, stronger, lighter, and, relative to our current economy, cheaper. The study of materials sciences is one of the fastest growing and most profitable technologies today, and rope rescuers will certainly benefit from that technology. At the very least, someone will design a rope grab that holds/slips a rope at a preset tension regardless of the variables.

As long as we have gravity, ground-based space launches will maintain the necessity for lightweight payloads. That means that as we speak someone is developing the lightest and strongest metals ever known. Plastics will probably make their next evolutionary jump and will even exceed metals in their ability to withstand omnidirectional tensile and compressive forces. A string of thirty plastic/composite carabiners in the 10,000 lbf breaking strength range will probably weigh 3 pounds and be simultaneously capable to bear the markings of NFPA, ANSI, NIOSH, OSHA, VIOSH, EIEIO-OSH, UL, SEI, FM, UIAA, CE, ASTM, ESPN, and TRTRT, et al.

The motorized ascending devices (MADs) used first in the early 1970s will finally see miniaturization and become practical. Gasoline-fueled power heads for chainsaws are already about the size of a softball and are capable of lifting several hundred pounds with ease when coupled with the right reduction gear transmission and personnel attachment. Expect personal ascending systems in the 5-pound range including fuel to be available and affordable in the very near future. Hydraulically powered winches will be available that will adapt to any local hydraulic power unit. Anchored to tension any rope system, they will be able to raise any load anywhere the pump can be hauled.

Electromagnetic levitation (mag-lev) pulleys will eventually make pulleys virtually frictionless. With a mag-lev pulley system, one person pulling on a 4:1 could easily raise a 600-pound load. A mag-lev trolley riding on a highline would make a trip with gravity on the ride down one side of the catenary and momentum most of the way back up the other side. A MAD would handle all of the vertical access jobs once the mag-lev trolley was positioned over the target.

Software is destined to be the most remarkably different of all of the new technologies. Ropes may have "dynotapes" imbedded in their sheath fibers to indicate when certain tensions have been reached or exceeded—green fibers for less than 1,000 lbf, yellow fibers for 1,000 lbf to 1,500 lbf, and red fibers for 1,500 lbf to 2,000 lbf. As these tensions are reached, the dynotape will break and bubble up on the surface of the sheath indicating a potentially dangerous situation. Eventually, the end of the ropes may have a receiver that accepts a force link indicator connected to a transrope-transducer. Tensions could be read accurately and digitally every second of an operation. The entire operation may be fed into a computer for real-time stress/strain analysis.

Tape or flat rope technology may take the round out of our ropes. Equipment could be built to accept tape more easily than round rope. Tape could be easier to store in greater quantities and promises almost unbelievable tensile properties. Pulleys could be smaller with flat sheaves, while DCDs could be the size of matchboxes.

Polyester and various exotic and exciting fiber technologies will be refined in the immediate future. Ropes will glow and/or float on water as needed. Chemical contact will no longer mean certain failure of a rope. Personal emergency escape ropes will become lighter, stronger, and almost impervious to the effects of fire, chemicals, and sharp edges. Cable and wire rope technologies will probably come out from their slumber and become improved and versatile enough to build into highline systems and others for almost instantaneous emergency egress in multivictim incidents. Exotic fiber technologies will find the perfect blend of carbon fiber and others to make extraordinarily strong and durable ropes of smaller and smaller diameter. A carbon and beryllium composite rope fiber may someday reach 10,000 lbf, be only 5 mm in diameter, and be able to have a knot tied in it. Fiber performance of that kind will probably be just the beginning of the next fiber revolution.

Systems of the future will take advantage of equipment improvements to reach far beyond today's limitations. Smart systems will replace much of the human component and will make rope systems safer and much easier to use. A laser will accurately track rope elongation and report readings to an operations center or a laptop for continuous analysis. Systems can then safely be pushed closer to their breaking strength when necessary to effect human rescue.

Highlines will venture into the third dimension, depth, by articulating an x,y,z profile with a parallel system or by suspending a counter-weighted boom beneath a trolley and rotating the length of the boom into the third dimension. Using third-dimension tensioned rope systems will free the rope rescue system engineer from using anchors that are way out of position.

Cultural changes in our environment can play havoc on the development of special operations teams and on technical rescue equipment and innovation. The next important training arena for fire and rescue services is in terrorism and domestic preparedness. There are several possibilities. All-out war could divert the attention of talented people away from advancing rope rescue technique and, at the same time, become the motivator that propels us into the next technological era of rescue. Local teams can be drawn off to fight a war, while a national and coordinated program like FEMA's USAR gets the opportunity to become a really powerful force for responding to incidents of man-made and natural domestic disaster. USAR, municipal fire and rescue teams, and military teams could further standardize training and equipment, making rope rescue and other disciplines as safe as a walk in the park. There could be national or perhaps worldwide standards for the techniques, training, and equipment used for mountain rescue, cave and wilderness rescue, and urban rescue.

■ SUMMARY

The look into the future of rope rescue is limited only by individual paradigms, friction from those who slow progress but raise awareness and strengthen resolve, and the limits of imagination. Allow yourselves to think outside of the parameters of your mentors and instructors, or you will become stagnant and stale. Work to keep rope rescue on the edge of technology and thought. Learn not to be critical of your brother and sister rope rescuers. Do not fall under the specter of commercial competitiveness and rival conflict. It is only marginally good for the business. Do not put one another down because of the NIH factor—not invented here or we somehow feel a little bigger by stepping up on someone else. In the name of safety, some say that other providers are dangerous and uninformed. The latest backyard test, often with subconscious predetermined conclusions, is offered up as fact, instead of food for thought.

Discussion is good, but controversy is destructive to the family of technicians. The future is an opportunity to unite in common goals, and in service to citizens, targets, patients, and customers. There are many emergency services providers who refuse to accept rope rescuers as a viable entity and who think rope rescuers are strange and new. Controversies need to stay in-house and successes shared humbly with all who will listen. Like the changing family structure of this newest millennium, technical rope rescuers are changing, learning, owning, and doing for the good of our business. With a spirit of unparalleled cooperation, coupled with incredible materials advances, the future for rope rescue is indeed bright.

Glossary

actual mechanical advantage (AMA) This is found by calculating TMA and subtracting from it the effects of friction.

alloy Combination of two or more elements where the majority is the *base* element and the other are the *alloying* elements.

aluminum The most common metal used today in the construction of rope rescue tools.

anchor slings Synthetic-fiber flat or tubular webbing configurations that wrap around anchors for the purpose of attaching system components.

anchor systems Groups of anchors that work in unison to provide either more conveniently located anchors or backup anchors if one of the anchors fails.

anchored pulleys They are directional only and provide no mechanical advantage.

anchors The items to which rescuers attach rope system components.

angle of approach The angle of the rope from the anchor to the edge relative to the ground. The higher the angle of approach, the easier it is to load the rope.

angled highline A tensioned rope system with a substantially higher anchor on one side than the other.

anodization A process whereby aluminum is subjected to the electrolytic action of the coating material in an energized solution, causing a strong molecular bond.

anti-torque devices (ATDs) Another name for swivels and snap links.

Authority Having Jurisdiction (AHJ) The organization, office, or individual responsible for approving equipment, an installation, or a procedure.

auxiliary equipment Any equipment that is not a rope, a harness, or a belt. For purposes of NFPA 1983, it refers to hardware and webbing.

belt Harness-like device without webbing that fastens under the thighs or buttocks.

bend A configuration where two ends of a rope are joined.

block creel As defined in NFPA 1983, "rope constructed without knots, or splices in the yarns, ply yarns, strands or braids, or rope. Unavoidable knots might be present in individual fibers as received from the fiber producer."

bombproof, bombing, or bombing off Terms that have been applied to guaranteed foolproof anchors.

bottom belay Pulling down on a rope, when required or requested, by a person positioned below a rappeler.

brake bar rack A versatile descent control device made popular by caver John Cole in the mid-1960s.

butterfly knot Used instead of eight on a bight where a strong midline, rather than a terminal end attachment, is desired.

carabiners Devices that cause rope system components to become interrelated and useful.

catenary angle The angle created in a flexible linear medium by the effects of span noise.

change of direction (COD) A system component that alters the direction of the rope.

cheater prusik Single prusik used in a variety of ways to enhance the flexibility of a system.

Class I lever Mechanism where the fulcrum is placed between the object being lifted and the applied force. The force is then applied in the opposite direction of the object being moved.

Class II lever The fulcrum is at the end of the lever, the object being moved is between the applied force and the fulcrum, and the applied force is at the end opposite the fulcrum.

Class III lever Applies force between the fulcrum and the object being moved.

closed-end rack (or closed frame rack) Modification of the John Cole brake bar rack.

compounding pulley systems Applying the output force from one pulley system to the input force of another.

compression Pulling the pulley system down to its smallest stroke.

conservation of mechanical energy This law states that the total mechanical energy in an isolated system is always constant.

contextual training Involves developing students' ability to handle anxieties related to the practical applications of a given discipline, and entering the environment in a progressive sequence to enhance skill levels.

control lines Used to move the trolley along the length of the trackline(s), and they act as the belay in the event of tracking failure.

counterbalance A rope system that uses gravity to perform a raise. A heavy object pulls down on one end of a rope that has been reeved over an anchored pulley. The object connected to the other end of the rope is raised by the countereffect of this object.

descent control device (DCD) Metal configurations (friction producers) that rub the rope and slow its movement.

de-set When the system is compressed so it will expand again when using the MA system as the DCD to lower a load rather than raise it.

double bight figure eight An inherently tight and enormously strong rope configuration that provides two bights in the end of the rope for connecting people or equipment components.

double fisherman's bend Two opposing fisherman's knots tightened against one another to make the common prusik loop.

double-line dynamic rappel system Two-rope rappeling system in which the rappeler's descent control device slides on a fixed rope and the second rope, or belay rope, is attached to an anchored DCD on the top and fed through the DCD as a backup in the unlikely event of main line failure.

dressing Manipulating the parts of a knot so they are all parallel, neat, and not crossing.

Dulfersitz Body rappel, a method of descending in which the body becomes the DCD and the rope is anchored at the top.

dynamic roll out Rare but dangerous condition that occurs when there is torque in a system and the components twist against themselves.

eight-plates Relatively inexpensive and the most common of descent control devices.

emergency egress Any means of escaping a structure, container, or vessel under emergency conditions.

emergency egress improvised harness A harness tied onto a person in an emergency using rope or webbing to allow escape from an untenable environment.

extension To pull out a pulley system to its longest strike.

faking A nautical term meaning to lay rope out for the most efficient deployment and to prevent kinks and entanglements.

fall factor The severity of a fall that is expressed as a ratio calculated by dividing the distance fallen by the length of the rope used to arrest the fall.

figure eight bend Quick and simple method of connecting two ropes of equal diameter.

figure eight on a bight A preferred, inherently tight and relatively strong knot that makes a bight in the end of a rope for connecting people or equipment components.

fisherman's knots A series of inherently tight and very strong rope configurations used to secure equipment to the end of the rope and as safety knots for inherently loose knots.

force vectors Any angle or curve in a rope system that is not in a straight line between the anchor and the load. All vectors add to overall system forces.

friction The resistance to relative motion between two bodies in contact.

friction coefficient (FC) A measure of the loss of advantage created by the effects of friction on a machine.

Golden Angle A force vector with the angle of 120 degrees, or one-third of a circle (360 degrees). The Golden Angle effectively triangulates system forces.

gravity The gravitational attraction of the mass of the Earth, the moon, or a planet of bodies at or near its surface.

half hitch A simple rope configuration that is made by passing the running end around an object, around the standing end of the rope, and then back through the resulting loop. It is inherently loose and is often misused as an adequate safety knot for inherently loose knots.

hardware According to NFPA 1983, "a type of auxiliary equipment that includes but is not limited to, ascent devices, carabiners, descent control devices, pulleys, rings, and snap links."

harness A webbing or rope configuration tied onto people to connect them to a safety tether or other component of a rope system.

hasty hitch harness An improvised webbing harness used when the victim has no reasonable harness, is conscious and in good medical condition, and time is of the essence.

hazard curve paradox Compares funding and other allocated resources with the hazard profile of a given operation.

high-stretch, or dynamic, kernmantle Rope with high elongation characteristics such as stretch of between 30% and 75%.

hitch A configuration that is tied around an object.

hokie hitch A load-releasing hitch devised to transfer a load safely from one part of a system to another.

horizontal or near-horizontal highline Rope transportation systems anchored at or near the same height.

hyper-bar A longer than normal top bar, extending an inch or so outside of the rack frame.

improvised harnesses Harnesses made from webbing, usually 2-inch tubular webbing, which are tied onto a person to be lowered.

inclined plane The simplest machine, which is a ramp.

inherently tight or inherently loose configurations Two categories of knotted rope and webbing configurations. Inherently tight knots, such as the eight on a bight, double bight eight, and the butterfly, stay in position, do not come untied by themselves or during use, and do not require safety knots as backups. Inherently loose knots, such as the clove, girth, and münter hitches, have a history of coming loose or capsizing (changing shape) during some uses and require backup safety knots to make them hold their shape.

Integrated Harness System (IHS) Combines two modified webbing harnesses integrated with old-fashioned "diamond" lashing techniques.

internal water bend Variation of the water bend.

knot A rope or webbing configuration comprised of bights, loops, and round turns that maintains its shape independent of outside factors.

leading edge The part of the edge of a high place that can be considered a danger to fall from and the immediate area surrounding the edge, usually considered to be about 10 feet back from the edge; the area where fall protection equipment is mandatory; the loading zone for loading a rope rescue system.

lever A combination of four parts: the lever, fulcrum, applied force, and the object being moved.

life safety rope Generally designated as a one- or two-person rope with a factor of not less than 15.

line A rope that is in use.

line hangers, or festoons Cordage and carabiner rope riders that lift the weight of the control lines up onto the trackline.

litter bridle The attachment device that connects the stretcher to the system. The bridle spreads and balances the weight of the litter over four or more legs.

load-releasing device (LRD) A system component that will allow the release of another loaded component, used to transfer forces from one component to another or when passing knots.

load sharing anchor system components Sections of rope or webbing configured to transfer the rope system load to two or more anchors.

lowering systems Rope systems engineered to use the effects of gravity to transport people and equipment from a high point to a low point.

mechanical advantage (MA) The advantage that is gained by using a machine to transmit force.

mechanical rope grabs and ascenders Devices constructed to grip the rope for raising or holding loads or for climbing.

messenger cord Stronger and somewhat heavier than the pilot cord (3 to 7 mm).

metallurgy The study of the science and technology of metals.

middle of the rope technique Most pulley systems start with the terminal end of the rope; this technique starts in the middle of the rope.

miscellaneous DCD Myriad devices sometimes used for special applications or for recreational use.

modified Swiss seat A safer harness than the emergency egress harness but takes more time to tie.

multi-trackline highlines Highlines with more than one trackline.

münter hitch A friction hitch that can be used for rappeling in a pinch.

newton One newton is roughly 0.225 lbf, or about the weight of an apple.

network A group of people spanning a specific geographic area (local, state, national, or international) with common interests that work together to complement educational and practical experiences.

nonanchored, or moving, pulleys Move in the same direction as the load toward the anchor provided no CODs have been added to the main line. They provide a mechanical advantage by sharing loads among multiple ropes in the pulley system and by the rolling action of the sheave that facilitates the transfer of applied force to the load.

overhand knot The most basic knot; a good safety backup to inherently loose knots.

overhand on a bight Makes a connection point in the terminal ends of webbing.

PEERS Pak A double bag system containing the equipment required for emergency escape: escape cordage, DCD, two carabiners, a harness (if one is not a part of the protective ensemble), and edge protection.

Personnel Accountability Report (PAR) A roll call.

personal emergency escape rope system (PEERS) A combination of equipment and training that gives trapped firefighters a last ditch means of egress after all other means have been considered.

personal escape rope Used for emergency self-rescue and escape.

pilot cord A strong but lightweight piece of cordage (1 or 2 mm diameter) that can be projected some distance to the other side.

playing An ancient nautical term meaning to direct and work the line to a desired location.

pounds A term used to measure an amount of weight or mass.

pounds force (lbf) The force created on an object by the gravity on the mass.

presewn harnesses Manufactured harnesses that are designed for quick donning, eliminating worries about knots and connections.

pretension To pull hard on the knot components in the direction of the intended load to work out any looseness that could cause the knot to be dressed differently when carrying a rescue load.

primary anchor The main anchor that a load depends on.

prusik loop A loop tied using the double fisherman's bend to connect the ends of a single piece of cordage.

pseudo-anchors Items that can be used as anchors but may not always be considered bombproof.

pulley A simple machine that transfers applied force from one location to another when used in conjunction with a flexible medium, such as rope, chains, belts, and so forth.

qualified person Defined by the Authority Having Jurisdiction (AHJ) generally as personnel employed by the local jurisdiction who are recognized as trained and proficient at performing a given task.

rappeling The act of sliding down a rope in a controlled fashion.

rescue quality (R/Q) pulley system components Combinations of R/Q pulleys, ropes, and carabiners that develop mechanical advantage in a neat and orderly fashion.

re-set When the system is extended again during raising operations and another haul segment is made on the main line.

rigging plates Flat pieces of aluminum or steel that have been manufactured with holes in them for the attachment of other pieces of equipment.

risk-benefit analysis An incident management technique where the factors of an emergency situation are studied to determine the risks and compare them to the benefits of attempting a particular action. For example, the benefits of committing live rescuers to extremely hazardous environments for the benefit of recovering dead bodies is negligible.

rope calls Standardized words used as communications tools to eliminate mistakes and confusion while a team is operating a rope system.

rope system The sum of all the equipment components that are engineered into a rescue tool for the purpose of saving lives and transporting people and equipment to areas that are otherwise not accessible.

ropebotics An esoteric term sometimes used to describe rope rescue instructors that are like robots in their delivery and rope use philosophy, highlighted by their inflexible view toward other types of equipment and different techniques.

running part The working end of the rope.

secondary anchor The backup, alternate, or belay anchor.

self rescue Ability of a rescuer to escape unaided from a given situation.

sensory deprivation An advanced training technique where one or more senses is removed or inhibited.

simple machines Machines that accomplish a task in a single movement. They translate input by people into multiplied force. The exchange is defined in terms of time and distance.

single figure eight A knot that is a turn and a half around itself.

single-line dynamic rappel system A single-rope system that has been rigged through a descent control device so the rope can be lowered or moved while a rappeler's weight is on the rope.

single-line static rappel line A single rappel rope that has been tied off to anchors (static) and that cannot be moved while a rappeler's weight is on the rope.

situation A position with respect to the condition and surroundings of an occurrence: a real-time "live" rescue has different operational factors from a training scenario.

slackline highline Involves releasing or increasing the tension in the trackline of a tensioned rope system to raise and/or lower the load on the line.

span The distance from anchor point to anchor point.

span of control The number of people that can efficiently and safely be managed and accounted for.

span noise All the factors that create variables in the ropes from anchor point to anchor point, including rope weight, trees, rain, waterfalls, wind, convergent volunteers, etc.

square bend Recognized as a square knot; can also be used to join webbing.

stainless steel A type of steel that is very strong, has a relatively low iron content, and is alloyed to be almost rust free and nonmagnetic.

standby DCD method Easiest way to pass a knot through a DCD and other system components.

standing part The anchored side of the rope.

steel Metal primarily composed of iron that has been refined from naturally occurring iron ore.

stroke The distance from the start of lever movement to the finish of lever movement.

super bombproof anchor (SBA) A bombproof anchor in which the rig master feels so confident that no other anchors are used in the system.

swivels and snap links Devices designed as attachment points for other system components that pivot on an axis to eliminate unwanted torque and twisting.

t3w Tandem three-wrap prusik loop on a rope.

team A group of people organized to accomplish a common goal.

team efficiency concept (TEC) Refining team member interaction to be productive without waste in every team endeavor.

tensioned rope systems, or highlines Rope rescue transportation systems where a trackline rope is anchored, crosses a span, and is pulled tight enough to safely support the weight of a rescuer, any patients, all ancillary system components, and any associated span noise.

terminal end A designation for the very last part of the rope. The terminal end is used to make connections to equipment and anchors.

tertiary anchor The third anchor.

theoretical mechanical advantage (TMA) The ratio of output force compared with input force generated by a machine with a complete lack of friction.

tie backs These tension two in-line anchors together to incorporate the strength from both anchors simultaneously.

training to fundamental correctness Sticking to the rules by not cutting corners when building rope rescue systems. Building a system using a series of rules and philosophies one day and then bending those rules the next day makes for irregularities that are confusing to a team and can cause a general lack of focus on complicated and stressful rescue calls.

trolley The wheeled device that rolls along the trackline carrying the weight of the load vertically and the force of the control lines horizontally.

tube type DCD Anchored pipe-like mechanisms with rope spiraled around them like stripes on a barber pole.

two in two out rule OSHA's rule that if two firefighters enter a building with an IDLH environment, there must be two firefighters suited up and waiting outside for possible rescue.

Urban Search and Rescue (USAR) Search and rescue teams whose primary service area is in the urban, semiurban, and industrial environments.

vertical function A term describing the raising and lowering of the rescue load while on a highline.

water bend Joins the ends of the same piece of webbing to make a sling or connects two different sections of webbing.

Index

Made in the USA
Lexington, KY
03 February 2012